W9-CKR-285

THE FIRST FOSSIL HUNTERS

THE FIRST FOSSIL HUNTERS

PALEONTOLOGY IN GREEK AND ROMAN TIMES

Adrienne Mayor

PRINCETON UNIVERSITY PRESS

PRINCETON, NEW JERSEY

Published by Princeton University Press,
41 William Street, Princeton, New Jersey 08540
In the United Kingdom: Princeton University Press, Chichester, West Sussex

Mayor, Adrienne, 1946–
The first fossil hunters : paleontology in Greek and Roman times /
Adrienne Mayor.
p. cm.
Includes bibliographical references and index.
ISBN 0-691-05863-6 (cloth : alk. paper)
1. Paleontology—Greece—History. 2. Paleontology—Rome—History.
3. Science, Ancient. I. Title.
QE705.G8M39 2000
560′.938–dc21 99-43073 CIP

This book has been composed in Galliard Typeface

The paper used in this publication meets the minimum requirements of
ANSI/NISO Z39.48-1992 (R1997) (*Permanence of Paper*)

www.pup.princeton.edu
Printed in the United States of America

1 3 5 7 9 10 8 6 4 2

For my parents

BARBARA AND JOHN MAYOR

CONTENTS

ILLUSTRATIONS

Maps

Figures

FOREWORD

Stones, Bones, and Exotic Creatures of the Past

AS A CHILD, I was powerfully drawn to stones and bones. By age eleven I told my parents I wanted to be a paleontologist. Pursuing my dream of becoming a paleontologist, I studied geology in college. Here I learned rocks, I learned minerals, I learned invertebrate fossils. My head was stuffed with facts. Later in graduate school I learned vertebrate fossils, and I learned bones, and anatomy, and evolutionary theory. I handled bones, and my wish to study fossils themselves became a reality. It has been a joy and a delight to pursue my career. But there was something missing: a historical perspective. History matters. What is almost as important as what we learn is how we have come to know. Paleontology is a historical science. Even though knowledge is advancing rapidly today (far more rapidly than when I was a graduate student), there is a component of paleontology that does not change and that can never be outdated: that is, the descriptions of the fossils themselves. In bygone days, descriptions and illustrations were often quite elegant. The old monographs from the early days of the twentieth century or even the nineteenth century must still be consulted. Rarely does the elegance of descriptive work today match that of former times. Remember how Darwin spent eight years producing his monographs on barnacles! The world moves too fast today even to permit consideration of such a project. Darwin would never have gotten tenure at a modern university! What I have gained from reading nineteenth-century works in paleontology is enormous respect for the intellectual giants of the past, the founders of the field of paleontology. I have also found that there is little new under the sun—many ideas about dinosaur biology that have been claimed as new in the 1970s or even today were widely discussed before the advent of the twentieth century!

In my training, nobody pointed out to me that history is important. I was at a bit of a disadvantage in not having the benefit of a liberal education. I attended a Canadian university at which I studied geology and other sciences, but little else, least of all history. I remember encountering during my graduate studies at Yale a book by M.J.S. Rudwick, entitled *The Meaning of Fossils* (1972). I excitedly opened it, but quickly closed it again as I discovered that it was not at all about fossils, but only about the *history* of understanding of fossils. So intent was I then on learning about the fossils that I was too focused to consider anything else. I often think that my real education began only after my formal training was completed and I no longer had to take courses or write a thesis. Perhaps ten years later I returned to Rudwick's book. By now I had matured enough to be receptive to a historical treatment, and I found it to be of surpassing interest. I learned the story Rudwick had to tell—indeed, perhaps I learned it too well!

Rudwick traced the history of attempts to understand the meaning of fossils, beginning with Conrad Gesner, who published a book, *On Fossil Objects* (*De Rerum Fossilium*), in 1565, which in Rudwick's judgment represented the beginning of the modern attempt to understand fossils. From the Latin word *fossilis*, meaning "dug up," the term *fossil* initially included minerals, interesting concretions and curiosities, as well as organic remains. Rudwick emphasized the difficulty of the task of interpreting ancient remains. Nothing in nature comes with a label attached—all interpretation is the result of hard-won knowledge. In some cases, the remains in the rock are of familiar form, such as seashells, snails, and clams, including remains of species still alive today. When such remains are found close to the sea, they present no insuperable problem of interpretation. But what of seashells found at alpine heights, far from the sea and high above sea level? Such fossils presented legitimate problems of interpretation. Moreover, what were remains of living creatures doing inside of solid rock? How did they get there? Certain triangular tongue-shaped objects inside of rocks were well-known curiosities in the medieval world; they were called *glossopetrae* ("tongue-stones"). Gessner suspected that they were the teeth of sharks but could not explain their existence inside of rocks. A century later, in 1666, a great shark was brought ashore in Tuscany. It was presented to Niels Stensen, better known by his Latin name Steno. Not only was Steno able to confirm that the tongue-stones were indeed the teeth of sharks, but he was able to discern the mechanism whereby the teeth had predated the formation of the surrounding rock. An inkling of the process of fossilization was born. One thing that greatly impeded interpretation of fossils back then is the fact that there was so little understanding of the natural world, of life in the seas beyond the shoreline, and especially of the tropical regions where so much of biological diversity resides. When we learn that seventeenth-century natural philosophers (the word *scientist* was a nineteenth-century invention) subscribed to the idea that earth was created in 4004 B.C., we moderns tend to snicker about the ignorance of our predecessors, but such mirth is entirely gratuitous. Once again we repeat our theme, that natural objects do not come with labels. It is not fair to use the knowledge we have at our disposal to judge investigators from other historical periods. The most mediocre scientist today has an understanding of the universe that vastly exceeds that of Isaac Newton, one of the great geniuses of the human race. Today we stand in the debt of generations of investigators who came before us. The concept of deep geological time was formulated at the end of the eighteenth century, particularly owing to the efforts of the Scottish savant James Hutton and his popularizer John Playfair. The stratigraphic system consisting of successive periods of geological time (Cambrian, Ordovician, Silurian, etc.) grouped into eras (Paleozoic, Mesozoic, Cenozoic) is an essential tool for our understanding of geological time and the age of fossils. This tool was developed during the nineteenth century, principally but not entirely before the publication of *The Origin of Species* in 1859. The stratigraphic system is not intrinsically an evolutionary system but was based simply on the observed change of faunas through time. Evolution is the explanation of the change, not the a priori assumption. In the twentieth century, radiometric dating techniques were developed, allowing the depth of geological time to be quantified. Thus knowledge of the meaning of fossils is hard-won, the result of a painstaking process of growth of awareness of nature. A correct understanding requires an understanding of the biology of the world, the age of the earth, and geological processes such as sedimentation and the formation of fossils. Knowledge of evolution is helpful but not essential.

Martin Rudwick made an extremely convincing case. I felt that I now had a grasp of the history of paleontology. Clearly an interpretation of fossils was the domain of the learned and the sophisticated. The ancients were unlikely to have achieved much understanding of fossils. Once again my history failed me. I forgot Alexander Pope's dictum, "A little learning is a dangerous thing." There was no room in my worldview for fossils in the ancient world. My complacency was quite shattered by Adrienne Mayor. Several years ago Mayor consulted me concerning a truly novel idea. She sought my advice because she knew of my experience with horned dinosaurs in general and with *Protoceratops*, a small horned dinosaur from Mongolia, in particular. She outlined a startling hypothesis about the identity of the griffin. Griffins are typically described as a race of four-footed birds having the beaks of eagles and the claws of lions, probably not flying but leaping in the air and digging in the ground, living in desert wilderness ferociously guarding hoards of gold. The legend of the griffin dates from at least 675 B.C., when the Greek adventurer and writer Aristeas met the nomadic Scythians near the foot of the Tien Shan (Heavenly) Mountains. For the next thousand years legends flourished of the gold of the Scythians and of the fierce warriors and creatures guarding it. Griffins became a highly popular subject of art and drama in the classical world, appearing, for example, in *Prometheus Bound* by Aeschylus (460 B.C.). No writer from classical time claims to have seen a live griffin.

Was the griffin a totally fictitious creature, or did it have a basis in fact? It is Mayor's truly arresting thesis that the source of the griffin legend was none other than Scythian gold-miners passing through the wastes of the Gobi Desert on their way to the Altai Mountains in search of the precious metal. The Gobi Desert is justly famous in paleontology as one of the richest sources of dinosaur fossils on the planet. Fossils from the Gobi caused great excitement in the 1920s when expeditions from the American Museum of Natural History brought back extraordinary fossil treasures to New York. Fossils from the Gobi Desert cause great excitement today as expeditions from the American Museum of Natural History continue to bring back extraordinary treasures, with each year's haul seeming to exceed the previous year's. At certain locations in the Gobi, dinosaur skeletons are abundant, often lying exposed on the surface of the ground. The most common dinosaur there is *Protoceratops*, a herbivore about six or eight feet long including its tail; many specimens of smaller size are also known. Its physical characteristics include a beak, an elevated bony frill at the back of the skull, and four prominently clawed feet. With the benefit of nearly 175 years of dinosaur study, today we know very clearly that *Protoceratops* is (1) a reptile and (2) a dinosaur. However, to an untrained observer the beak could easily have been seen as birdlike, and the elevated frill might have been seen as some sort of avian feature, possibly giving rise to the idea that this was a bird, albeit an abortive flier. Descriptions of the essential physical characteristics of the griffin evidently match *Protoceratops* very well indeed (pay no attention to the luxuriant artistic treatment on Greek vases and friezes). But this alone is insufficient to carry Mayor's point. Even if Scythian miners passed by, why would they notice bones in the ground? After all, a visit to the dinosaur hall of any great museum, including my own at the Academy of Natural Sciences in Philadelphia, reveals that fossil bones are usually dark colored and inconspicuous in the ground. Here I was able to provide Mayor with a valuable clue. The fossil bones of the Gobi are highly distinctive in their coloration. The bones of *Protoceratops* are creamy white; moreover, the sediments of the Djadochta Formation in which they are found are typically bright red. The color contrast is striking—the bones stand out and are unmistakable even to the casual passerby! Recollect that no one reported seeing a live griffin. In a certain sense, the Gobi was a land of danger and death. Travelers would have seen ample signs of death—carcasses of horses, camels, even bodies of their unfortunate riders, dead of thirst in the desert, and subsequently ripped apart by ubiquitous scavengers, all evidence to the imaginative mind of the ever-present danger of the desert and its frightful guardian, the griffin. In short, I find the case for the identity of the griffin both startling and convincing.

I have read *The First Fossil Hunters* with the greatest interest. As I child I greatly enjoyed Greek mythology (always in preference to its more derivative Roman counterpart). I might also mention that my father, a biologist, majored in ancient Greek in college. I devoured Edith Hamilton and Bulfinch and D'Aulaire. Mayor has caused scales to drop from my eyes. She has brought me to a fresh appreciation of the wellspring of the culture in which I live. The ancient Hellenic world was actually very rich in vertebrate fossils. No, these were not dinosaurs fossils for the most part, but rather the bones of impressive large mammals from the more recent geological past, primarily from the Miocene, Pliocene, and Pleistocene epochs: i.e., from about 25 million years ago to almost the present. The island of Samos off the coast of Turkey is particularly rich in such remains. It stands to reason that fossils would have been noticed by the ancient residents. How surprising can it be that these stones and bones were woven into tales, legends, and myths? Ours is a storytelling species. We delight in understanding our past and weaving available data into compelling narratives. So powerful is oral tradition that, as Jaroslav Pelikan tells us, a completely recognizable version of Homer's *Odyssey* is still recounted orally in the Balkans. No wonder that Greek myths tell us of heroes and giants of the past—bones of unfamiliar creatures rendered into familiar forms at unfamiliar sizes! A wonderful example of this are Pleistocene elephant tusks conjured into the image of the mythical Calydonian Boar.

It is an inexcusable conceit of the modern intellect to think that we alone have access to knowledge of the past. Paleontologists, classicists, and historians as well as natural history buffs will read this book with the greatest delight—surprises abound. I look forward to having this book on my bookshelf. I salute its author!

PETER DODSON
University of Pennsylvania

ACKNOWLEDGMENTS

As an independent researcher of classical legends about natural history, I explore the borders of ancient and modern knowledge, collecting unclassifiable passages in Greek and Latin texts, searching for meaningful patterns, and relating the results to modern science. My investigations would be impossible without the help of experts in a wide range of disciplines. Like the ancient Greek investigators Herodotus and Pausanias, I freely admit that over the years I have pestered many authorities in ancient history, classical literature, archaeology, geology, and paleontology with questions that may have seemed bizarre. In recovering the story of ancient Greek and Roman encounters with prehistoric fossils and synthesizing the ancient experiences with recent paleontological discoveries, I am deeply indebted to the generosity and enthusiasm of many individuals in the humanities and the sciences.

Lowell Edmunds, Richard Greenwell, and Dale Russell gave early and sustaining encouragement. From the beginning, Jack Repcheck's zeal was an inspiration. This project would not have been possible without the expertise of Eric Buffetaut, David Reese, and Nikos Solounias. I'm especially conscious of valuable criticism from Paul Cartledge, Peter Dodson, William Hansen, Geoffrey Lloyd, Michelle Maskiell, Barry Strauss, and Norton Wise who read chapters in draft. The following people went out of their way to assist in various ways: Carla Antonaccio, Filippo Barattolo, John Barry, John Boardman, Phil Curry, Kris Ellingsen, Sue Frary, Neil Greenberg, Arthur A. Harris, Ed Heck, Jenny Herdman, Jack Horner, George Huxley, Brad Inwood, Christine Janis, Sheldon Judson, Robert Kaster, George Koufos, Helmut Kyrieleis, Kenneth Lapatin, Adrian Lister, Beauvais Lyons, John Oakley, John Ostrom,

Ana Pinto, Rip Rapp, Bonnie Robertson, William Sanders, Alessandro Schiesaro, Gérard Seiterle, Sevket Sen, Janet Stern, Angela P. Thomas, Dorothy Thompson, Evangelia Tsoukala, David Weishampel, William Willers, and Elaine Zampini. Thanks also go to the many members of the international Vertebrate Paleontology and Aegeanet Internet discussion groups who replied to specific queries.

I am grateful to my sister Michele Mayor Angel for creating the maps. I thank my editor, Kristin Gager, for guiding the book through production, Debbie Felton for proofreading, and Lauren Lepow for her nimble copyediting.

I have appreciated the invitations to present my research on paleontological legends to the International Society for Cryptozoology/Folklore Society joint meeting, the Princeton Program in the Ancient World, the Biology-Classics group at Villanova University, and the American Museum of Natural History.

I am lucky to have Josiah Ober as my companion of ideas.

THE FIRST FOSSIL HUNTERS

GEOLOGICAL TIME SCALE

Era	*Period*	*Epoch*	*Years Ago*
		Holocene	Recent
	Quaternary	Pleistocene	1.7 million to 10,000
		Pliocene	5–1.7 million
Cenozoic		Miocene	23–5 million
	Tertiary	Oligocene	34–23 million
		Eocene	55–34 million
		Paleocene	65–55 million
	Cretaceous		145–65 million
Mesozoic	Jurassic		215–145 million
	Triassic		250–215 million
Paleozoic to Precambrian			250–4,500 million

INTRODUCTION

THE CLASSICAL Greek landscape evokes many images—heroes and Amazons, gods and goddesses, painted vases and bronze statues, marble columns and temple ruins. The enormous fossil bones of mastodons and mammoths are not likely to appear in anyone's mental picture of classical antiquity. But immense skeletons of creatures from past eons indeed lie buried all around the lands known to the Greeks and Romans. And for the ancient Greeks and Romans themselves, vestiges of giants and monsters of the distant past were important features of their natural and cultural landscape. This book explores the relationship between two simple but surprising facts: the Mediterranean world was once populated by giant creatures, and the ancients were continually confronted by their remarkable petrified remains.

The ancients collected, measured, displayed, and pondered the bones of extinct beasts, and they recorded their discoveries and imaginative interpretations of the fossil remains in numerous writings that survive today. Yet "paleontology" is missing in the standard lists of the great cultural inventions of the Greeks and Romans. How did modern science and history come to lose the significant paleontological discoveries, thoughts, and activities of classical antiquity? That paradox inspired my project, to recover the long-neglected evidence of human encounters with fossils from the time of Homer to the late Roman empire (ca. 750 B.C. to A.D. 500).

The history of the ancient engagement with fossils has languished in shadow for several reasons. In the first place, few people are aware that millions of years ago, huge mammals of the Miocene, Pliocene, and Pleistocene eras roamed what would become the lands of the Greeks and Romans. Those who do study Mediterranean vertebrate paleontology are not acquainted with the detailed descriptions of immense bones and teeth that figure in vivid ancient writings about giants and monsters. Most classical scholars have no idea that huge remains of prehistoric mastodons and mammoths, woolly rhinoceroses, giant giraffes, cave bears, and saber-toothed tigers continually erode out of the earth in Mediterranean lands. Fewer still realize that those fossil exposures exist precisely where ancient Greek myths located the destruction of giants or monsters, and where the ancients claimed to have observed gigantic bones. So it's not surprising that classicists tend to read the ancient allusions to the bones of giants or monsters as mere poetic fantasies or as evidence of popular superstition. I will argue that these allusions are evidence for a native natural history of the prehistoric Mediterranean. Tracing that natural history will lead us down some rarely traveled pathways of classical studies, since many of the best-known ancient thinkers, such as Thucydides and Aristotle, failed to mention remarkable remains.

We can recover the lost fossil knowledge of antiquity only by a new reading of the neglected classical material about extraordinary skeletons in the light of little-publicized modern fossil discoveries in the lands once inhabited by the ancient Greeks and Romans. That means bridging the gulf in communications between modern humanities and the sciences, in order to restore a vital missing chapter in the early history of paleontology.[1]

The typical history of paleontology begins with praise for the ancient Greek philosophers who recognized that small fossil shells found far from the seashore represented evidence of former oceans. Next, historians of paleontology remark on the lack of evidence for discoveries of large vertebrate fossils in antiquity. Here, some modern historians do mention one notable insight about animal fossils attributed (erroneously, it turns out) to a Greek philosopher named Empedocles. To explain the supposed disinterest in large prehistoric bones in the classical era, scientists assume that Aristotle's notion of unchanging species was a dogmatic principle that suppressed paleontological speculation in antiquity, just as it did in the Middle Ages. A few amusing medieval fossil misunderstandings about toadstones and unicorns come next, followed by a nod to Renaissance thinkers. We quickly arrive at the official starting point of paleontological history, the scientific discoveries of the eighteenth and nineteenth centuries by Georges Cuvier, Richard Owen, and Charles Darwin.

Four errors about ancient experiences with fossils recur in paleontological histories. First, for the reasons outlined above, it's taken for granted that even though the early Greeks grasped the meaning of tiny marine fossils, somehow the ancients never noticed the huge fossil remains of dinosaurs, mammoths, and other extinct vertebrate species. To account for this mysterious lapse, historians of paleontology speculate that large, mineralized bones were not recognized as bones, and some even suggest that they were "just too big to be noticed."[2] The extensive ancient evidence, gathered together for the first time here, proves how wrong such assumptions are.

Big vertebrate fossils, even isolated bones and teeth, were objects of intense curiosity and speculation in Greco-Roman times. This fact was well known to the founder of modern paleontology, Georges Cuvier (1769–1832), the French naturalist who first proposed that mammoth bones belonged to extinct elephants. Cuvier was a scientist, but his eighteenth-century classical education meant that he was also familiar with Greek and Latin literature. In his monograph on living and extinct elephants published in Paris in 1806, Cuvier summarized the history of fossil mammoth discoveries around the world up to his time. He traced the earliest finds to classical antiquity, citing several ancient accounts of giant skeletons and tusks that came to light in Greece, Italy, Crete, Asia Minor, and North Africa between the fifth century B.C. and the fifth century A.D. In a sense, then, I am pursuing a historical path first broached by Cuvier, but subsequently forgotten amid the exciting scientific discoveries of his day. It's daunting to embark on a topic initiated by the great Cuvier, but advances in classical studies and paleontological science—and the new information made available to me by experts in both fields—now make it possible to restore the ancient fossil investigations to their rightful place in the history of science.[3]

The recovery of this long-forgotten evidence contradicts another institutional myth of modern paleontology, that no serious consideration of vertebrate fossils could occur in classical antiquity because the scientific theories of evolution and extinction had not yet been invented. Meaningful interpretation of fossils as organic remains of the past requires an understanding of natural history that the ancients could not have possessed—that is the assumption evident, for example, in Martin J. S. Rudwick's influential book *The Meaning of Fossils: Episodes in the History of Paleontology.*

It's time to rethink those assumptions. It may be true that no natural philosopher—not even Aristotle—articulated a formal theory to explain vertebrate fossils, and that famous writers like Plato and Thucydides made no mention of giant bones. But that should not mislead us into thinking that the ancient Greeks and Romans had no concepts or paradigms to explain the observed phenomena of unexpectedly large, petrified bones that matched no living creatures. In fact, the earliest recorded paleontological speculations are preserved where no one since Cuvier has thought to look: in the Greco-Roman myths of nature's past and scattered throughout the lesser-known writings of geographers, travelers, ethnographers, natural historians, and compilers of natural wonders. Reading about the fossil discoveries described in these less-polished texts humanizes the ordinary Greeks and Romans as never before and gives a new immediacy to ancient life. It also reveals the wealth of natural knowledge that lies hidden in the rarely studied popular literature of antiquity.[4]

The third error concerns a "fact" often asserted about large vertebrate fossils in antiquity. Since the early twentieth century, numerous reliable international historians of paleontology have perpetuated a myth: that the Greek philosopher Empedocles studied fossil skulls of elephants in caves in Sicily. This modern myth holds that Empedocles, writing in the fifth century B.C., was the first to relate prehistoric elephant skulls to the ancient legend of the Cyclops, the one-eyed giant killed in a cave by Odysseus in Homer's *Odyssey.* It is sometimes further claimed that Giovanni Boccaccio was the first to publicize Empedocles' finds. Here is a typical recent example of this modern myth: "When fossil elephant bones turned up [in Sicily] in the 5th century B.C., Empedocles interpreted them as cyclops-bones. . . . In the 14th century A.D., Boccaccio repeated the identification, citing Empedocles."[5]

It is true that Boccaccio was present when peasants discovered a giant skeleton in a cave in Sicily in about 1371. He was among the crowd that gathered, daring one another to touch the giant. When someone finally poked it, the skeleton instantly crumbled into dust, leaving only three huge teeth, parts of the skull, and a vast thigh bone. Boccaccio identified the giant as the Cyclops from the *Odyssey*. But Boccaccio never mentioned Empedocles in his account. And in the surviving fragments of Empedocles' writings, the philosopher never referred to skulls, caves, giants, or the Cyclops—much less to elephants, which were unknown to Greeks until a hundred years after his death.[6]

So how did this fake fact arise? With the help of other historians who had also been duped by the myth, I traced its origin to statements made by the eminent Austrian paleontologist Othenio Abel in 1914 and 1939. What led Abel to make his unfounded claim about Empedocles? Writing about fossil folklore in 1914, Abel hit on the idea that ancient sailors mistook the large nasal opening in unfamiliar fossil elephant skulls for the eye socket of a one-eyed giant. To support his own ingenious speculation, Abel attributed the idea back in time to Empedocles, an ancient philosopher who pondered the origins of life. With no basis in the surviving record, Abel declared that "Empedocles reported such finds in Sicilian caves and believed these to be unassailable proof of the existence of an extinct race of giants." In the 1940s, Willy Ley, one of the first historians of paleontology to repeat Abel's Empedocles myth, added the false claim that Boccaccio had cited Empedocles as his authority when he announced the discovery of the Cyclops. In the manner of folk legends, Abel's and Ley's plausible-sounding assertions were taken up and elaborated by successive writers who never bothered to check what Empedocles and Boccaccio had really said.[7]

Just as Boccaccio's giant crumbled to dust when touched, this generally accepted "fact" about ancient paleontology collapses as soon as it is tested against the evidence. It would be easy to see

Abel's misinformation about Empedocles as a deliberate hoax, but I think Abel elaborated on Empedocles' genuine insights about primeval life-forms out of a desire to fill in that disturbing blank in the ancient record mentioned above, namely, the absence of any surviving philosophical theories to explain big fossil bones. Similar impulses have motivated other paleontological fictions both ancient and modern, as we will see in the final chapter.

The fourth error is an unexamined commonplace among paleontologists. It's often suggested that Aristotle's "fixity of species" idea was a deathblow to rational speculation about evolution and extinction in classical antiquity and the Middle Ages. This misleading view unfairly conflates two very different cultures and eras. The notion of immutable species created in one fell swoop was not a monolithic principle in classical antiquity—it only became so in the Middle Ages when Aristotelian thought was merged with biblical dogma in Europe. Instead, for a millennium *before* the Middle Ages, the Greeks and Romans identified large prehistoric remains as vestiges of gigantic, unfamiliar creatures that had appeared over time, reproduced, and transmuted, and then were destroyed by catastrophe or died out long before current human beings appeared on the earth.[8]

The adventurous Greeks and Romans observed prehistoric fossils all around the Mediterranean and as far afield as India. The nature of the ancient sources and the wide-ranging topography of ancient and modern fossil discoveries mean that we will need to jump around a bit in geography and chronology. The word "ancient" has different meanings in different contexts: we are asking how people who lived *thousands* of years ago interpreted the remnants of creatures that lived *millions* of years ago. Timetables and maps will provide us with guideposts as we undertake the temporal and spatial travel required by the synthesis of ancient discoveries with modern paleontological knowledge. The literary and archaeological evidence ranges from the eighth century B.C. to the fifth century A.D. (see Historical Time Line); the geological ages are shown in the Geological Time Scale.

The first chapter traces the paleontological origins of one exotic creature of ancient oral folklore, the griffin. By following clues fortuitously preserved by literate Greeks and Romans, I propose that the griffin image was based on illiterate nomads' observations of dinosaur skeletons in the deserts of Central Asia. But how did the Greeks and Romans interpret the very different kinds of prehistoric fossils they saw firsthand in their own lands? What species once roamed the prehistoric Mediterranean basin? Who really owns important fossil relics, and who should interpret them? These questions emerge in chapter 2, which surveys the violent geological history of the Mediterranean lands and the modern discoveries of fossil deposits there.

Armed with a working knowledge of the various species of mastodons, mammoths, and other large, extinct mammals whose remains are now known to exist around the classical world (listed by region in appendix 1), we turn in chapter 3 to the narratives describing the exciting discoveries of colossal skeletons from the time of the Trojan War (ca. 1250 B.C.) to the end of the Roman Empire (about A.D. 500). Many sensational firsts in the history of paleontology are registered here: the earliest recorded measurements of prehistoric fossil skeletons, the first descriptions of the natural and human conditions that exposed fossils, the first paleontological museum, the earliest recognition of Miocene mastodons as elephants, the first reconstruction of a prehistoric creature from its remains, the oldest illustration of a fossil discovery, and the earliest-ever descriptions of fossil deposits in Greece and the Aegean islands, Italy, France, North Africa, Egypt, Turkey, the Black Sea, and India. (Appendix 2 gathers a wide sampling of literary evidence to supplement the passages cited in chapter 3.) The bones of gigantic beings were treasured as relics of the mythic past and displayed as natural wonders in temples and other public places. Close readings of the ancient accounts and consultations with modern paleontologists familiar with Mediterranean and Eurasian fossils allow us to determine the true identity of the giant skeletons unearthed in antiquity.

How did classical artists visualize the giants and monsters whose bones were buried in the earth? Have any fossil bone relics collected in antiquity been excavated from ancient sites? Chapter 4 uncovers the little-known artistic and archaeological evidence for the ancient interest in fossils.

Chapter 5 delves into Greco-Roman myth to find the concepts

that helped ordinary people interpret the mysterious, enormous remains that emerged from the earth. Popular lore contained scenarios of life-forms changing over time and their destruction in the deep past. An important theme of this book, the tension between official scientific certainty and popular belief, comes to the fore when we plumb the strange silence surrounding giant bones within the established circle of ancient natural philosophers, including Aristotle. The chapter concludes with a summary of the impressive insights of classical paleontology.

In chapter 6, we learn that paleontological hoaxes originated in the Roman era as a response to the tension between popular and scientific beliefs. Today the gulf between popular superstition and scientific knowledge seems unbridgeable, and scientists often bemoan their failure to communicate their own excitement in the enterprise of separating truth from specious fiction. Some scientific skeptics conclude that it "insults" both myth and science to try to reconcile such different worldviews.[9] Yet when we compare ancient paleontological "fictions" to some startlingly similar modern examples, it becomes clear that creative scientific curiosity and the exercise of the mythic imagination are more closely related than one might suppose.

"What are fossils, after all," asks French paleontologist Pascal Tassy, "if not vestiges both destroyed and preserved by time?" The same definition applies to the fragmentary ancient literary and archaeological evidence. Just as a fossil is "petrified time," so is an ancient artifact or text. The tasks of paleontologists and classical historians and archaeologists are remarkably similar—to excavate, decipher, and bring to life the tantalizing remnants of a time we will never see.[10] This book represents the first attempt to integrate those efforts. I hope the results will encourage further investigations into the earliest stirrings of paleontological inquiry.

Historical Time Line

ca. 70,000 years ago	Humans appear in Greece
2300–1100 B.C.	Bronze Age Greece
1300–1200 B.C.	Tons of fossils enshrined at Qau, Egypt
1250 B.C.	Trojan War Pelops's giant bone lost at sea
800 B.C.	Homer, ca. 750 Hesiod, ca. 700 Hero cult in Greece, 700–500 Monster of Troy myth first written down
700 B.C.	Aristeas travels to Scythia, ca. 675 Griffins become popular motif in Greece Large fossil femur dedicated at Samos
600 B.C.	Anaximander, d. 547 Xenophanes, ca. 560 Sparta finds Orestes' giant bones, 560 Monster of Troy vase painted in Corinth
500 B.C.	Empedocles, ca. 492–432 Pindar Kimon discovers giant bones of Theseus Aeschylus, ca. 460 Euagon records Neades legend of Samos Scythians tattoo themselves with griffins Herodotus, ca. 430 Peloponnesian War, 431–404 Plato, 429–347
400 B.C.	Aristotle, 384–322 Theophrastus, 372–287 Palaephatus Alexander the Great, 356–323 Greeks first learn about elephants
300 B.C.	
200 B.C.	Euphorion says giant bones shown in Samos

100 B.C.	Anteaus's giant bones seen in Morocco Giant bones exposed in Crete Virgil, b. 70 Strabo, b. 64 Lucretius Scaurus displays Joppa monster in Rome Diodorus of Sicily, ca. 30 Manilius, ca. 10 Calydonian Boar tusks looted from Tegea Roman Empire, 31 B.C.–ca. A.D. 450
A.D. 1	Ovid, 43 B.C.–A.D. 17 Augustus's reign, 31 B.C.–A.D. 14 Augustus's museum of giant bones in Capri Tiberius's reign, A.D. 14–37 Model of giant constructed from tooth Sea monster remains reported in Gaul Orontes' giant bones discovered in Syria Josephus, b. 37 Huge bones displayed in Palestine Pliny the Elder, 23–79 Claudius's reign, 41–54 Centaur remains displayed in Rome Skulls of dragons displayed in India Apollonius of Tyana travels to India
A.D. 100	Plutarch, ca. 100 Fossil mastodons of Samos recognized as elephant remains Hadrian's reign, 117–138 Suetonius Ajax's giant bones discovered at Rhoeteum Phlegon of Tralles, ca. 130 Giant bones displayed in Nitria, Egypt Bones of giants Hyllos and Asterios in Anatolia Pausanias, ca. 150 Achilles' giant bones seen at Sigeum Huge skull measured on Lemnos Lucian, ca. 180
A.D. 200	Solinus describes giant bones in Pallene and Crete Philostratus, ca. 230 Aelian, d. ca. 230

A.D. 300	Quintus Smyrnaeus
A.D. 400	Augustine, 354–430, finds huge tooth at Utica Claudian, ca. 370–425
A.D. 500	Giant bones displayed in Constantinople Procopius, ca. 540, views mammoth tusks identified as Calydonian Boar's

CHAPTER

1

The Gold-Guarding Griffin: A Paleontological Legend

I BOARDED the overnight ferry from Athens to Samos, a Greek island just off the coast of Turkey, in the late summer. My destination was a small museum in Mytilini, a village in the mountainous interior. Intrigued by a note in an old guidebook, I hoped to see a collection of colossal bones that had been found north of the village, dug out of a dry streambed in a place known to locals as the "Elephants' Cemetery." The guidebook mentioned that significant prehistoric fossils had been stored in a room above the village post office since the 1880s. One of the skeletons was named *Samotherium*, the "Monster of Samos."

The gigantic fossil bones of Samos had been pointed out to curious travelers in classical antiquity, according to ancient Greek authors. I was drawn to coincidences linking modern paleontological discoveries and ancient stories of extraordinary bones, because I was interested in the way legends of fantastic creatures can arise from observations of the remains of unfamiliar extinct animals. I was keen to see whether any of the Samos fossils might resemble the features of the griffin of classical legend. Tracing the identity of those mysterious gold-guarding creatures with bodies like lions' and beaks like eagles' had become an obsession for me that year in Athens. I had already learned that since the seventeenth century, classical scholars, ancient historians, art experts, historians of science, archaeologists, and zoologists had all insisted the griffin was simply an imaginary composite of a lion and an eagle, a symbol created to represent vigilance, greed, or the difficulties of mining gold. I suspected otherwise.[1]

To me, the griffin seemed a prime candidate for a paleontological legend. This animal was no simple composite; it didn't seem to belong with the obviously imaginary hybrids of Greek tradition like Pegasus (a horse with wings), the Sphinx (a winged lion with a woman's head), the Minotaur (a man with a bull's head), and the half-man, half-horse Centaurs. Indeed, the griffin played no role in Greek mythology. It was a creature of folklore grounded in naturalistic details.[2]

Unlike the other monsters who dwelled in the mythical past, the griffin was not the offspring of gods and was not associated with the adventures of Greek gods or heroes. Instead, griffins were generic animals believed to exist in the present day; they were encountered by ordinary people who prospected for gold in distant Asia. Modern historians have judged the ancient Greek and Roman authors who spoke of griffins as either gullible fools or perpetrators of fantasy. But I noticed that the writers who described the griffin avoided sensational language. Griffins were simply said to roam in pairs or packs, nesting on the ground, defending gold from intruders, and preying on horses, stags, and perhaps even humans. They had no supernatural powers. The most striking thing about griffins remained consistent over many centuries: this animal went about on four legs but also had a powerful beak. That odd combination of bird and mammalian features was what I hoped to find in the fossil *Samotherium* skeleton (fig. 1.1).

When our ferry arrived in Samos harbor that morning, I noticed a modest museum of the island's archaeological finds, tucked behind the public garden. I decided to visit it before heading for Mytilini. Inside the museum, I was thrilled to find a trove of bronze griffins, hundreds of them, all retrieved by German and Greek archaeologists from the ruins of the sanctuary of the goddess Hera, the Heraion. The work of Ionian craftsmen (in western Anatolia), most of the statuettes were busts of griffins, made to decorate the rims of big bronze bowls. The bronzes had been ded-

1.1. Griffin and nomad. Red-figure cup, late sixth-early fifth century B.C. Drawing by author, after R. A. Valotaire, "Vases peints du Cabinet Turpin de Crissé," *Revue Archéologique* 17 (1923): 51.

icated to the Heraion between the eighth and sixth centuries B.C.—about the time that the first written accounts of griffins appeared in Greece. I spent all morning sketching rank upon rank of griffins, trying to capture with my pen the strong, hooked beaks; the enormous, staring eyes; the menacing, predatory intent. I noticed that some griffins' necks were either feathered or scaled like lizards', with peculiar folds or ruffs; many had distinctive long "ears" or "horns" and forehead knobs, and most had stiff, stylized wings.

As I drew the individual griffins, all the while anticipating what I might find in that bone room over the post office in Mytilini, I noticed that the bronzes fell into two general types. I found myself hurrying past the elegantly streamlined griffins whose long, graceful necks were decorated with swirls. I lingered before the bulky, sturdy, leathery-looking beasts, mostly the products of artists of earlier dates. The contrast seemed to be that of idealizing portraiture versus homely taxidermy. These heavier, short-necked, earthbound brutes were not as aesthically pleasing as the classical griffins so beloved by art historians. But I was struck by their vitality and by the convincing realism of the gnarly details. Something about these early models was so lifelike, so real. . . . As I stared at their powerful beaks, empty eye sockets, leathery necks, and bumpy skulls, I was struck by a sense of *déjà vu*—they seemed so *prehistoric*—

I rushed outside. I had to rent a motorbike at the quay immediately, reach that mountain village, and then find someone with the key to the bone room! It was a hot, drowsy island afternoon. I passed no traffic on the steep dirt track out of town except for some listless goats in the dust and a long, low-riding, maroon Dodge of the long-finned era. Foot-long black-and-white beaded lizards basked on great stone blocks; beyond them lay the gullies where immense petrified skeletons have been emerging over millennia.

The road twists up the switchbacks. Finally in Mytilini, the plane trees cast pools of late shade. At the tiny post office I make my request. The postmaster hurries to find the mayor, who brings the old key to unlock the upstairs room, the paleontological museum of Samos. Sunlight slanting through dusty glass reveals a jumble of great fossil skulls, vertebrae, femurs. The old postmaster shows me yellowed newsclippings of sweaty workers in undershirts and men in suits posing next to the big bones poking out of ravines. He locates the femur and skull of the *Samotherium*, awesome and stony in glass cases. The samothere's thigh bone is twice the size of a human femur. The impressive skull is about two feet long, with two bony horns, large eye sockets, and big teeth, but, alas, no beak (figs. 1.2, 1.3).

It was a wondrous thing to imagine this giant ancestor of the giraffe alive and grazing seven million years ago where goats now browse. These formidable fossils had indeed amazed the ancient farmers of Samos, but that was another story. I realized that the inspiration for the ancient griffin legend must lie in more distant lands.[3] I returned to Athens to undertake more research. There under the slow fans of the Library of the American School of Classical Studies, the sound of doves and cicadas monotonous in the fig trees outside, I delved deeper into ancient griffin lore.

1.2. *Samotherium* skull, about 2 feet long, excavated in Samos, Greece, 1923–24. Photo courtesy of Nikos Solounias.

One paleontological legend collected in Siberia in 1827 by a German geologist seemed at first glance to reveal the identity of the ancient griffin. Georg Adolph Erman (1806–1877), whose surveys of Siberia won a Royal Geographical Society medal, also gathered ethnographical material. Erman believed that the Siberian customs, language, and oral history corresponded to fifth-century B.C. Greek descriptions of the ancient Scythian nomads, the people who had first told the Greeks about griffins. Like those ancient nomads, the natives of the northern Urals mined gold sand. They told Erman that they often came across the remains of enormous "bird-monsters," which had been slaughtered in great numbers by their brave ancestors.

Erman identified these monsters as the remains of Ice Age rhinoceroses and mammoths, which were embedded in the layers of peat overlying the gold-bearing sand along rivers flowing into the Arctic Ocean. The Siberians called the rhinoceros horns "birds'

1.3. Author with display of fossil limb bones in the old Paleontological Museum, Samos, Greece, 1988. Photo by Josiah Ober.

claws," even though they readily admitted that they knew the beasts were not really birds. It is "just our custom to call them that," they told Erman. "We know all about rhinoceroses."[4]

The Siberians' use of a conventional name for the fossils parallels the way the Chinese referred to all fossils of extinct animals as "dragon bones" even though they recognized that some of the remains belonged to deer and horses. Consider our name for *Dinosauria* ("terrible lizards"). Even though we now know that the name conjures up an outdated vision of all dinosaurs as "cold-blooded, sluggish, [and] dim-witted," we go on calling them dinosaurs. In discussing the history of the word *dinosaur*, Montana paleontologist Jack Horner points out that the "saurian" misnomer perpetuated "unexamined assumptions" about the image and behavior of dinosaurs. That name shaped the "search image" of paleontologists, leading them to "overlook, misinterpret, or dismiss" any evidence that contradicted their image of dinosaurs as cold-blooded lizards. Horner's concept of the "search image"—the conscious or unconscious scanning of evidence to concentrate on specific patterns or constellations of features—will be a handy one in this book. Narrow search images can help refine scientific research, but they can also lead investigators to ignore potentially significant information.[5] As we will see, search images influenced the ancient Greeks and Romans too. It was their custom to refer to oversize fossil skeletons as the bones of "heroes" or "giants."

In 1848, Erman declared that he had discovered the "prototype of the Greek story of the Grifons" in the Siberian gold-miners' legend of bird-monsters. Erman's theory fell into obscurity, revived in 1962 only to be rejected by classicist J.D.P. Bolton. Erman's Siberian informants lived above the Arctic Circle, noted Bolton, some 2,000 miles (3,200 km) northwest of the origin of the griffin tale in the Altai Mountains of Central Asia.[6] Besides, the rhinoceros and mammoth skulls had no beaks, the essential hallmark of griffins. Like my own early hopes pinned on the *Samotherium*, the Monster of Samos, Erman's identification of the griffin focused on the wrong fossils in the wrong place.

But Erman's notion that the griffins were based on observations of prehistoric remains was on the right track. And Bolton was correct to stress the geographical origin of the original tales of griffins. All the descriptions between 700 B.C. and A.D. 400 pointed to a specific homeland for griffins: the desolate wastes of Central Asia, where nomads called Scythians (now known as Saka-Scythians) had prospected for gold in antiquity. I found that this was a desert region so obscure that the designers of atlases typically stitch page bindings directly over that very latitude and longitude, obliterating the map's topography as surely as any sandstorm.

The following pages piece together the puzzle of the Scythian griffin. I argue that the griffin is the earliest documented attempt to visualize a prehistoric animal from its fossil remains. But it is important to keep in mind that the griffin story is known to us only because it was taken up by literate Greeks and Romans. Their preservation of scraps of oral folklore will allow us to solve a dramatic 2,700-year-old paleontological enigma. Then, in the chapters to come, we'll see how the Greeks and Romans struggled to interpret for themselves the extraordinary fossil skeletons that emerged in their own lands.

The Griffin Legend

Nearly 3,000 years ago, Saka-Scythian nomads prospected for gold in the western reaches of the Gobi Desert between the Tien Shan and Altai Mountains. Saka-Scythian culture stretched across the steppes from the Black Sea to the Altai; their nomadic lifestyle flourished between about 800 B.C. and A.D. 300. The Scythians were not a literate culture—they had fine art, but their stories were preserved only by accident, when a literate culture became interested in them.

Sometime in the seventh century B.C., Greeks first made contact with Scythian nomads. Along with gold and other exotic goods from the east came folklore about the remote land and its inhabitants. One such tale, about gold-guarding griffins, first appeared in writing in an epic poem about Scythia by a traveler named Aristeas, a Greek from an island in the Sea of Marmara (southwest of the Black Sea). Aristeas visited the easternmost tribe of Scythian nomads, the Issedonians, at the base of the Altai Mountains in about 675 B.C. These people told him about the vast wilderness beyond Issedonia, where gold was defended by fierce "griffins." (The word *griffin* comes from the ancient Greek word *gryps*, which means "hooked," as in a beak; *gryps* is related to the ancient Persian verb *giriften*, which means "to grip or seize.") Aristeas recorded that gold-seeking nomads on horseback battled griffins; the Issedonians described griffins as lion-size predators with strong, wickedly curved beaks like those of eagles.[7]

Scythia was an important source of gold in antiquity, and archaeologists have excavated spectacular gold treasures from Saka-Scythian tombs across southern Russia. Using a distinctive decoration known to art historians as "animal-style," the nomads covered their goods with exuberant zoological designs. Gold and bronze artifacts teem with realistically detailed rams, deer, stags, horses, asses, and eagles. Among these real animals are intertwined unknown creatures, especially griffin-type animals.

I was particularly fascinated by the discoveries of the Soviet archaeologist Sergei Rudenko. In the 1940s, Rudenko excavated several fifth-century B.C. tombs near Pazyryk on the northern slopes of the Altai Mountains, in the old Issedonian territory visited by Aristeas. Besides many gold artifacts featuring griffins, Rudenko was astonished to find some mummified nomads preserved in the permafrost for 2,500 years. The skin of one of the male warriors was covered with dark-blue animal tattoos. In the welter of recognizable beasts inscribed on his body were several unknown creatures, including griffins (fig. 1.4). Decades later, in 1993 and 1995, Russian archaeologists unearthed two more tattooed mummies of the same era near Pazyryk, a young man with a large elk figure on his shoulder and a young woman with a flamboyantly antlered deer and a griffin-like creature on her shoulder and wrist.[8] These tattooed images closely matched the earliest literary records and the bronze Greek griffins from Samos. The conclusion was clear: These nomadic people had known the griffin lore collected by Aristeas!

Strange creatures combining the features of birds and mammals had appeared in Near Eastern art as early as 3000 B.C., and peacock-

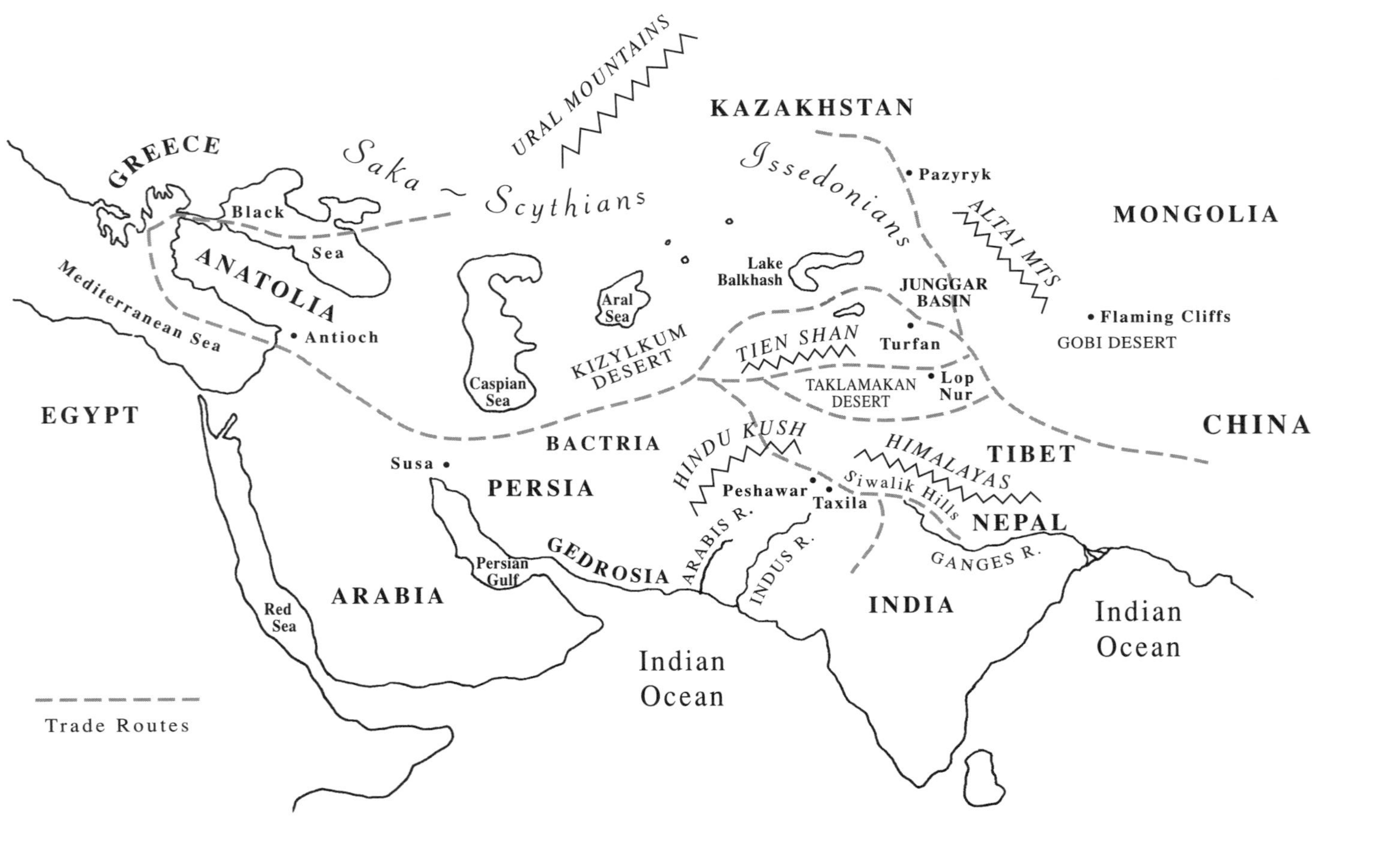

Map 1.1. Trade routes between the Mediterranean and Asia. Map by Michele Mayor Angel.

1.4. Tattoos of unknown animal (*top*) and griffin (*bottom*) on a mummified Saka-Scythian nomad discovered by Sergei Rudenko in a fifth-century B.C. tomb at Pazyryk, northern slopes of the Altai Mountains. Drawing by author.

headed griffins can be seen in Mycenaean art of the Greek Bronze Age (ca. 1200 B.C.). As a student of ancient folklore, I regretted that there was no way to know what kind of stories, if any, corresponded to those images. But with Aristeas and the Greek and Roman writers who followed him, I could relate the snippets of folk knowledge about the griffins of Issedonia to the artistic images of the same era. A careful reading of the literary traditions that began with the fragments of Aristeas in the seventh century B.C. and thrived until the third century A.D., and a close look at the artistic representations of the same thousand-year period, might allow one to uncover the real identity of the mysterious griffin.

There may well have been tales about griffins told in Mesopotamia and Greece before Aristeas, but his expedition and epic poem about Scythia seem to coincide with an explosion of interest in griffin art—and the interest in griffins continued solidly through the next thousand years. As exciting details about Scythia began to circulate in Mediterranean lands, griffins became favorite motifs of Greek and then Roman artists. On vases, artisans painted exotic scenes of nomads fighting griffins (fig. 1.1), and in mosaics and sculptures ferocious griffins attacked stags and horses. In the seventh and sixth centuries B.C., lifelike bronze griffin busts with large, staring eyes and threatening, open beaks became wildly popular, especially as decorations on the mammoth bronze bowls the Greeks used for mixing wine.

The griffin was much more than a static decorative motif; it was imagined and depicted as a real animal with recognizable behavioral traits. When Judith Binder, a lifelong scholar at the American School of Classical Studies at Athens, heard about my quest, she urged me to go see the griffin "pup" in the museum at Olympia (the ancient home of the Olympic Games, in the Peloponnese). I sought out that unique bronze *metope*, a decorated plaque created for the Temple of Zeus in about 630 B.C. The artist had portrayed a fierce mother griffin with a baby griffin nestled under her ribs (fig. 1.5). Sarah Morris, now a professor of classical archaeology at UCLA, showed me another griffin family scene on a Mycenaean vase of about 1150 B.C., painted well before the first known written accounts of the griffin. In that vignette, a griffin pair tends two nestlings. What inspired such naturalistic images of griffin life? The imagery of griffins did not follow any standard mythological narratives—instead, the artists seemed to be imagining the behavior of an unusual animal they had heard described but had never seen.

The territory of the Issedonian Scythians where Aristeas learned about the griffin in about 675 B.C. is a wedge bounded by the Tien Shan and Altai ranges, in an area that straddles present-day northwestern Mongolia, northwestern China, southern Siberia, and

1.5. Griffin and baby. Hammered bronze relief, ca. 630 B.C., found at Olympia, Greece, Olympia Archaeological Museum no. B 104. Drawing by author.

southeastern Kazakhstan. The second-century A.D. Alexandrian geographer Ptolemy and ancient Chinese sources agree in locating the Issedonians along the old trade routes from China to the West, from the western Gobi Desert to the Dzungarian (or Junggarian) Gate, the mountain pass between modern Kazakhstan and northwestern China. Recent linguistic and archaeological studies confirm that Greek and Roman trade with Saka-Scythian nomads flourished in that region from Aristeas's day to about A.D. 300—exactly the period during which griffins were most prominently featured in Greco-Roman art and literature.

The nomads left no written records. Even Aristeas's poem itself

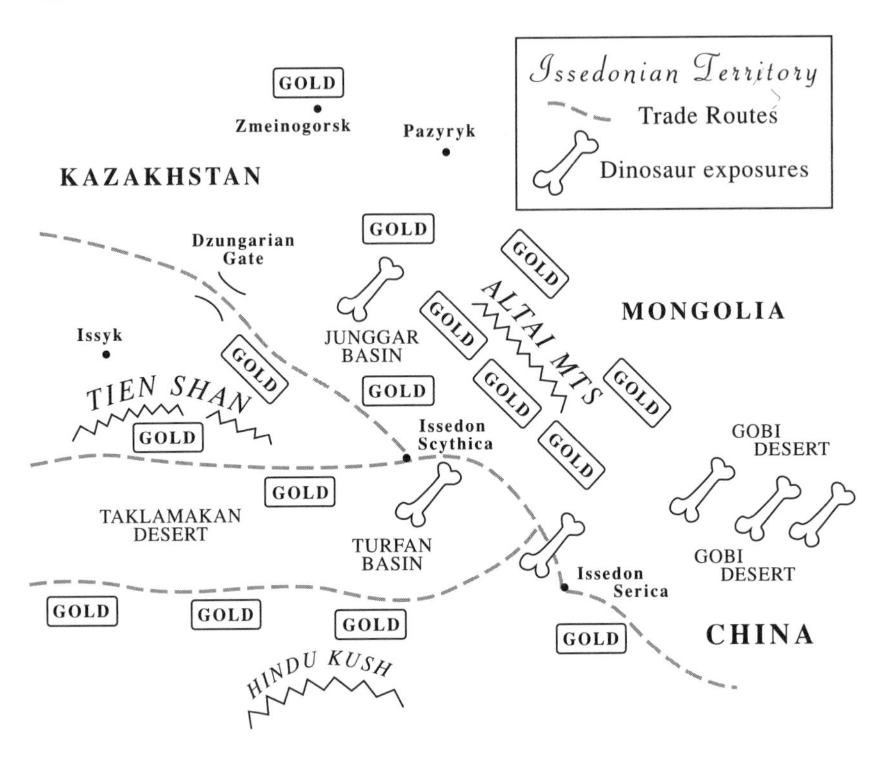

Map 1.2. Issedonian territory: trade routes, gold deposits, and dinosaur fossil exposures. Map by Michele Mayor Angel.

no longer exists, but his epic was so famous in antiquity that quotations from it are preserved in works by several ancient authors. Those authors also refer to other, now lost, works of Greco-Roman writers who collected information about Scythia, gold mining, and griffins. So, in piecing together the natural history of the griffin of Scythian lore, we must rely on terse, fragmentary passages that derive from a much fuller and richer tradition. Moreover, many of the travelers, historians, geographers, and other writers who mention griffins not only were dismissed as "mythographers" by their colleagues in antiquity, but they have been discounted by modern scholars for recording hearsay instead of "objective facts." Yet it is their very interest in curious natural phenomena of their day that makes those writers so valuable to the recovery of paleontological legends.

The Natural History of the Griffin

The Athenian playwright Aeschylus (b. ca. 525 B.C.) was fascinated by the geography and customs of strange lands. He was the earliest writer to use the information about Scythia's landscape and folklore gathered by the traveler Aristeas. In his tragedy *Prometheus Bound* (460 B.C.), the gods punish the Titan Prometheus by chaining him to a cliff in remotest Scythia. Relying on Aristeas's material, Aeschylus sets the scene. He describes the lonely caravan trails leading to a desolate country where nomads prospect for gold, a desert inhabited by monstrous Gorgons who magically turn living things to stone, and by fearsome griffins. The griffins Aeschylus likens to "silent hounds with cruel, sharp beaks." Despite the poetic license granted by tragedy and myth, Aeschylus was a careful zoologist—he takes pains to distinguish the *wingless* eagle-beaked griffins from actual *winged* eagles.[9]

Around the same time that Aeschylus was writing *Prometheus Bound*, the widely read and well-traveled historian (and the world's first anthropologist) Herodotus was visiting the westernmost of the far-flung Scythians, just beyond the Black Sea. He had read Aristeas's poem, and he interviewed Black Sea Scythians about the lives of their nomadic brethren who lived much farther to the east. Remarking that some of his information had passed through a chain of seven translators stretching eastward to the Altai Mountains, Herodotus transcribed demonstrably authentic ancient vocabulary from Issedonia. His descriptions are the oldest comprehensive picture we have of the lifestyle, language, and legends of the steppe nomads, and many of the cultural features he described in his *Histories* (ca. 430 B.C.) continue to be confirmed by artifacts excavated from Saka-Scythian graves found by Rudenko and others in south Russia and Kazakhstan. Linguistic analysis of the nomads' Indo-Iranian vocabulary, otherwise unknown to the Greeks, confirms that Herodotus had access to genuine information from Central Asia.

Herodotus was born in Halicarnassus, Caria (modern Bodrum, on the Turkish coast across from the island of Kos) in about 484

B.C. He is a controversial but key figure in the history of paleontological legends. In antiquity, elite historians considered him a storyteller who reported useless hearsay or made up entertaining tales from whole cloth—and his old reputation as the "Father of Lies" has lingered to this day. Some even question whether he undertook the travels he claimed. But as scholars and archaeologists discover more about the Saka-Scythians and other non-Greek cultures discussed by Herodotus, he is beginning to be appreciated as a faithful recorder of historical reality as well as popular beliefs. The "extraordinary quality of Herotodus" as a reporter, writes Neal Ascherson in *Black Sea*, "is that his information grows in importance from year to year as archaeology confirms it." As an explorer and sympathetic listener who believed that legends preserved traces of real history, Herodotus was assiduous in ferreting out facts, oral traditions, and local opinions. He invited his readers to consider alternative versions of events, often adding his own comments.[10]

As for the dangers of mining gold beyond Issedonia, Herodotus had heard that prospectors made long expeditions to remote deserts marked by extreme heat and cold. "I cannot say for sure how the gold is obtained there," he commented, "but some say that one-eyed men called Arimaspeans steal it from griffins." Herodotus scoffed at the idea of a race of one-eyed men, but he expressed no doubts about the existence of griffins. In a typical griffin scene on a vase from Caria about a century after Herodotus, the artist painted a nomad on horseback encountering a griffin in a stark landscape indicated by a dead tree and rocky ground. A nugget of gold is shown just above the griffin, which almost seems to be part of the ground, an outline emerging from the rock (fig. 1.6).[11]

A few decades after Herodotus, in about 400 B.C., another Greek from Caria, a physician named Ctesias, settled in the Persian city of Susa (in modern Iran). His writings about the exotic lands east of Persia, based on his own experiences and on reports from Persian sources, now exist only in fragments. They have been considered fantastic and untrustworthy by modern scholars, but his comments about griffins seem quite down-to-earth. Ctesias ex-

1.6. Griffin and mounted nomad. Red-figure vase from Mylasa (Caria, Turkey), fourth century B.C. Note the nugget of gold above the griffin, and the barren landscape with rocky ground and dead tree (petrified trees are found in the Gobi desert). Drawing by author, after F. Winter, "Vase aus Mylasa," *Mitteilungen des deutschen archäologischen Instituts* (1887), plate 11.

plained why Asian gold was so notoriously hard to obtain. It came, he said, from "high mountains in an area inhabited by griffins, a race of *four-footed birds*, almost as large as wolves and with legs and claws like lions."[12]

By the time the Latin author Pliny the Elder compiled his encyclopedic *Natural History* based on more than two thousand sources (A.D. 77), Romans were trading extensively over the spice routes to Asia. Griffinology kept pace. In A.D. 43, the geographer Pomponius Mela had reported that fierce griffins guarded gold in a sunbaked wilderness of Scythia. Summarizing the work of Aristeas, Herodotus, Pomponius Mela, and "many other authorities" (now lost), Pliny agreed that griffins were encountered in the vicinity of Scythian gold mines. In addition to noting the "terrible hooked beak," he was the first writer to allude to the peculiar "ears" and "wings" that had long appeared in paintings and sculpture. But

Pliny was skeptical; he rejected the idea that the griffin could be like any known bird, even though it was alleged to have some avian features. Nevertheless, Pliny added an intriguing new detail: "The griffins toss up gold when they make their burrows." This is the first mention of griffin nests!

At about the time of Pliny, a traveling sage named Apollonius of Tyana (in Cappadocia, Turkey) journeyed widely in Asia. His biographer, Philostratus (ca. A.D. 230), has been maligned by modern historians as a fabulist, and yet the remarks he records about griffins are notably naturalistic. According to Philostratus, Apollonius reported that rocks in the region of the griffins were "flecked with drops of gold like sparks." Apollonius speculated that their strong beaks might enable them to quarry this gold. He estimated their size and strength as about equal to that of lions, and he supposed that griffins would be powerful enough to overcome lions, elephants, and dragons but not tigers. Like Pliny, he denied that griffins—despite their beaks—were anything like real birds. They cannot fly because they lack true bird-wings, said Apollonius. He suggested that they had webbed membranes, which might help them make short hops during combat.[13]

By the second century A.D., some people were claiming that griffins had spotted fur like leopards. Pausanias, a knowledgeable traveler from Asia Minor who described the antiquities of Greece in about A.D. 150, disagreed. "Those who love to hear marvelous stories cannot resist adding details, thus ruining the truth by mixing in lies," he remarked. Pausanias briefly quoted Aristeas about nomads fighting griffins and affirmed that the animals resembled "lions but with the beak and wings of an eagle." In the land of the griffins, Pausanias tells us, gold "emerges near or on the surface of the earth."

Within a generation of Pausanias, griffin reports began to accumulate even more realistic details. Aelian, a learned compiler of facts and popular knowledge about natural history in the early third century A.D., is another author often dismissed as a perpetrator of fantasy. He drew on travelers' tales and written texts no longer available to us to write the most complete narrative we have about griffins.

> I hear that the griffin is a quadruped like a lion, with talons of enormous strength that resemble the claws of a lion. It is reputed to have black plumage on its back with a red chest and white wings. Ctesias says the neck is variegated with dark-blue feathers and it has an eagle-like head and beak just as artists portray. Griffins make nests near mountains, and although it is impossible to take a full-grown griffin, people sometimes capture the chicks. The Bactrians say that griffins guard the gold of those parts, which they dig up and weave into their nests. . . . However, [others] sensibly deny that the creatures would *intentionally* guard the gold. The truth is that when the prospectors approach, the griffins fear for their young, and so give battle to the intruders.

Aelian tells how miners journey, in caravans of one or two thousand men, to the wilderness of the gold deposits. "I am informed," he says, "that they return home after three or four years." "Dreading the strength of griffins, the men avoid hunting for gold in the day. They approach at night when they are less likely to be detected. Now, the place where the griffins live and the gold is found is a grim and terrible desert. Waiting for a moonless night, the treasure-seekers come with shovels and sacks and dig. If they manage to elude the griffins, the men reap a double reward, for they escape with their lives and bring home a cargo of gold—rich profit for the dangers they face."

Aelian was the last ancient author to record any new information about gold-guarding griffins. About a century after Aelian, as the Roman Empire's lines to Asia were fraying, a huge mosaic floor was constructed at a villa in Sicily. The celebrated "Great Hunt" panorama at Piazza Armerina shows the methods of capturing exotic wild beasts (lions, tigers, elephants, ostriches, and so on) from the far reaches of the empire. In the last panel, the villa's owner, perhaps recalling Aelian's discussion of capturing griffins, commissioned the artist to portray a griffin being lured into a wooden trap baited with a nervous-looking man (fig. 1.7). I take this mosaic and Aelian's text, created as the Roman Empire was disintegrating, to mark the end point of a thousand years of interrelated writing and art about what was believed to be a real animal of Asia.[14]

1.7. Griffin lured to trap baited with a man, Roman mosiac, ca. A.D. 310. Piazza Armerina, Sicily. Photo by Barbara Mayor.

Geology of Griffin Territory

Even though no commentator had ever claimed to have sighted a live griffin, the salient features remained consistent over a millennium. And the new details were consistent with the original framework of quadruped bird. Something real must have continued to confirm the most remarkable features about griffins: They had four legs but also a beak; they were found in deserts near gold. What kind of physical evidence might have verified their existence for so many people over so many centuries?

I studied two other legendary creatures that had been inspired by observations of the fossils of extinct mammals. The statue of a dragon created in 1590 at Klagenfurt, Austria, is often cited as the earliest known reconstruction of an extinct animal; it was modeled

1.8. Dragon statue at Klagenfurt, Austria, sculpted by Ulrich Vogelsang in 1590. The head was modeled on the skull of an Ice Age woolly rhinoceros discovered by quarrymen in about 1335. Photo by Josiah Ober.

on the skull of an Ice Age woolly rhinoceros unearthed in a local quarry in about 1335 (fig. 1.8). In 1914, Othenio Abel, an Austrian paleontologist, suggested that the Cyclops of Homeric legend was based on fossil elephant finds in antiquity. Abel, who excavated many Mediterranean fossil beds, related the image of one-eyed giant cavemen to the remains of Pleistocene dwarf elephants, common in coastal caves of Italy and Greece. Shipwrecked sailors unfamiliar with elephants might easily mistake the skull's large nasal cavity for a central eye socket (fig. 1.9).

The small elephants ranged from 3 to 6 feet (1–1.8 m) high at the shoulder, and the skulls and teeth are much larger than men's. In profile, elephant skulls do resemble grotesque human faces, and the vertebrae and limb bones could be laid out to resemble a giant man. Anyone who saw such an assemblage would try to visualize how such a creature looked and behaved when alive. Since Cyclopes lived in caves, the ancient Greeks imagined them as primi-

1.9. Woman contemplating Cyclops head and elephant skull. According to Othenio Abel's theory, the one-eyed giant of Homeric legend was based on observations of fossil elephant skulls. Drawing by author; Cyclops based on Greek sculptures of the fourth and second centuries B.C.

tive troglodytes who used rocks and clubs as weapons. The great piles of bones on the cave floors might be the remains of shipwrecked sailors—the savage Cyclopes were probably cannibals! Human occupation of Sicily and other islands where dwarf elephant bones abound occurred long before Homer, and descriptions of them probably circulated among sailors from Mycenaean times onward. The Cyclops story was assimilated into the epic poetry tradition and made famous in Homer's *Odyssey.*[15]

The Klagenfurt dragon and Abel's interpretation of the ancient Cyclops legend spurred me on to inquire what kind of distinctive fossil remains might have inspired the image of the griffin. But first, I needed to locate the secret gold deposits exploited by the nomads of Central Asia.

Grave looters and archaeologists have carted off prodigious quantities of golden artifacts (with griffins as a favorite motif) from Saka-Scythian tombs along the slopes of the Altai and Tien Shan. These mountains (the word *Altai* means "gold" in the local dialect) were the traditional source of famed Scythian gold, but many ancient writers described prospecting expeditions in the deserts below. In trying to locate the goldfields in this region, I became aware that Russian and Chinese gold deposits have long been closely guarded secrets. Yet by consulting three sources of evidence—the writings of nineteenth- and early-twentieth-century travelers, notes made by early archaeologists, and even declassified Western intelligence documents—I was eventually able to unravel the mystery. I determined that gold sand had been mined since antiquity in the foothills skirting the old caravan routes. Just as Pliny and Pausanias claimed, the gold here *does* emerge as particles on the surface of the desert, in the form called placer gold. And Ctesias was right about its mountainous origin—the gold from the massifs continually erodes down into the gravel basins below.

Like Russians of the 1800s who were guided by old place-names to exploit the gold-bearing detritus of ancient Scythian mines, I pored over old maps and journals to locate along the caravan routes from the Gobi to Lop Nur and the Dzungarian Gate places whose names meant "gold." Between 1860 and 1950, this technique had allowed Russian archaeologists to rediscover more than one hundred Altai gold mines worked since about 1500 B.C. Their most astounding find was the skeleton of a Bronze Age miner at Zmeinogorsk, whose leather bag held gold nuggets. The man had apparently died a sudden death, and since the gold was still in his bag, he had not been felled by a thief. An ancient traveler coming across his skeleton might imagine that he'd been attacked by a griffin guarding its nest.

Aelian and Herodotus's accounts of parties of gold-seekers setting out for years at a time in hope of great profit offer a realistic scenario. The risks the men faced were real: extreme heat and cold, thirst, shifting dunes, blinding winds, violent sandstorms. The miners would have to leave the main caravan trail and disappear into the lonely wilderness of eerie red sandstone formations to reach the gullies where they would sift the gravel for flakes of gold. They had to know exactly where to find water, and they were forced by hellish heat to travel and work at night. At night, moon- and starlight would cast bizarre shadows and illuminate strange shapes. We can be sure that—like other long, dangerous tracks across wastelands, such as the Oregon Trail and the sinister Skeleton Coast of Namibia—the desert floor was littered with the bleached bones of animals and travelers who had died of thirst, sandstorm, or other misadventure. A creature's remains might be alternately hidden and revealed by landslides, torrential rains, flash floods, perpetual whirlwinds, and blizzards of sand and gravel. The sandstorms themselves were legendary. During the Roman Empire, it was rumored that an entire legion of soldiers marching in these deserts had vanished forever in a cloud of swirling sand.[16]

Such hallucinatory, and perilous, landscapes make powerful impressions on travelers, judging by the diaries of pioneers in American badlands and modern travelers in African and Asian wastelands. As they left the main routes, then, ancient prospectors, reminded of their own vulnerability by the parched, vulture-pecked corpses and scattered bones of earlier seekers and their beasts of burden, would remain sharply vigilant against every possible danger in the desert—especially if they had heard tales of fierce griffins "guarding" the approaches to the gold. We now know, of course, that no aggressive, wolf-size, four-legged flightless "birds" ever coinhabited the earth with humans. So what could account for the consistent reports of just such an animal?

Stone Bones

As I combed the literature for fossil deposits near the Central Asian goldfields, I learned that thirteenth-century Chinese travelers had feared the ominous "fields of white bones" and "heaps of hard, bright white stones like bones" in the deserts around Turfan and Lop Nur, old Issedonian lands rumored to be haunted by terrifying demons and dragons. It is generally believed that the earliest written descriptions of "dragon" bones appeared in a Chinese chronicle of the second century B.C. During the digging of a canal in north-central China, "dragon bones were found and therefore the canal was named Dragon-Head Waterway."[17]

But even older dragon lore appears in the *I Ching*, a collection of divinations probably compiled just after the time of Homer. Like Homer, the authors drew on older traditions as far back as

1000 B.C. The omens and advice in the *I Ching* originally centered on agricultural matters. One of the "peasant" omens is "dragons encountered in the fields." I think that this omen refers to farmers plowing up prehistoric bones in cultivated fields, a common occurrence in fossiliferous regions. According to the *I Ching*, this was a good omen. Indeed, such a find was a cash crop, since "dragon" bones and teeth had been harvested from the earth as folk remedies for millennia in China. Farmers counted on the extra income they received for dragon bones, and many worked through the winter months in long-established fossil-digging operations. The suppliers of Chinese apothecaries hid their sources, although they themselves kept careful records of provenience. As the French paleontologist Eric Buffetaut comments, nineteenth-century Europeans made many "important paleontological discoveries in the drugstores of large Chinese cities." According to imperial customs records of 1885, twenty tons of dragon bones were *exported* that year, with untold tons consumed in China.[18]

In 1919, a Swedish geologist, Johann Andersson, was the first Westerner to discover a major dragon bone works in northern China, and the Austrian paleontologist Otto Zdansky studied the operations in 1923–25. Over centuries the fossil miners had honeycombed galleries throughout the red clay formation. Working like modern paleontologists, they extracted heavy, calcified bones with pulleys and sifted sediments in baskets, cleaning, labeling, and storing the remains. Zdansky identified many extinct Pliocene species, including the equine *Hipparion* and deer *Cervocervus.* He learned that the workers recognized the bones, teeth, and antlers as belonging to strange versions of familiar species, such as horses and deer, although by convention everyone called all petrified remains "dragon bones." The British paleontologist Kenneth Oakley has shown that certain features of the traditional Chinese dragon, such as the distinctive antlers resembling those of fossil deer, replicated the lineaments of Pliocene and Pleistocene prehistoric mammals of northern China and Mongolia.[19]

In 1920, the interest of American adventurer Roy Chapman Andrews was piqued by reports of dragon teeth and bones from the Gobi Desert. He knew that a Russian geologist had collected a fossil rhinoceros tooth on the old caravan route in 1892, and he examined specimens purchased by paleontologist Walter Granger in 1921 from peasant diggers at Zdansky's dragon bone pit. On the basis of these finds, and hoping to find evidence of early hominids, in 1922 Andrews led an expedition, sponsored by the American Museum of Natural History, to search for prehistoric remains in the southern Gobi Desert.

Following ancient caravan trails from China through the desolate landscape in southwestern Mongolia, the team discovered—on the surface or only partially embedded—a multitude of fossils. The men were incredulous at the sheer numbers: in Andrews's words, remains appeared to be "strewn over the surface almost as thickly as stones," and the ground seemed "paved with bones." Many of the species were dinosaurs new to science, and Andrews's team was the first to recognize dinosaur nests and eggs. The spectacular finds were shipped to New York with great fanfare.[20]

On their very first afternoon at the site called Flaming Cliffs, about 30 miles (48 km) from the Altai Mountains, Andrews and each member of the Central Asia Expedition had located a dinosaur skull. In two weeks they gathered over a ton of fossils from the red sediments; and in two summers they excavated the bones of more than one hundred *Protoceratops* and *Psittacosaurus* skeletons, related denizens of the Cretaceous period (ca. 100–65 million years ago). Andrews's original photographs show dinosaurs emerging from the ground that combine the features of birds and mammals in a very striking way. The body of the hatchet-faced *Protoceratops* is about 6 to 8 feet (about 2 m) long, roughly the size of a lion, and has four limbs, but the head has a nasty-looking beak, large eye sockets, and a thin, bony frill at the back of the skull (figs. 1.10, 1.11). The smaller (4 to 6 feet long, about 1.5 m) *Psittacosaurus* ("parrot-beaked") has very prominent jugals, or cheek projections. In these deserts, the exquisitely preserved skeletons are frequently fully articulated, with the beaked skulls still attached. "Tiny surface features—grooves and pits that mark the routes of blood vessels and nerves"—are still evident.[21]

The beak and the hip structures of the *Protoceratops* and its relatives do resemble those of birds, something the ancient Asian

1.10. *Protoceratops* skeleton from the Gobi Desert, Mongolia, excavated by Roy Chapman Andrews. Neg. 312326, photo R. C. Andrews. Courtesy Department of Library Services, American Museum of Natural History.

1.11. *Protoceratops* skull embedded in the ground, excavated by Roy Chapman Andrews, Gobi Desert, Mongolia. Neg. 410737, photo R. C. Andrews. Courtesy Department of Library Services, American Museum of Natural History.

1.12. Clutch of fossilized dinosaur eggs, found by Roy Chapman Andrews, Gobi Desert, Mongolia. Neg. 258267, photo R. C. Andrews. Courtesy Department of Library Services, American Museum of Natural History.

nomads—as falconers familiar with large raptors—could hardly fail to notice. Indeed, Ctesias's description of 400 B.C. is not far off: this really *is* "a race of four-legged birds"! Moreover, in shallow depressions scattered over the desert, numerous clutches of fossilized eggs and even newly hatched young are clearly visible. The eggs—which we now know belonged to several dinosaur species—were arranged the way living birds arrange their eggs (fig. 1.12).

Andrews's highly publicized shipments of Gobi fossils to the United States and his auction of dinosaur eggs (for five thousand dollars each) angered the Mongolians and the Chinese, who forbade further Western expeditions in the Gobi. In the 1960s, Polish-Mongolian teams undertook significant Gobi excavations, unearthing numerous protoceratopsids and other species. In 1986, in the remote Xinjiang province of northwestern China, the Canada-Chinese Dinosaur Project, led by paleontologists Dale Russell, Philip Currie, and Dong Zhiming, began to explore the western extension of the Gobi, in the heart of ancient Issedonia. The red rock formations of the Junggar Basin badlands and the foothills of the Tien Shan yield an equally rich array of dinosaurs, from the Triassic to the Cretaceous. Nests and eggs are also common. Here, psittacosaurs are very plentiful—indeed, the parrot-beaks are the most abundant dinosaurs known. Beaked dinosaurs have also been found even farther west along the old caravan routes, in the Kizylkum (Red) Desert in Uzbekistan.[22]

In the wind-scoured dunes, alluvial basins, and red sediments along the old caravan trails, prehistoric remains are continually revealed by the very same forces of erosion that bring the gold down from the mountains. The desert is extremely arid with little vegetation, so it's possible to spot fossils on the ground. The shapes of the skulls and skeletons are quite obvious, even to amateurs, and the surrounding rock is soft and crumbly, making it easy to uncover partially embedded bones. And the white bones stand out against the red matrix. In 1992, I learned that a pair of American amateur fossil hunters in the Gobi had spotted a beak poking out of a sandstone cliff. By nightfall they'd uncovered enough to reveal an entire *Protoceratops* skeleton in a standing position. In 1993, many more articulated *Protoceratops* skeletons were found standing on their hind legs, in poses indicating that they died during sandstorms. "Superb specimens" of beaked dinosaurs, often preserved in lifelike poses, are the most ubiquitous remains in the Gobi, says Peter Dodson in *The Horned Dinosaurs.* He calls them "ridiculously easy to find" and confides that some paleontologists consider them a "nuisance"![23]

When I contacted Russell and Currie to see what they thought of my theory about the origins of the griffin legend, they agreed that ancient nomads certainly would have observed constantly emerging, fully articulated skeletons of beaked dinosaurs. The protoceratopsids are about the size of wolves or lions, and they resemble large, flightless four-legged raptors. The fossil beds' proximity to gold deposits led to the notion that they "guarded" the approaches to gold in the Altai foothills. The mystery that had begun for me on the Greek island of Samos was solved: the long-lost griffin was found at last in its original Central Asian homeland.

Like modern paleontologists, the ancient observers of "griffin

1.13. Artist's conception of the Scythian nomads' reconstruction of the fossil *Protoceratops* skeleton (*top*) as a griffin (*bottom*). Drawings by Ed Heck.

bones" relied on what they knew about familiar animals to visualize what the creatures would be like in life. As hunters, falconers, and herdspeople, they were knowledgeable about the anatomy and behavior of raptors and mammals—and they were keen observers, as evidenced in the realistic details of Scythian animal art (fig. 1.13).

In the ancient natural history of the griffin, its birdlike features—the beak and egg laying—included the collection of shiny objects like gold for its nest. This habit was well known to ancient bird-watchers such as Pliny, who related that precious gems were found in birds' nests. But how could gold turn up in a dinosaur's nest? Gold in the Altai comes from igneous rocks millions of years older than the Cretaceous sediments that hold dinosaur bones. But gold sand is continually washed down from the mountains by rain and streams. Gravity on slopes and strong winds then scatter the gold-bearing sand over the geologically younger sediments. A sand blizzard in the Gobi can transport pebbles the size of silver dollars! In classical antiquity, the fourth-century writer Theophrastus knew that mounted nomads prospected in the deserts after high winds shifted the dunes and exposed minerals. Pliny reported that the desert-dwellers rushed out after violent storms to gather precious stones glinting in the dunes or caught between rocks. Modern travelers confirm that minerals are exposed after hellacious windstorms in the Gobi. A chance find of a gold particle lodged in among petrified dinosaur eggs might well have sparked the ancient idea that griffins had gathered the gold.[24]

What is so striking about the ancient griffin narrative is the way it anticipates some of our most advanced—and contested—theories about dinosaurs. Most experts now visualize dinosaurs as agile, warm-blooded *birdlike* quadrupeds, instead of sluggish reptiles. To reach these conclusions, paleontologists infer the animals' behavior and physiology from the fossil record and from what we know about living species. The ancients, looking at the same kinds of physical evidence some 2,500 years ago, envisioned griffins as energetic, warm-blooded, egg-laying, four-legged, flightless *birds.* Paralleling today's movement in depictions of dinosaurs away from heavy-bodied, ponderous lizards to leaner, sleeker models, the ancient illustrations of griffins also appear to progress from bulkier, reptilian forms to more streamlined, agile creatures.

Jack Horner has given some thought to the implications of this "earliest encounter on record" between humans and dinosaur remains. "It now seems that the winged griffin of the Gobi Desert, that fanciful hybrid of avian and mammalian features, struck closer to the truth than anyone could have guessed." The nomads "could only guess at the origin of the unfamiliar animal, its relationship to other creatures, its current whereabouts," Horner notes, and yet their imagination and careful observations resulted in a model that "more closely resembles" the actual "protoceratopsian skeletons of the Gobi" than do many nineteenth- and twentieth-century attempts to reconstruct dinosaurs. "Why," asks Horner, "were prehistoric Mongolian nomads, supposedly backward and superstitious, more adept at interpreting paleontological evidence than post-Enlightenment scientists?"

Horner's answer is that the nomads' "conceptual toolbox" was unencumbered by a rigid Linnaean-style system of categories of genus, species, and so on. "If the nomads had viewed the natural world" in Linnaean terms, "they would have felt compelled to place *Protoceratops* in one category or another, bird or mammal. They would not have been so open-minded about what their eyes told them—that the creature they found in the Gobi Desert possessed features from both categories and so belonged to both, or neither."[25]

Indeed, that tension between empirical observation and reliance on authoritative texts arose long before Linnaeus devised his classification system in the 1750s. Going by what they heard and the pictures they saw of griffins, the Greeks and Romans joined the nomads in visualizing the animal as a flightless, four-legged bird that nested on the ground. An exception was the skeptical Pliny, who modeled his zoology on Aristotle's classifications of species. As chapter 5 will show, the ancient natural philosophers were more nervous about ambiguous categories than were the Saka-Scythian nomads and open-minded writers like Herodotus. The philosophers' theoretical concerns apparently prevented them from even taking notice of unusual remains that excited the curiosity of everyone around them. It's interesting that the great philosopher-biologist Aristotle did remark on the ambiguous nature of one exotic *living* creature, the ostrich. In *Parts of Animals*, he detailed the way an ostrich combines the features of a bird and a large mammal, and he noted that despite wings ostriches are flightless. The existence of this passage, and another one on sea monsters, makes it all the more striking that Aristotle never mentioned the problematic griffin, which many in his day believed to be as real as the ostrich.[26]

In 1860, the ancient disagreements over whether the griffin was really a bird were reflected in the nickname that the zoologist Andreas Wagner gave to the 150-million-year-old fossil *Archaeopteryx*, a feathered reptile "disguised as a bird." Wagner dubbed it *Griphosaurus problematicus* ("the problematic griffin-lizard"). Today, controversy boils over the relationship between birds and dinosaurs, the development of feathers, and the origin of flight. The controversy has been sharpened by the discovery in 1994 of feathered dinosaurs (*Sinosauropteryx*, *Confuciusornis*, *Liaoningornis*, and *Protoarchaeopteryx*) in a trove of new dinosaur species discovered by a farmer in China. John Ostrom, the Yale paleontologist who championed the idea that birds orginated from warm-blooded dinosaurs, calls these new discoveries "mind-boggling." Phil Currie's excited response gives us a sense of how some ancient observers may have reacted to seeing "griffin" remains: "Suddenly you're dealing with an animal that isn't extinct any more. The dinosaurs are still alive," in the form of living birds![27]

The question of whether these Chinese specimens are true primitive birds, or transitional forms between dinosaurs and birds, was foreshadowed in the ancient disagreements over the birdlike griffin's appearance, flight capabilities, and behavior. Recall that Aeschylus had emphasized in his play that the griffin had a beak *like* that of an eagle, but no wings. Was it to complement the beak that Greek artists added stylized wings to ground-dwelling griffins? The wings they supplied are the same standard issue given to mythological creatures such as the Sphinx, another flightless being. Despite the wings, however, ancient writers and artists alike agreed that griffins were earthbound: the animal did not fly, but stalked prey and nested on the ground.

Just as paleontologists guess at the outward appearance of dinosaurs, no one in antiquity was sure whether griffins were furry, leathery, or feathery. Ctesias had heard that the four-legged birds had black and red feathers. Apollonius of Tyana speculated that some kind of webbed appendages might help griffins hop around when fighting—an idea that prefigures Ostrom's notion of dinosaurs whose feathered arms allowed them to make prolonged leaps in the air. Pausanias pooh-poohed the popular belief of his day that griffins were furry, with spots, rather than feathered. And Aelian recounted differences of opinion over the color of the plumage. Sculptors and painters reflected this ancient uncertainty and used their imaginations to indicate either feathered, scaly, rolled, or crosshatched leathery skin on the necks of griffins. And, as I had noticed in the Samos archaeological museum, griffin body types ranged from gracile to robust. Postures varied too: some artists showed griffins standing sturdily on four legs, while others depicted them balanced on hind legs as they grasped prey with their forelimbs. Interestingly, many *Protoceratops* skeletons are found standing in both positions.

Besides wings, odd "ears" and crests appear in artistic representations. In the gallery of bronze griffins in Samos, forehead gear ranged from nonexistent, to bulges and bumps, to tall or squat horns or knobs. Some griffins had long rabbit ears while others had elfin ears. A few artists gave griffins crested necks that resemble those of sea serpents or plated dinosaurs. Artistic license, or attempts to illustrate garbled descriptions of eyewitnesses? A few griffin artifacts show strong similarities to the blunter parrot-beaked skulls of psittacosaurs, and many have indications of teeth inside the beaks, just like the protoceratopsid skulls (fig. 1.14).

Maybe the complex head structures were intended to represent the prominent jugals of *Psittacosaurus* skulls. In his cautious endorsement of my idea that griffins were inspired by beaked dinosaur remains, Peter Dodson noted that the wings and ears might reflect "bafflement" resulting from "attempts to interpret the bony

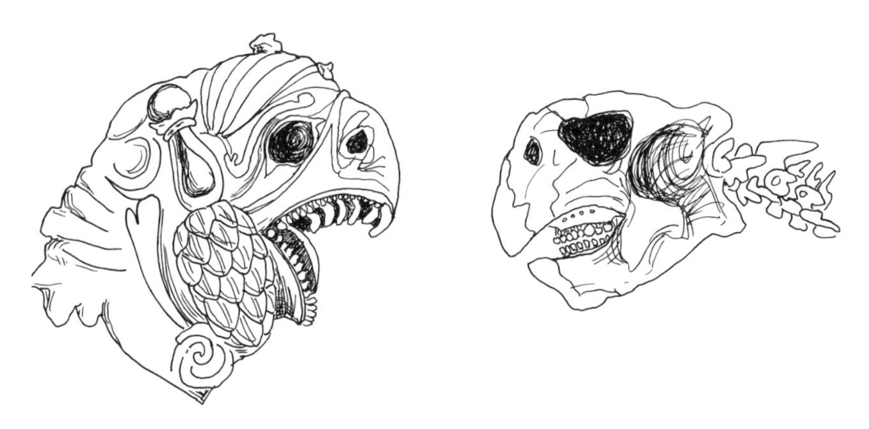

1.14. *Left*: gold griffin head, Saka-Scythian, seventh century B.C., found in northwest Iran. *Right*, fossil *Psittacosaurus* skull, common in the Junggar Basin, in Uzbekistan, and the western Gobi. Drawings (not to scale) by author; griffin head in Teheran Archaeological Museum, after K. Jeffmar, *Art of the Steppes* (New York: Crown, 1964), plate 47.

frills at the back of the skull" of the *Protoceratops.* Using taphonomy, the study of what happens to animal remains after death, Jack Horner proposes that the ears, knobs, and wings may be attempts to represent the appearance of weathered *Protoceratops* specimens. The translucent bony neck frill is eggshell thin, and Horner notes that "more often than not it breaks off and disappears long before anyone finds the skull, leaving behind a structure resembling a horn, very much like the horn depicted in Greek and Roman pictures of the griffin." Horner also points out that the dinosaur's "elongated shoulder blade" is "located in exactly the same place as the griffin's wing." That long, slender scapula may have been interpreted by ancient observers familiar with raptors as an anatomical anchor for wings, and then artists added stylized wings as the story was passed along.[28]

Features of different dinosaur species may have contributed to the griffin's image, too, such as the 28-inch (70-cm) isolated *Therizinosaurus* claws found in Kazakhstan and the western Gobi and the gigantic *Deinocheirus* claws found by the Polish-Mongolian team in southern Mongolia. The bony, ridged, frilled, knobbed, plated, and spiked skulls and spines of other Asian species may have been conflated with the beaked dinosaurs. The sharp-beaked, crested Dzungarian pterosaur, whose remains are found west of the Altai Mountains in Issedonian territory, for example, sports a 10-foot wingspan.[29] Was Apollonius's suggestion that griffins had webbed membranes based on observations of pterosaur remains?

Protoceratopsian bone beds typically contain hatchlings and adults, leading some paleontologists to conclude that the dinosaurs nurtured their young, another hotly disputed notion. Ancient artists portrayed griffins defending their young, a scenario that was described by Aelian. It is possible that ancient nomads came across dinosaur skeletons brooding atop clutches of eggs. Just such an assemblage was found in the Gobi by Roy Chapman Andrews in 1923. But that and similar scenes were misinterpreted as egg stealing rather than nesting until Mark Norell and Michael Novacek of the American Museum of Natural History discovered an assemblage of a dinosaur protecting its own eggs in 1993. In *Dinosaur Lives*, Jack Horner calls attention to modern misunderstandings about dinosaur aggression and defensive behavior.[30] Interestingly, aggressive behavior of nesting griffins was a question addressed by Aelian. Some gold-seekers claimed that griffins attacked intruders to safeguard the gold in their burrows, while others thought they preyed on humans. But Aelian reasoned that if griffins attacked, it would be to defend their young.

It has been suggested that the Issedonians circulated the tale of rapacious griffins to scare away rival gold-seekers. Once the story was known, the sight of the fearsome skeletons with powerful beaks in the desert would convince skeptics and provide physical evidence that matched the griffin's reputed appearance. At any rate, and at whatever remove, the griffins of ancient Greco-Roman narratives and art must certainly reflect reports of fossil remains of the beaked dinosaurs of Central Asia, dinosaurs whose abundant and well-preserved skeletons have contributed so much to modern paleobiological knowledge.

The gold-guarding griffin is the earliest known example of a legendary monster that can be traced to dinosaur remains. That picture of the last dinosaurs to walk the earth was developed nearly three thousand years ago by nomads who had no idea of the vast geological forces and awesome time spans involved, and no formal concepts of evolution or extinction. Horner imagines us in their place: "If you came across a *Protoceratops* skeleton, or any other unusual skeleton, there was every reason to believe that similar animals existed, if not in the immediate area, then somewhere else. What's more, there was nothing to suggest that the group to which an unfamiliar animal belonged might have died out."[31] Except for the fact of extinction, then, the reconstruction of the griffin by the Saka nomads and their literate Greco-Roman reporters came very close to our most up-to-date knowledge about protoceratopsids. Drawing on minute observation and zoological knowledge, the two groups imagined an extraordinary—but reasonable and consistent—natural history for the beaked dinosaurs whose physical traces haunted the wilderness where nomads searched for gold.

This chapter shows that the mind-boggling fossil remains of alien creatures have always demanded answers, and that curious observers will always strive to provide them. Pre-Darwinian interpretations are worthy of our attention because they tell us something about the human imagination and about ourselves. As John Ostrom recently remarked to me, "Ancient populations perceived much more about their local scenes than we modern, intelligent, educated descendants like to admit." To the Saka-Scythians, "the griffin was as much of the real world as the *Sinosauropteryx* is to us today." The effort to recapture open-minded, keen observation combined with unfettered imagination is highly valued among creative scientific thinkers. As Horner puts it, "an ever-restless imagination" plays "a crucial role in my work—helping me fill in gaps, recognize patterns, and make guesses about where I should look for additional clues." He believes that the "interplay between fact and imagination" is the key to understanding the wondrous creatures that we will never see alive—and it might even help us comprehend our own "tenuous" destiny on earth.[32]

Piecing together the griffin legend demonstrated to me how rewarding close attention to even fragmentary and time-ravaged evidence could be. If the ancient Greco-Roman writers preserved enough evidence to show that illiterate Asian nomads created an impressive natural history for beaked dinosaur fossils, then, I thought, we should try to recover more evidence of paleontological discoveries by the Greeks and Romans themselves. I returned to the island of Samos to pursue the ancient Greek legends that grew up there to explain the extraordinary *Samotherium* bones—and that story led me to search out other little-known ancient descriptions of remarkable fossil remains all around the Mediterranean.

Griffin lore was transported by travelers over great distances by word of mouth. Then, over miles and centuries, Greek and Roman writers elaborated on the hearsay and debated the details about griffins—without ever realizing that the exotic creature had been brought to life from mere bones by the Scythian nomads. What is unique about the Greco-Roman narratives in the chapters that follow is that—unlike Scythian lore transmitted by cultural outsiders—these accounts detail a literate ancient culture's *direct* experience with fossils. Even though they are fragmentary and never as complete as we might hope, the Greek and Roman texts contain some of the world's *oldest written descriptions of fossil finds*, many of them firsthand. Writers like Herodotus, Pausanias, and Aelian tell us what they and their contemporaries thought, said, and did when they came upon bones of startling magnitude. Gathered together for the first time and never before examined in light of modern paleontological discoveries, these narratives allow us to rediscover the earliest episodes in the history of paleontology.

The story of ancient encounters with fossils is full of twists and coincidences, frustrating lapses and unexpected insights, and many paleontological firsts. To recover that story, we first need to know something of the terrain. The next chapter looks at the violent geophysical upheavals that created the Mediterranean landscape. Most of the land is too young to have been inhabited by the Cretaceous dinosaurs of Scythia. Instead, the perplexing bones observed by Herodotus, Pausanias, and other ancients belonged to extinct megafaunas from the Miocene, Pliocene, and Pleistocene epochs. Modern paleontology in Greece began with the now largely forgotten international "bone rush" of 1839 and the revolutionary discoveries of mastodons, mammoths, rhinoceroses, grotesque chalicotheres, giant giraffes like the *Samotherium*, cave bears, saber-toothed tigers, and other huge paleomammals around the Mediterranean. These important, but relatively little-known, modern discoveries allow us to picture the dimensions and appearance of the stupendous bones that caused consternation from ancient Samos to Rome.

CHAPTER

2

Earthquakes and Elephants: Prehistoric Remains in Mediterranean Lands

The Big Bones of Samos

INQUIRING MINDS wanted to know: Whose enormous bones littered the island of Samos? The ancient historian Plutarch took on this mystery in his work *Greek Questions*, a compilation of curious facts about Greece written in about A.D. 100. Best known for his biographies of ancient celebrities (the *Parallel Lives*), Plutarch studied philosophy, lectured in Rome and Egypt, and served as a priest at the oracle of Delphi. A prolific writer of tireless curiosity and staggering erudition, Plutarch left a treasure trove of antiquarian information about the topics that fascinated him, from live Centaur sightings to Persian magic. In his Greek Question no. 56, he offers a glimpse of the popular and learned debate that surrounded the discovery of remarkable bones in the ancient world.

Huge bones were displayed to travelers at two places in Samos: Panaima ("Blood-Soaked Field") and Phloion ("Crust of the Earth"). To explain the name Blood-Soaked Field, Plutarch alluded to the notion, widespread in folklore, that red-colored earth had been stained by bloodshed in a battle of great magnitude. The greatest mythological conflict on Samian soil was the battle between the god Dionysus and the mighty Amazons. According to a Hellenistic version of the myth, Dionysus came to Greece from India with a train of war elephants. This myth must have arisen sometime after the fourth century B.C., after the Greeks first learned about elephants from Alexander the Great's campaigns in India. In the myth, Dionysus encountered the Amazons around Ephesus on the Turkish coast across from Samos. Ephesus was the legendary Amazonian stronghold; Amazon "graves" had been pointed out in western Anatolia since the time of Homer. According to the story related by Plutarch, Dionysus and his elephants pursued the warrior women to Samos and defeated them in battle there.

The massive prehistoric bones that emerged from the ground in Samos were believed to be the remnants of that battle. Some classical-era tourist guides may have identified certain bones as those of fallen Amazons, in keeping with the ancient belief that men and women of the mythic past were much larger than present-day people (see chapter 5). But Plutarch clearly states that some of the immense bones of Samos were displayed as the remains of Dionysus's war elephants. This is an astonishing moment in the history of paleontology, because the remains of mastodons (prehistoric elephants) do exist in the bone beds of Samos. Plutarch's statement means that, some 1,700 years before Cuvier, fossil mastodons were correctly recognized as a species of elephant! The legend of a great battle between the Amazons and Dionysus's Indian elephants in the distant past was a rational attempt to explain how in the world *elephants* came to be buried on an Aegean island (figs. 2.1, 2.2).[1]

Long before the Greeks knew about elephants, however, a different interpretation of the huge bones of Samos had prevailed. "Some also said that at Phloion the very earth had cracked open and collapsed upon certain huge beasts as they uttered great and piercing cries." Here Plutarch refers to a much older legend about monsters unique to Samos called the Neades. These Neades supposedly inhabited Samos in primordial times, before human beings

2.1. Typical fossil exposure in Greece, with scattered mastodon bones. Jawbone, tibia, and ulna of *Zygolophodon (M.) borsoni* excavated in northern Greece in 1998. Photo courtesy of Evangelia Tsoukala, Aristotle University, Thessaloniki.

2.2. Excavators with mastodon mandible and limb bone, northern Greece. Photo courtesy of Evangelia Tsoukala, Aristotle University, Thessaloniki.

arrived. Our earliest source for that legend is Euagon, a historian of Samos who lived in the fifth century B.C., some five hundred years before Plutarch. According to surviving fragments of Euagon's lost work, the Neades' shrieking raised such a din that the ground was torn open and swallowed them. In this legend, the people of Samos attributed the mass death of the titanic creatures whose bones they saw trapped under solid rock to the severe earthquakes (with their deafening noise) that periodically wrack Samos.[2]

The Neades were proverbial throughout the Greek world because of a humorous saying: "So-and-so shouts louder than the Neades!" In about 330 B.C., that proverb caught the attention of the great natural philosopher-historian Aristotle, who studied old sayings for germs of historical facts. Aristotle knew the complete writings of Euagon, but regrettably the work (*Constitutions*) in which Aristotle quoted Euagon about the Neades is itself only a fragment. Even so, Aristotle's indirect acknowledgment of the legendary Neades is noteworthy because it is the *only* allusion to unusual animal remains in his extensive writings.

"According to Aristotle," writes classical scholar George Huxley, "Samos had been an empty waste until the enormous beasts invaded the island." We know that Aristotle began his zoological inquiries when he lived at Assos (on the north-central coast of Turkey), and on Lesbos, an island north of Samos. Fossil bone beds similar to those of Samos exist in both locales. Huxley is certain that, like Euagon before him, Aristotle "knew of the deposit of large fossil bones in Samos." But Aristotle's interest in the proverbial Neades was political, not zoological. He says nothing about the famous big bones of Samos. (We will return in chapter 5 to the puzzling matter of Aristotle's silence on the subject of conspicuous petrified bones.)[3]

Aelian, the natural historian who recorded new information about griffins (chapter 1), also discussed the Neades. Like griffins, the Neades were thought of as real animals of a specific landscape where people saw curious prehistoric remains. But, unlike the Scythian nomads with their fear of lurking griffins, no Greeks expected to see *live* Neades; the monsters of Samos had gone extinct in the remote past. Aelian quoted Euphorion, the Greek librarian at Antioch (in ancient Syria, now Turkey) who collected sensational geographical legends in about 200 B.C. All of Euphorion's works have disappeared except for a few scraps of papyrus and tantalizing quotations like Aelian's.

"Euphorion says that in primeval times, Samos was uninhabited except for dangerous wild animals of gigantic size, called Neades. The mere roar of these awesome beasts could split the ground." Aelian continues, "Euphorion says that their huge bones are displayed in Samos." That statement was confirmed in 1988, when German archaeologists discovered a very large thigh bone of an extinct animal in the ruins of the Temple of Hera on Samos. The big fossil had been brought to the temple and dedicated to the goddess Hera by a pious Samian in the seventh century B.C. It's likely that Euagon and Euphorion saw for themselves the great

Map 2.1. Greece, the Aegean, and Anatolia. Map by Michele Mayor Angel.

fossil skeletons of Samos, either in the ground or displayed in the temple.[4]

Though fragmentary, the Neades legend is a remarkable paleontological event, because it means that as early as the fifth century B.C., people correctly recognized that the enormous fossilized remains of extinct mammals on Samos belonged to strange beasts that no longer existed. And the legend proposed a naturalistic scenario—an earthquake—for their demise sometime in the prehuman past. If only the complete version of the Neades tradition had survived, we might have an ancient Greek paleontological legend to surpass that of the griffins!

Embedded in the ancient beliefs about the remains of Neades and Dionysus's Indian elephants are some significant geological and paleozoological truths. The extinct creatures whose bones stud Samos indeed died out there long before the arrival of humans. Earthquakes played a part in their extinction and in burying their remains. Notably, once the Greeks learned about elephants, the new zoological information was used to create a new explanation of the fossils, that they were elephants from India. Very early cousins of Indian elephants are among the five prehistoric elephant species buried on the island—as we will see later in this chapter when we return to the modern paleontology of Samos. Even the story of Dionysus and his Asian elephants chasing the Amazons from Ephesus to Samos is paleogeographically apt, since the island was connected to Asia Minor until about fifteen thousand years ago (a shallow strait, less than two miles wide, now separates Samos from Turkey).

The following pages give a brief overview of the geological history of the Mediterranean, along with some surprising ancient insights about geophysical processes. Next we look at the conditions of fossilization and the natural forces that exposed prehistoric remains to view in antiquity. Pausanias, the learned Greek traveler of the second century A.D. who wrote about griffins, also recorded some striking personal experiences with giant skeletons that weathered out of the ground in Asia Minor. We'll consider the logic that led the ancients to perceive many of those and other extinct animal bones as vestiges of giant or monstrous humans of

Greek myth. The real identity of the "giants" becomes clear when we look at the history of modern paleontological discoveries in the circum-Mediterranean, in the same locales where the ancient Greeks and Romans observed enormous skeletons. That history also shows that the questions of who assigns meaning to remarkable fossils and who claims ownership of them are timeless themes. A quick tour of typical Mediterranean fossil deposits will acquaint us with the various extinct species whose immense bones captured attention in antiquity.

Complex Mediterranean Geology

The turbulent forces that forged the topography of the Mediterranean are not fully understood by geologists. The terrain once occupied by the old Greco-Roman world is a dynamic "crunch zone" of colliding continental plates, violent earthquakes, upwarping seismic activity, and volcanicity. The geophysical history and ongoing processes shaping the eastern Mediterranean are "extremely complex," in the words of Turkish geologist Oguz Erol. Comparing the Eurasian and African continents to "two converging rugby forwards," geologist Derek Ager remarks on the many "unsolved complexities" of the Mediterranean region. Those complexities and upheavals were remarked upon in antiquity, sometimes with amazing perception.[5]

Ancient Geological Knowledge

Asia Minor's advancing plate and severe silting by rivers have moved the Turkish coast westward by 20 miles or so (over 30 km) since the fourth century B.C., stranding many classical Greek seaports far inland and burying or drowning other towns. The ancients were well aware that Greece and the neighboring lands were in flux, with shifting coastlines and emerging land masses. Herodotus, for example, described the silting of the Nile over the course of thousands of years and speculated that if the Nile

Map 2.2. The ancient Greco-Roman world. Map by Michele Mayor Angel.

emptied into the Arabian Gulf for the same time span, the gulf would be completely silted in. The Greek historian Polybius (b. 204 B.C.) predicted that over the ages the Black Sea would become land.

The natural harbor of Troy, site of the Trojan War (which may have taken place about 1250 B.C.), was a landlocked plain by the time of the geographer Strabo (first century B.C.). The deep harbor of Ephesus disappeared over a few centuries, despite strenuous dredging in the Roman era. Pliny the Elder remarked that the sea there no longer lapped the steps of the Temple of Diana as it once did. The ancient towns around present-day Izmir saw headlands transformed into islands and vice versa, marching shores, transgressive seas, and vanishing harbors. Priene, founded in about 1000 B.C. on the coast, had to abandon its silt-choked port in about 350 B.C. Ancient Sardis is now buried by almost 33 feet (10 m) of sedimentation and landslides. The ruins of Miletus and Herakleia, thriving port cities in antiquity, now overlook fields and lakes.[6]

Besides these gradual changes, catastrophes were also recorded. During an earthquake in Crete in the first century A.D., for example, Philostratus wrote that 7 *stadia* (over 350 feet; 107 m) of the shore suddenly collapsed into the sea near the southern promontory of Lebena, and on the same day a new island appeared north of Crete. Pausanias detailed the earthquake and tsunami that utterly destroyed the Greek town of Helike on the Gulf of Corinth in 373 B.C. As a native of Magnesia in Asia Minor, he also wrote about the sudden inrush of the sea that destroyed a city near Mount Sipylos. Archaeologists have located some of the drowned towns mentioned by Pausanias and others. Submerged ruins of several cities can be seen along the Turkish coast, and other examples exist in the Black Sea and the Aegean.[7]

Many ancient authors expressed surprising insights about volcanoes, earthquakes, land creation, and alluvial deposits. The fifth-century B.C. poet Pindar, for example, knew that the island of Rhodes had emerged from the sea long after other lands already existed. So had Delos, noted Aristotle, who proposed an ongoing cycle of barely perceptible sea transgressions and land formations over eons. Pliny listed ten islands that had appeared within human memory, and he understood that Sicily was once attached to Italy, Euboea to Greece, Cyprus to Asia Minor, and so on.

In the fourth century B.C., Plato discussed the vast ocean that once covered Greece and the topographical transformations of Attica over the ages. Strabo wrote brilliant passages on volcanic island creation and the shifting crust of the earth; he extrapolated large-scale upheavals and depressions of strata from his observations of smaller-scale events. The Roman poet Ovid spoke of seashells stranded on high peaks by former oceans, and he told how several islands were joined or separated from mainlands. In the first century B.C., the Jewish philosopher Philo argued that one could deduce the age of the earth if one could measure the rate of erosion over eons.[8] In sum, the Greeks and Romans lived in a geologically tumultuous landscape—and they knew it.

The Sea of Tethys and the Aegean Land Bridge

The unique paleogeology of Eurasia and Africa determined the types of prehistoric animals whose remains would be found around the Mediterranean. Unlike Central Asia, where nomads observed skeletons of Cretaceous dinosaurs of the late Mesozoic era, the land around the Mediterranean Sea yields the skeletons of creatures of the late Tertiary and Quaternary periods (see Geological Time Scale). Most of the prehistoric bones that amazed the ancient Greeks and Romans belonged to the huge, strange mammals of the Miocene and Pliocene epochs (together called the Neogene, about 23 to 2 million years ago) and the Pleistocene epoch (about 2 million to 10,000 years ago). In the Gobi Desert, the 65-million-year-old dinosaurs are among the most exquisitely preserved prehistoric specimens on earth. Although the Mediterranean fossil mammals are millions of years *younger*, their bones, battered by violent tectonics and harsh erosion through the ages, usually emerge in a very different state of preservation.[9]

About 190 million years ago, the vast Sea of Tethys covered what would become Eurasia. Tethys existed for about 150 million years. Over this unimaginable period of time, masses of seashells and marine organisms settled in deep deposits on the sea floor, forming the limestones of future continents. While generations of *Protoceratops* dinosaurs were flourishing and then dying out in Central Asia, monumental upheavals buckled the Tethys sea floor into mountain ranges, forcing up blocks of Cretaceous and Eocene limestones, some 10,000 feet (over 3,000 m) thick, and stranding multitudes of marine fossils on their summits. Stone shells and fish on mountains and in deserts attracted attention very early in Greek history, and they were correctly perceived as evidence of former seas (chapter 5). "Shelly" limestone was a ubiquitous building material in antiquity—Pausanias described it as "soft and extremely white, with seashells all the way through it." The skeletons of larger marine denizens of Tethys, such as Eocene whales (*Zeuglodon* or *Basilosaurus*), some 70 feet (21 m) long, were also stranded on the vast sea's former shores, from North Africa to Pakistan. (Whale remains may account for some ancient reports of extremely long skeletons in the following chapter.)

About 15 million years ago, Tethys receded and the converging continental plates connected Europe, Asia, and Africa. During the Neogene, many species of large mammals migrated along this terrestrial corridor. These animals included the giant giraffe *Samotherium* and a great variety of prehistoric elephants and rhinoceroses whose huge bones emerge in Samos and other sites around the Mediterranean. Meanwhile, inexorable seismic activity continued to create more land masses, islands, and fluctuating sea levels. Africa and Spain jammed together, causing the early Mediterranean Sea to nearly dry up about 6 million years ago. Violent tectonic shifts allowed the Atlantic to gush back in some 5 million years ago, filling the present Mediterranean and Black Seas.[10]

Powerful seismic forces still wrenched the land in the Pleistocene epoch, and rising waters isolated more new islands. Only 20,000 years ago, Aegean islands like Samos and Lemnos were still peninsulas of Asia Minor. The great Ice Ages brought radical fluctuations of temperature and shorelines. During the Pleistocene, immense ancestral elephants and mammoths appeared; some mammals swam to or were isolated on islands, where they evolved new forms. Pleistocene mammal remains are abundant in deposits in Italy, Greece, some Aegean islands, and Asia Minor.

Crescents of volcanoes erupted across the Aegean, thrusting up yet more islands. On Lesbos, an entire Miocene forest of sequoia and palm trees was buried in volcanic ash 18 million years ago. On Thera (Santorini), where volcanoes began about 25,000 years ago, a spectacular eruption blew the entire core of the island (3,200 feet or 1,000 m high) to smithereens in 1638 B.C., destroying the great Minoan civilization centered on Crete.[11]

Half a dozen volcanoes are still active in Italy and Greece. Africa is still relentlessly advancing north as Asia Minor pushes southwest, warping the land with landslides, upthrusts, daily tremors, and severe earthquakes. In Greece, for example, about five hundred villages have been relocated since the 1960s owing to landslides alone. Since Roman times, some parts of Crete have been uplifted almost 30 feet (9 m). In Italy, the volcanic Phlegraean Fields on the Bay of Naples rose more than 2.5 feet (76 cm) over six months in 1970.

Such momentous events over the ages continually expose layers of very old rocks containing fossils deposited in earlier geological eras. As the stratigraphy is convoluted and sheared by faulting, the skeletons are disarticulated and jumbled. Rainstorms, floods, and subsiding coastal cliffs also expose and scatter fossil skeletons. The least durable bones, such as proboscidean (e.g., elephant and mammoth) skulls, are often destroyed, leaving only the strongest bones, such as the femur (thigh bone), patella (kneecap), scapula (shoulder blade), and teeth. Figures 2.3, 2.4, and 2.5 show how the huge bones of a prehistoric elephant appear in a typical fossil exposure in Greece. Whole skeletons are rare, and even when entire specimens are found, the bones may be crushed, like the fifteen complete mammoth skeletons discovered together in central Italy in the 1980s. If paleontology is already a "historical subject dealing with a complex, and quite literally, unimaginably complex chain of events," it's easy to appreciate how the Mediterranean's violent geology makes the fossil record even more difficult to decipher.[12]

Fossilization and Taphonomy

Fossils

All around the Mediterranean, perplexing bones of great size fired ancient imaginations. But before we look at some ancient narratives and identify the fossil remains that inspired them, let's focus on the definition of the term *fossil.* In antiquity a *fossil* (Latin *fossilis*) meant any curious or valuable item that emerged from the earth or could be dug up, such as crystals and gems. The exact connotations of "dug-up items" (Greek *ta orukta*) are obscure. Large petrified animal bones were *not* included in the ancient definition, but the word did apply to small fossils such as fish embedded in rock and to prehistoric ivory tusks buried in the ground.

Imported ivory (*elephas*) and local "*fossil*" or "buried" ivory were precious commodities in the Greek world long before living elephants became known. Large prehistoric tusks in the ground were also known to the ancients as marsh ivory (see fig. 2.6 for a pair of fossil mastodon tusks found in Greece). The fourth-century writer Theophrastus accurately described fossil or marsh ivory as "mottled brown and white." A longer account of marsh ivory by Apollonius of Tyana (first century A.D.) offers an impeccable description of the fossilized tusks of prehistoric elephants. They were much larger than ordinary tusks, Apollonius observed, but the ivory was "dark, porous, and pock-marked, and very difficult to carve." Pliny claimed that elephants deliberately buried their own tusks, thus providing the only surviving ancient explanation for the fact that "fossil ivory" could be dug from the ground.[13]

Nowadays, *fossil* refers to the preserved remains, impressions, or traces of life-forms of past geologic ages, including plants, animals, shells and marine organisms, and even footprints. For petrifaction to occur, groundwater containing minerals must gradually dissolve organic structures of bone, tooth, and shell, replacing them with crystalline calcite, gypsum, or silica and thereby slowly transforming them into stone. Mineral composition determines the weight, density, and coloration of fossils. For example, bones buried in

2.3, 2.4, 2.5. Fossil skeleton of large, straight-tusked elephant *Palaeoloxodon antiquus* of the Pleistocene, whose remains are found across southern Europe and Eurasia. The scapula, ribs, and vertebrae are recognizable in the jumble of bones excavated in Macedonia, Greece, 1992–94. Meter stick shows scale. Photos courtesy of Evangelia Tsoukala, Aristotle University, Thessaloniki.

lignite (soft coal or peat) are stained brown or black, while other fossil bones are chalky white (as in Samos and the Gobi Desert) or gray (as on Kos).

The processes of petrifaction inspired both myths and scientific inquiry in antiquity, as we'll see in chapter 5. In the first century

2.6. Fossil mastodon tusks, excavated in Macedonia, Greece. Photo courtesy of Evangelia Tsoukala, Aristotle University, Thessaloniki.

A.D., for example, Pliny noted that selenite crystals replaced the marrow of petrified animal bones in very deep selenite mines at Segobriga, near modern Toledo, Spain. (Selenite, a transparent form of gypsum, was sprinkled in the arena of the Circus Maximus to produce a sparkling surface for the Roman games.) Pliny assumed that the mineralized bones belonged to "wild animals" that had recently fallen into the shafts, but it's possible that the Roman-era miners observed the fossil remains of familiar-seeming Pleistocene rodents and carnivores in the mines, which were located in the vicinity of Spanish fossil deposits.[14]

The extraordinary dimensions of the remains of what we now know to have been giant giraffes, mastodons, mammoths, and other extinct animals captured the imagination of ancient Greeks and Romans and matched their search image for the corpses of primeval giants and monsters long ago buried in the earth. We can assume that normal-size bones and teeth of extinct deer, weasels, rodents, and the like went unnoticed because they looked like those of familiar living animals. But other categories of small fossils did attract interest. Fossilized plants, fish and other marine creatures, footprints, and the like were notable because they had curious shapes, seemed out of context, or were inexplicably petrified.

For example, several ancient writers remarked on footprints and hoofprints in bedrock in Sicily, in the heel of Italy, and along the Dniester River (Moldavia). These impressions were thought to have been left by the Greek hero Heracles and a herd of king-size cattle owned by a mythical giant named Geryon. I think that the mysterious hoofprints in stone can be explained as the distinctive fossil impressions of giant shellfish, megalodontid bivalves. These hoof-shaped fossils are embedded in late Triassic limestones of southern Europe and the Black Sea area, and they still figure in modern-day folklore about legendary cows or horses.[15]

Fossilized ammonites (cephalopod mollusks), amber, belemnites (fossil cuttlefish guards), shells, echinoids (sea urchins), crinoid stems (fossil "sea lilies"), shark teeth, and plants were curiosities in antiquity: they are mentioned in ancient literature (see appendix 2) and found in archaeological sites (chapter 4). As early as the fourth century B.C., amber was recognized as petrified tree sap, but some other fossils were misinterpreted in ways that persisted in medieval times: for example, Strabo recorded the popular notion that pebble-shaped nummulites found around the Pyramids in Egypt (single-celled *Camerina* fossils) were lithified lentils left over from ancient Pyramid-builders' meals.[16]

Taphonomy

Taphonomy (from the Greek, meaning "principles of the grave") looks at how animal remains are affected by geological and biological processes from the time of death until they emerge from the earth. Taphonomy explains the way *Protoceratops* frills break off in the ground, leaving structures that resemble griffin wings and ears. It explains why the skulls of elephants tend to crumble because of air cells in the bone. It includes the mineralization of bones, the breakage and scrambling of skeletons by earthquakes and erosion, and interference by animals and humans up through the present day. Among taphonomists the technical term for human activity is "grubbing." (No paleontologist sees his or her own work as grubbing, of course—the word is applied to amateurs and to scientists of the past whose methods are no longer judged scientific.) The processes of fossilization and the complex taphonomy of the Mediterranean determined the condition of the prehistoric remains that were seen by ancient observers.

In a remarkable passage in the fifth-century B.C. tragedy *Prometheus Bound*, set in the deep mythical past, the Greek playwright Aeschylus painted an evocative picture of taphonomic processes. To punish the Titan Prometheus for bringing fire to humans, the gods chained the giant to a mountain at the edge of the Asian desert populated by mute griffins. Aeschylus imagines that great landslides and torrential rains will bury the giant at the bottom of the gorge, where his body will be "trapped in stone for eons. Then, he must travel through vast tracts of time before he finally reemerges into sunlight as a carcass for eagles to ravage." Without realizing the true magnitude of the geological events involved, Aeschylus poetically conveys a vivid image of the forces that would hide and then reveal the body of a primeval giant creature.[17]

In keeping with the ancient interest in geology and geography, almost every writer who described the discoveries of giant skeletons tells us whether the bones were revealed by earthquake, landslide, sea or river erosion, rainstorm, animal activity, or human digging. Some even mention old artifacts found with enormous bones, suggesting previous human interference and reburial in earlier ages. These naturalistic details enhance the veracity of the accounts.

Pausanias as Paleontological Reporter

Pausanias traveled widely in Greece and Asia Minor in the second century A.D., viewing local antiquities and discussing their history and meaning with authorities and ordinary folk alike. His work, *A*

Guide to Greece, is a rich source of paleontological discoveries and interpretations made in his own day and in the more ancient past. Pausanias was a very learned man (possibly a doctor), steeped in Greek mythology, but his opinions are notably practical-minded. He reasonably concluded that so-called giants and heroes were genuine, mortal creatures of an earlier age, not supernatural or divine beings.

Pausanias described the exposure of several immense skeletons in his native Asia Minor and recorded the efforts of locals to explain them. The first incident occurred when the Roman emperor (probably Tiberius) diverted the Orontes River west of Antioch in Syria. In the clay of the dry riverbed, the workers found a skeleton 11 cubits long (about 15.5 feet, nearly 5 m). The huge bones seemed human, says Pausanias. How did the workers determine the size of the buried giant? The skeletons of several large extinct species could have been mingled; 11 cubits might refer to the length of the total fossil assemblage. Recognizable mammal bones may have been arranged end to end to conform to the mythical image of a two-legged giant. Or people may have estimated body length from one massive thigh bone.

Whose gigantic bones were buried in the riverbed? Based on modern paleontological discoveries along the Orontes River, we can guess that the bones belonged to a mastodon or even a steppe mammoth, *Mammuthus trogontherii*. (If it was a steppe mammoth, Pausanias's figure of 11 cubits is not far off, since these animals could reach 14 feet high at the shoulder; the femur would be about 4.5 feet long). According to Pausanias, the Syrians sent messengers to consult the great oracle of Apollo at Claros (in Ionia) about the identity of the giant. Oracles settled controversies and dispensed advice to individuals and city-states; the cryptic responses of the gods were interpreted by priests or seers. The Claros oracle determined that the big bones belonged to Orontes, the mythical giant-hero of India after whom the river was named. But Pausanias says that some people disagreed with the oracle's interpretation, arguing that it was the body of a different giant, from Africa, named Aryades. Others may have recalled that in archaic times the river had been named after Typhon, a fearsome monster killed by lightning during the mythical war between the giants and the gods (the Gigantomachy). According to that tradition, Typhon had tried to escape by burrowing underground, thus creating the riverbed.[18]

Pausanias reported another oversize skeleton at Miletus, on the Turkish coast southeast of Samos. This 15-foot (4.5-m) skeleton appeared on one of the "little islands that broke off from the island of Lade" in the harbor. The local people identified this skeleton as Asterios, the son of Anax, a giant offspring of Earth who was the legendary founder-hero of Miletus. The bones' dimensions and location, only a few miles from Samos, suggest that Asterios was probably a large Miocene mammal that eroded out of the cliffs of an unstable island. A similar skeleton may have inspired similar local lore in Ephesus, the city directly across from Samos: the founder-hero of Ephesus was said to have been another giant son of the Earth, named Koresos.

"I was surprised by yet another event in upper Lydia," says Pausanias. Near the river Hyllos (near modern Usak, Turkey) at Temenothyrae, "a storm broke open the ground, and some huge bones appeared. You would have thought they were human by their form, but the size was phenomenal!" Pausanias continues, "The story immediately went around everywhere that this was the body of Geryon." Geryon, the mythical monster-giant killed by Heracles in his Tenth Labor, had raised prodigious oxen. "Everyone in the area of Hyllos knew someone who had plowed up big horns of cattle," which they now realized must have belonged to Geryon's herd. But the learned Pausanias objected. Wasn't Geryon slain by Heracles in Cadiz, in Tartessus (Spain), and weren't his cattle driven from there to Greece? Pausanias knew that a gigantic skeleton had already been found at Cadiz, and he may have heard about the cattle's famous footprints in Italy. Now the Lydian priests and city fathers had to step in to resolve the impasse. They decided to say that the bones belonged to another giant, named Hyllos.[19]

Despite Pausanias's naturalistic details of the bones' magnitude and weathering, classicists have not thought to connect these anecdotes to actual fossil exposures in Turkey. I contacted paleontologist Sevket Sen (at the Paleontology Laboratory of the Museum of Natural History in Paris) about the presence of large fossil skeletons in Anatolia. Sen, who has excavated extensively in the Aegean and Turkey, confirmed that extensive Miocene sediments exist from the Gallipoli Peninsula (ancient Thracian Chersonese), the Troad (site of ancient Troy), and the island of Imroz (ancient Imbros), to the southern coast of Turkey. (These locales are all places where the ancients discovered enormous bones; see chapter 3.) Dense concentrations of large mammal fossils are "frequent and can be extremely rich," says Sen, who knows of more than fifty localities with mastodon, rhinoceros, and giraffid remains in western Turkey. "The Miocene is fabulously represented in Turkey," agrees William Sanders, a proboscidean specialist who excavates in Turkey. The Miocene-Pliocene was the heyday of prehistoric elephants, and Turkey in particular has "a rich proboscidean fauna of gomphotheres and deinotheres [very large Miocene mastodons]." In the region of Ozluce, Turkey (inland east of Samos), for example, exceptionally complete skeletons of Neogene elephants and rhinoceroses are abundant, while deinotheres are common in the lignite soils around Izmir (on the mainland east of Chios).[20]

Ancient Taxonomy

The impressive remains that the Lydians attributed to the ogre Geryon and his outsize cattle surely belonged to extinct mastodons and bovines whose fossils lie in western Turkey. National pride led the Lydians to identify their giant skeleton as a notorious mythical monster killed by Heracles, but Pausanias's geographical objection forced the local authorities to reassign the bones to a different giant, one who had given his name to the river Hyllos. The language here is mythological, yet all three of Pausanias's narratives demonstrate people's rational efforts to distinguish the geographical distribution of no-longer-extant creatures and to create a "taxonomy" of extinct species—which they visualized as giants and monsters—that was consistent with mythohistory, current events, and previous discoveries of big bones.

Taxonomy is an ordering process that uses a special vocabulary for identifying and classifying types of life-forms. Geoffrey Lloyd, historian of ancient science at Cambridge University, observes that Greek literature exhibits "complex classification systems" based on acute observation of natural phenomena similar to systems common in nonliterate societies, but he questions whether "such taxonomies" are the "products of deliberate research" motivated by a "desire to extend knowledge." Although formal, systematic paleontological research on big bones was not carried out in antiquity, the narratives gathered in this book show that people did compare historical examples of big bone finds, and they collected, measured, displayed, and even tried to reconstruct large skeletons, activities that seem to express a desire to extend their knowledge of giant species of the past.[21]

The ancient method of naming extinct giants based on geography and mythohistory parallels modern scientific nomenclature for extinct animals, in which paleontologists often choose Greek, Latin, and other names that reflect local legends about monsters. For example, the nineteenth-century fossil hunter Othniel C. Marsh named the immense *Brontotherium* ("Thunder Beast") of North America after the Sioux myth of the Thunder Beast, which the Indians associated with the big fossils exposed by thunderstorms in the Dakota badlands. The old Russian legend of the earth-shaking Indrik Beast was commemorated in the name *Indricotherium* for the largest-ever land mammal (discovered in 1911 in Central Asia). The Aztec feathered serpent god is honored in the name *Quetzalcoatlus* given to the pterosaur with a 35-foot (over 10-m) wingspan found in Texas in 1972. The recently discovered ceratopsian dinosaur *Achelousaurus horneri*, which appears to have a broken horn, is named after Achelous, the river monster of Greek myth whose horn was broken off by Heracles, and for Jack Horner, whose surname humorously restores the beast's lost horn.[22]

The urge to connect the remains of mysterious creatures with mythical and historical events, with monsters or heroes of the deep past, is widespread. In medieval Europe, the stupendous bones of prehistoric animals were believed to belong to giants, saints, and celebrities from antiquity. The big bones were placed in coffins and reburied by the hundreds in medieval churches as saints' relics. Mammoth and rhinoceros skeletons were arranged into upright positions and displayed as antediluvian giants, primitive cavemen, and even historical Visigoths and other barbarian warriors. The notorious case of King Teutobochus was thoroughly investigated by French paleontologist Léonard Ginsburg in 1984. Colossal bones and teeth were discovered in 1613 in southern France and widely exhibited as those of Teutobochus, the giant king of the Germanic tribe defeated in 105 B.C. by the Romans. Ginsburg examined "Teutobochus's" tooth and determined that the giant's remains actually belonged to an extinct elephant, the *Deinotherium*, one of the largest mammals that ever lived.[23]

Hominid Fever

In classical antiquity, the great bones of extinct mammals were frequently perceived as those of giant humans. The tendency to see the remains of elephants and other large mammals as human is not as odd as it may seem. As we have seen, because of geological disruption around the Mediterranean, the fragile skulls of prehistoric elephants often crumble away, leaving femurs, scapulas, and teeth. These surviving remains resemble enormous counterparts of our own anatomy (see figs. 2.7, 2.8, 2.9, 2.10). Who can resist comparing an oversize femur to one's own thigh? Moreover, people of antiquity visualized their ancestors and mythical heroes as superhuman giants, so they were primed to see larger-than-expected mammal bones as vestiges of huge humans (see chapter 5). In rural areas of Greece today, some people still imagine the ancients as larger than us moderns. In the 1990s, for example, a very large fossil femur was discovered by a resident of Luka, a village in Arcadia (Peloponnese), a region with Pleistocene mammoth remains. The doctor who examined it concluded that it be-

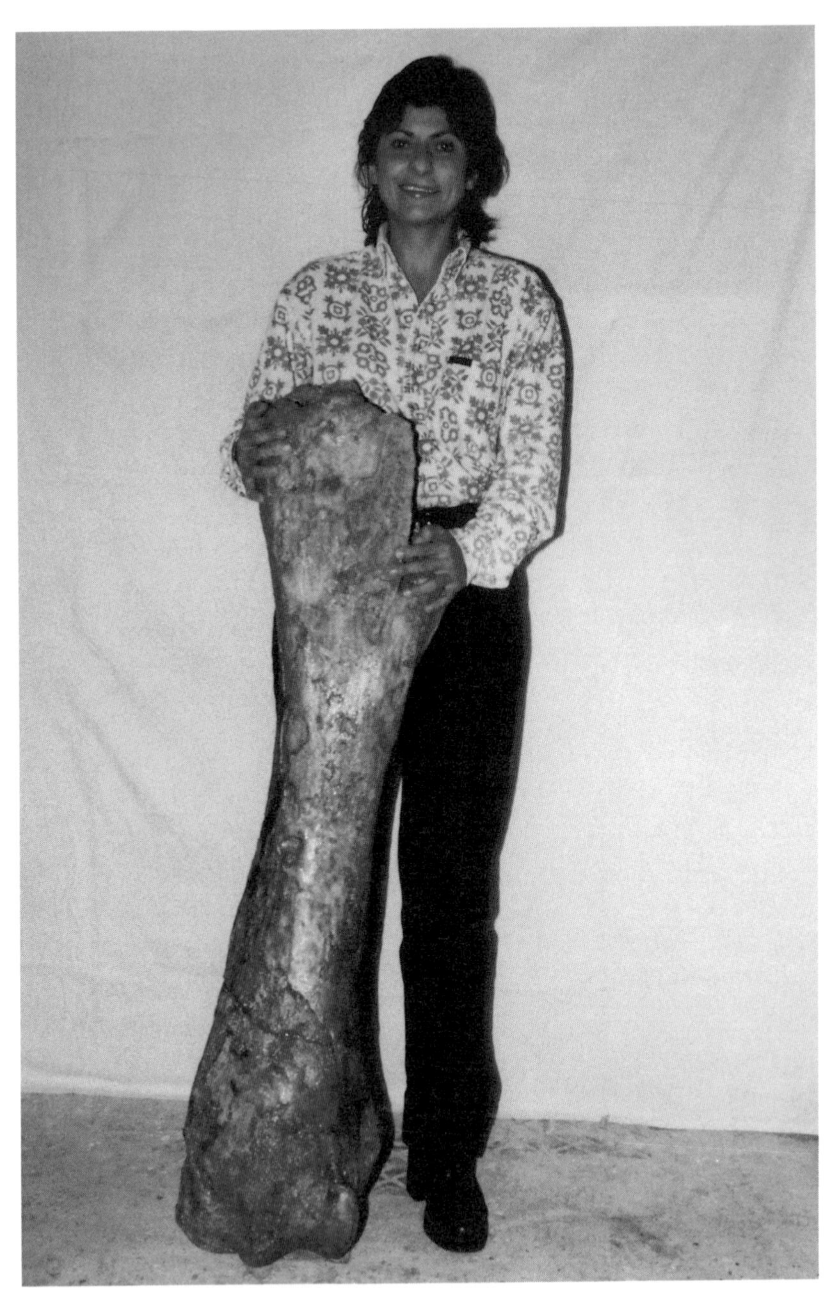

2.7. Paleontologist Ana C. Pinto with fossil femur of the Pleistocene elephant *Palaeoloxodon antiquus*, excavated in Asturias, Spain. Courtesy of Ana C. Pinto. Photo by M. Pajuelo.

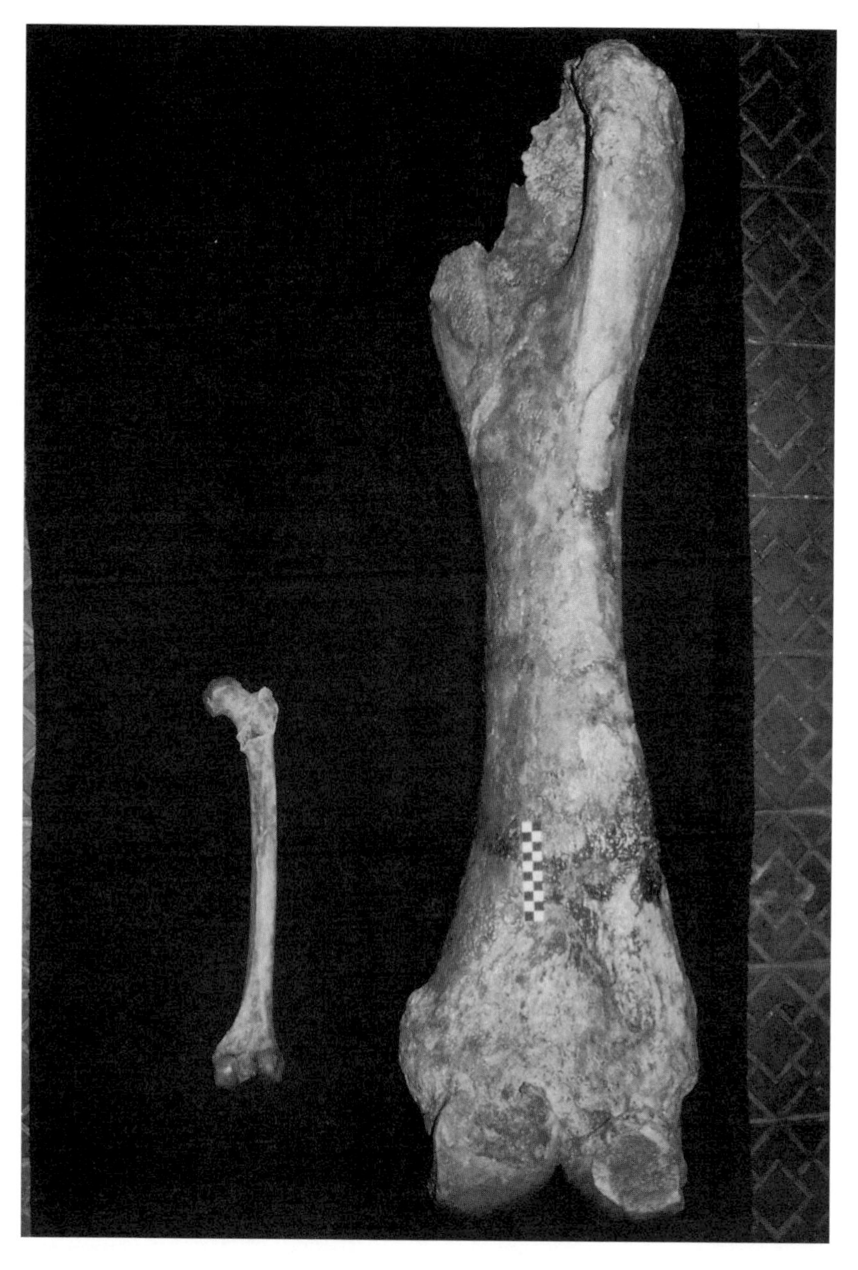

2.8. *Left*: human femur. *Right*: fossil femur of *Palaeoloxodon antiquus*. Courtesy of Ana C. Pinto. Photo by M. Pajuelo.

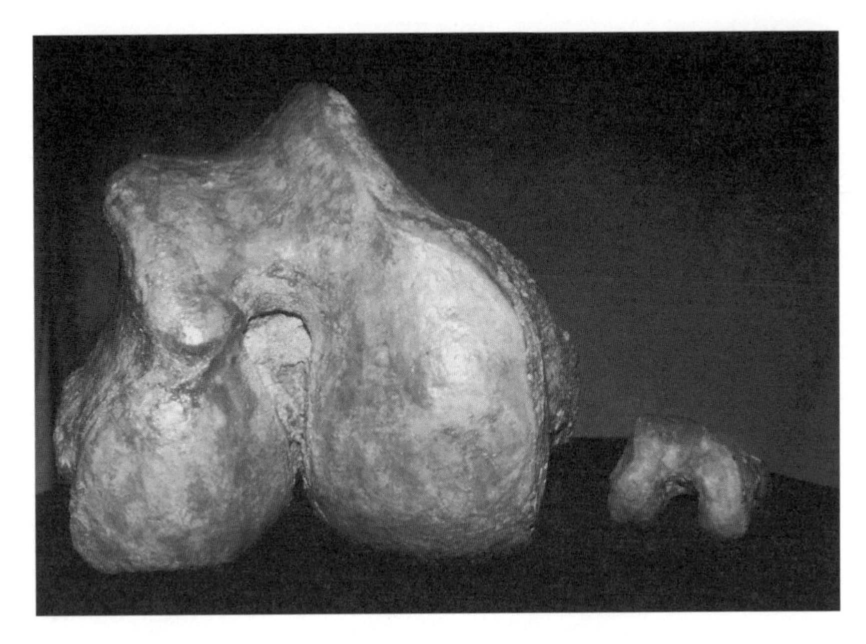

2.9. Left: distal end of *Palaeoloxodon antiquus* femur. Right: human femur. Courtesy of Ana C. Pinto. Photo by M. Pajuelo.

longed to a "very well-built human of antiquity, certainly bigger than any modern person."

According to forensic anthropologist Douglas Ubelaker, nonhuman bones, especially femurs, can fool even the most experienced medical experts. In his study of modern FBI files, he found that about 15 percent of "human" bones thought to be those of murder victims turn out to be animal bones. Ubelaker points out that similarities of mammal anatomy, the finder's expectations, and the context of discovery encourage the misidentification as human. Those same factors figured in antiquity.[24]

Are humans hardwired with an anthropomorphizing search image? That is the theory developed in 1993 by anthropologist Stewart Guthrie. The tendency was first noticed by the sixth-century B.C. Greek philosopher Xenophanes. For powerful psychological and evolutionary reasons, we scan the world for "what matters most," namely, "humanlike models." In Guthrie's theory, the ancients would have been predisposed to perceive oversize mammal

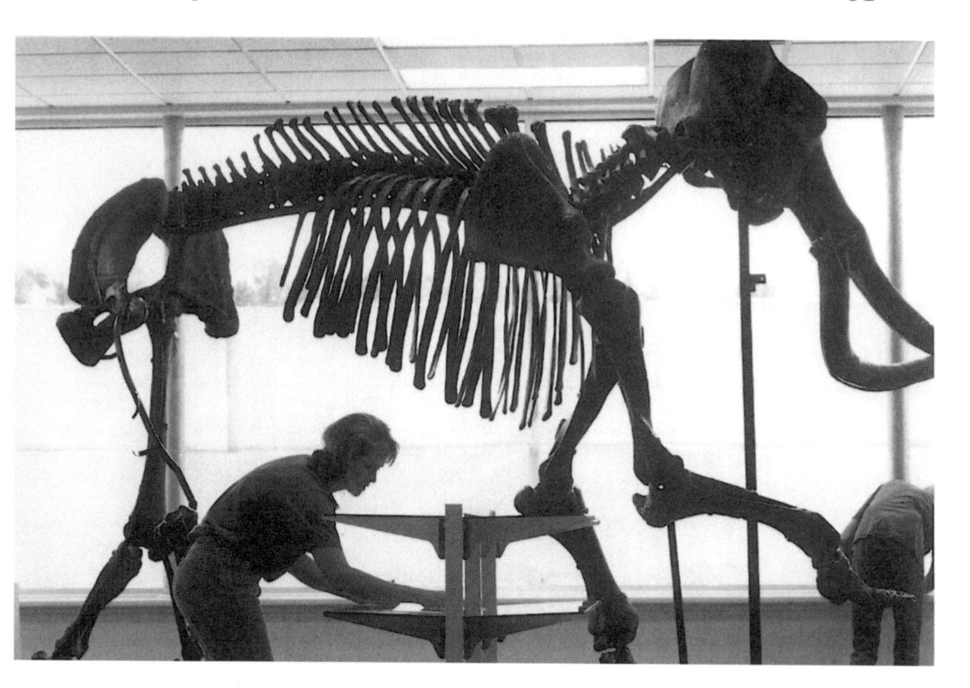

2.10. Mammoth skeleton (*Mammuthus primigenius*), a common fossil in Greece, Italy, and Anatolia. The position of the woman mirrors that of the mammoth, making it easy to see how the mammoth's bones might have been taken as giant counterparts of human anatomy in antiquity, especially if the skull were missing. Photo Thomas Lee, *Livingston* [Montana] *Enterprise.*

bones as belonging to highly organized creatures like themselves. Their expectations translated anatomical cues, such as mammalian thigh bones and large teeth, into human forms (fig. 2.11). As Guthrie notes, our anthropomorphizing drive creates illusions even as it ensures our survival.

Unusual remains are all the more exciting and meaningful if they are imagined as belonging to our own kind. Natural historian Leonard Krishtalka dubs this "driving passion to unearth our own evolutionary roots" *hominid fever.* The fever can be fruitful, in that it leads paleobiologists to search out human origins, but pride in "prehistoric pedigrees" can also spawn errors, hoaxes, and nation-

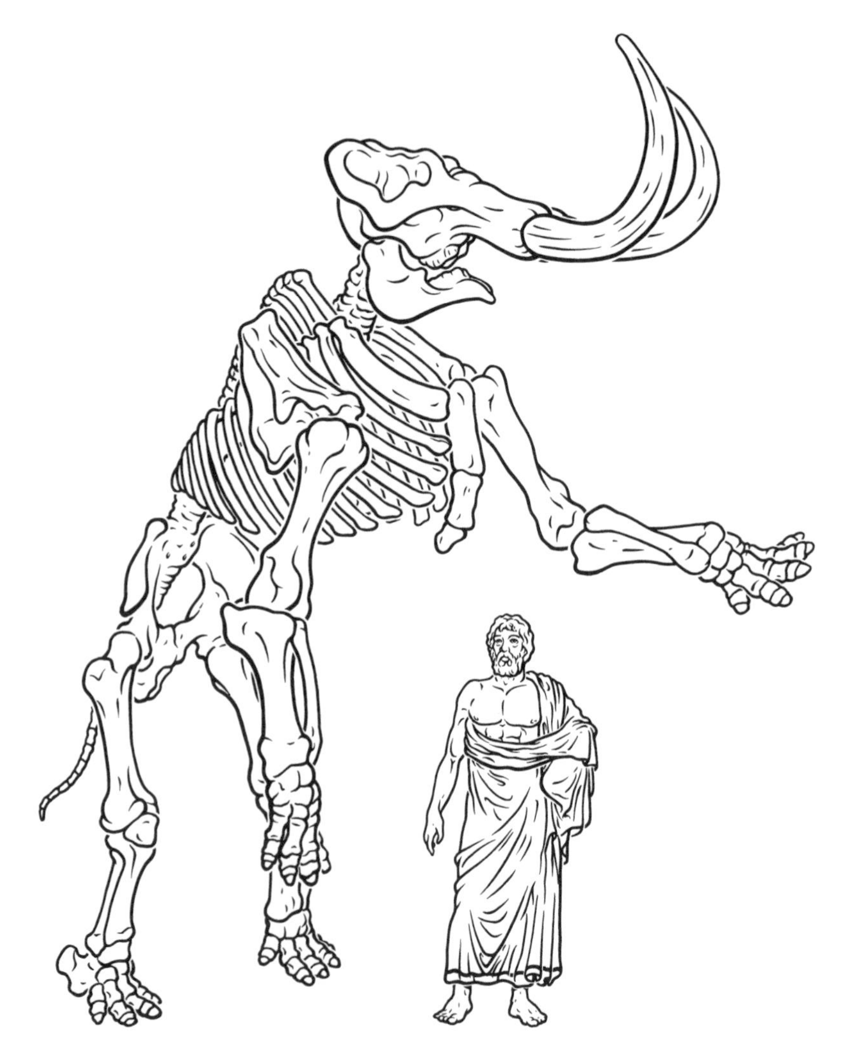

2.11. Mammoth skeleton arranged to conform to the ancient Greek image of a mythical giant whose remains were buried in the earth. Drawing by Ed Heck.

alistic propaganda. Nations "with fossil hominids feel somehow anointed," observes Krishtalka.[25]

In antiquity, hominid fever influenced the identification of big bones as relics of giants and heroes of the glorious Greek past, as we saw in Pausanias's experiences with big skeletons in Asia Minor.

In the next chapter, we'll see how Greek and Roman cities actively acquired impressive "heroes' bones" to enhance their religious and political power. Despite the strong tendency to anthropomorphize huge bones, however, some large prehistoric remains were correctly perceived as those of animals of the distant past rather than giant humans, as we saw in the legends of the Neades and war elephants of Samos.

Paleontology of the Mediterranean Lands

To illuminate what kinds of prehistoric remains were observed in antiquity, the following pages survey modern paleontological discoveries in the same places where Plutarch, Pausanias, and others said that giant bones emerged from the earth. (Appendix 1 gives examples of the largest, mostly Neogene and Pleistocene, fossil species reported in the paleontological literature, for the regions where ancient writers located remarkable remains.)

Three exceptionally rich fossil sites in Greece—Pikermi, Samos, and Megalopolis—are typical of the bone beds around the Mediterranean. Their colorful histories and contributions to paleontology are not widely known, even though the sites were scenes of feverish fossil hunting in the recent past. Modern paleontology in the Mediterranean began in 1839 after a chance find near Athens.

The Pikermi Bone Rush

In 1837, Greece was ruled by the German-born king Otto. One day, a Bavarian soldier in Otto's service was guarding the bridge over the Megalo Rhevma, a dry creek bed at Pikermi, a village few miles northeast of Athens. He noticed something glittering in the gully. Grubbing in the dirt, he uncovered what looked like a human skull encrusted with diamonds. He hid his treasure in the barracks near Pikermi.

Back in Munich, the soldier bragged about his jewel-encrusted ancient Greek skull. When he tried to sell it, he was arrested for

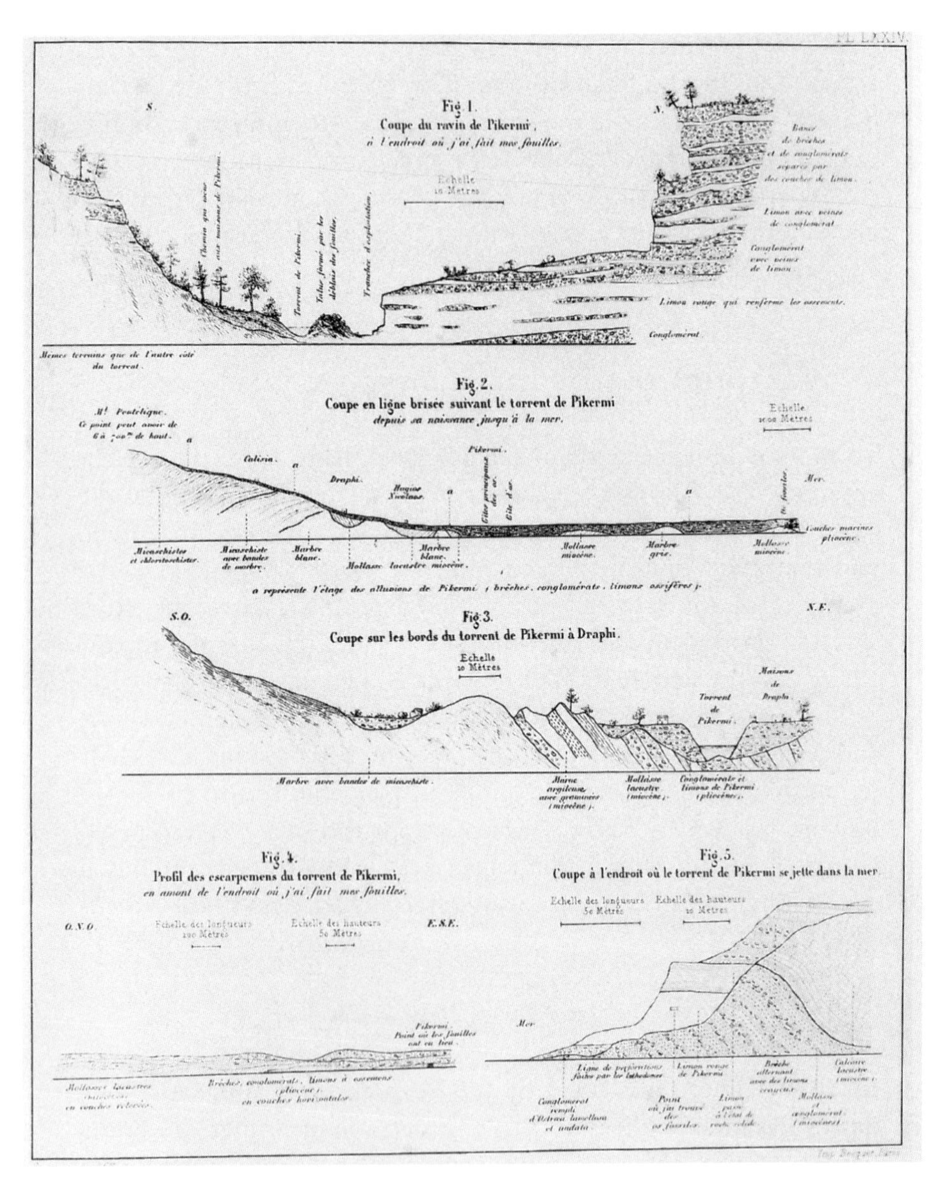

2.12. Cross-section drawings of the stratigraphy of the Megalo Rhevma bone beds, Pikermi, Greece. Gaudry 1862–67, *Atlas.*

grave robbery. The police called in the eminent zoologist Andreas Wagner as an expert witness. The "diamonds" turned out to be large calcite crystals. But to Wagner's amazement, the skull belonged not to an ancient Hellene but to a 13-million-year-old ape. The soldier's electrifying discovery sparked a bone rush in Greece.

German, Austrian, French, Swiss, British, and Greek scientists converged on Pikermi, where they commenced large-scale excavations between 1839 and 1912 (figs. 2.12, 2.13, 2.14). From the Megalo Rhevma came the remains of creatures at once oddly familiar and totally strange. Almost all of the Pikermi species turned out to be transitional types, "missing links" between successive epochs, which provided strong evidence for Darwin's new theory of evolution (first published in 1859). Today, more than 8,000 specimens from the Pikermi bone rush are stored in museums around the world.[26]

The paleontologists excavated a profusion of fifty-three different Neogene species (figs. 2.15, 2.16, 2.17, 2.18). Besides apes, ostriches, pigs, and three-toed horses, there were immense tortoises the size of a Volkswagen Beetle. The carnivores—hyenas, lions, bears, and saber-toothed tigers—loomed much larger than today's predators. The most massive bones belonged to great mastodons, the giant giraffe *Helladotherium*, rhinoceroses, and the grotesque chalicothere *Ancylotherium*, a lumbering herbivore with hooked claws instead of hooves. The most formidable behemoth found at Pikermi is the tremendous elephant *Deinotherium giganteum*, the second-largest land mammal ever to walk the earth (after the *Indricotherium*). Deinotheres stood about 15 feet (4.5 m) tall and sported tusks that arched down and back rather than up and forward (fig. 2.19). They ranged across Europe, Eurasia, and Africa until they died out in the Pleistocene epoch.

How did so many different types of animals end up jumbled together in the gully? Some paleontologists proposed that a cataclysm, say a vast landslide, caused a violent inrush of the sea, forcing the panicked beasts together before they were drowned and buried under mud. The French paleontologist Albert Gaudry imagined that the swiftly rising Aegean forced all the animals up Mount Penteli where they starved. Othenio Abel noticed that

2.13. Excavations at Pikermi, Greece, ca. 1900. Photo by T. Skoufos, in O. Abel, *Leibensbilder aus der Tierwelt der Vorzeit* (Jena, 1922), fig. 132.

2.14. Block from the bone beds at Pikermi, Greece. The fossils of numerous species are mixed together, in a manner typical of bone beds of the eastern Mediterranean. Note how the large limb bones and shoulder blade stand out. Photo A. Smith Woodward 1901, in O. Abel, *Leibensbilder aus der Tierwelt der Vorzeit* (Jena, 1922), fig. 130.

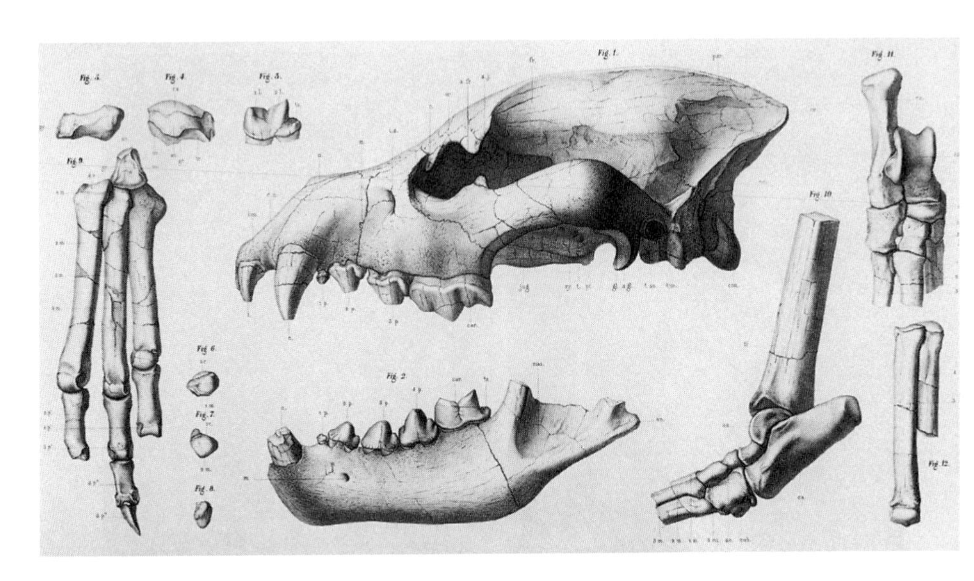

2.15. Skull of the giant hyena (*Hyaena eximia*) of the Neogene in Greece. It measures about 11 inches long; the incisor is about 1.5 inches long. Gaudry 1862–67, plate 13.

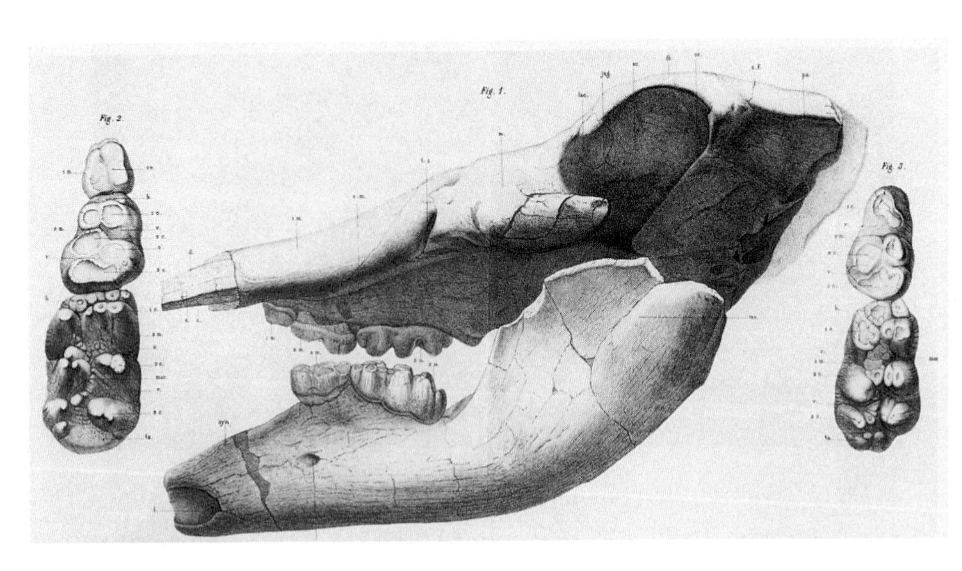

2.16. Mastodon skull and teeth from Pikermi, Greece. *Mastodon pentelici* is a common Neogene fossil of the eastern Mediterranean. The molar measures about 3 inches long. Gaudry 1862–67, plate 22.

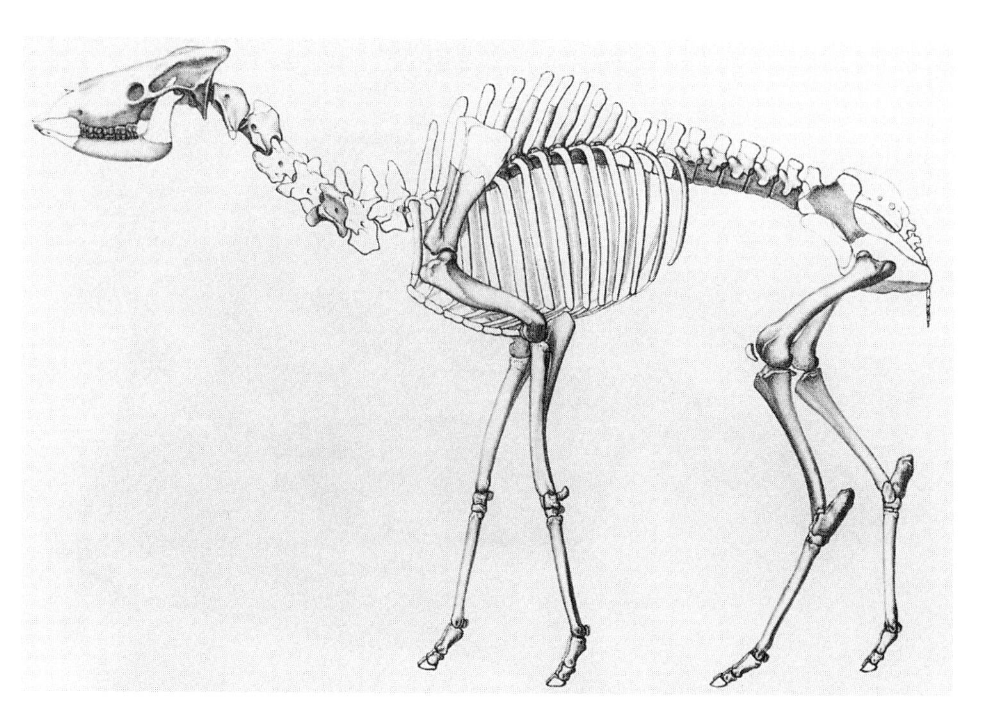

2.17. *Helladotherium* skeleton, Pikermi, Greece. The femur of this giant Miocene giraffe is about 28 inches long. Gaudry 1862–67, plate 44.

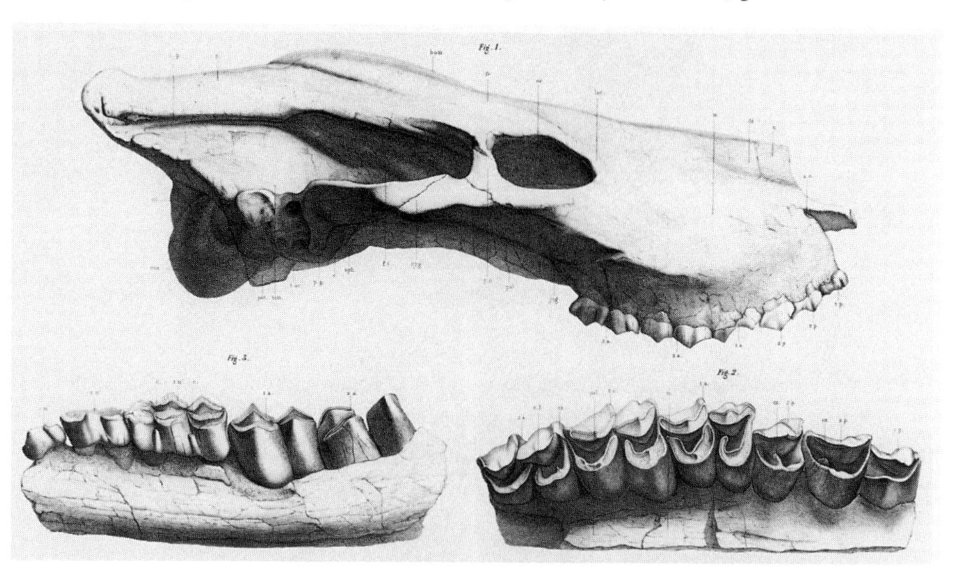

2.18. Giant giraffe skull, about 19 inches long. The *Helladotherium*, the ancestor of the living okapi of Africa, stood about 7 feet tall at the shoulder. Gaudry 1862–67, plate 41.

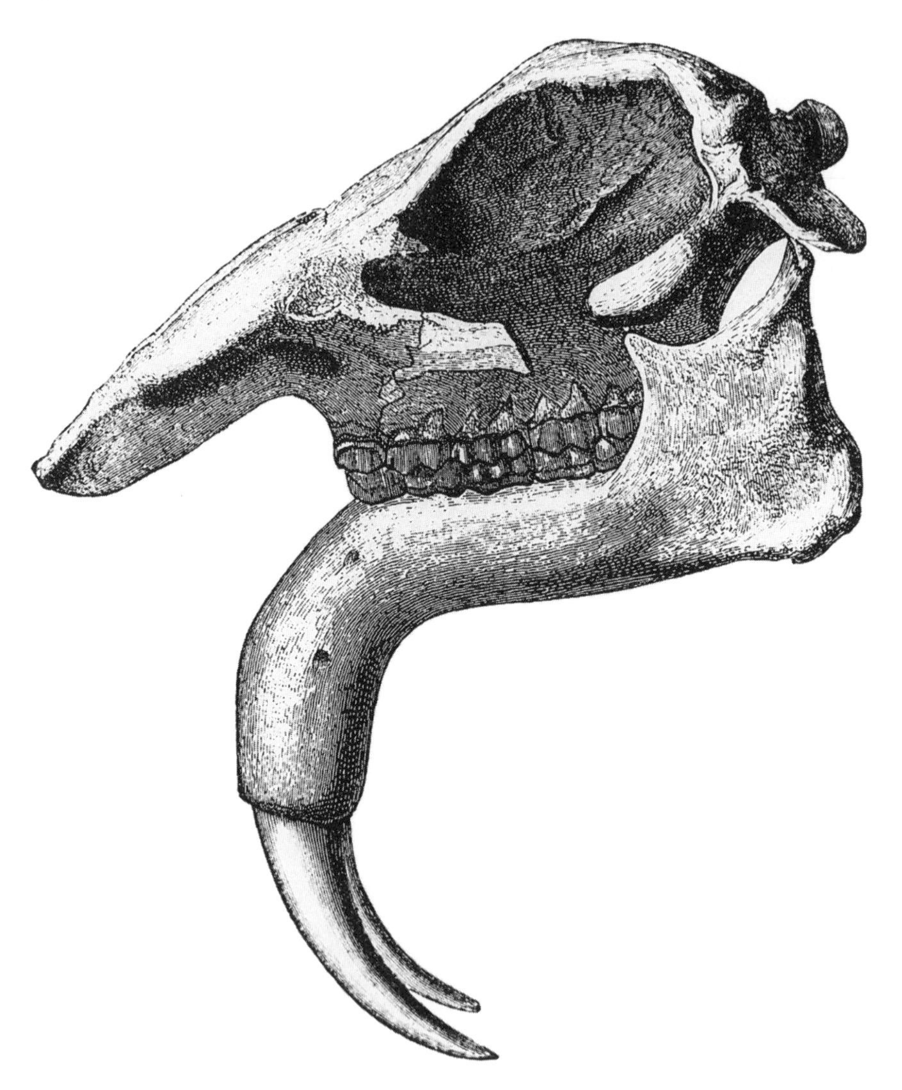

2.19. *Deinotherium giganteum* skull. This colossal elephant stood about 15 feet tall at the shoulder. It ranged across Europe, North Africa, and Asia during the Miocene to the Pleistocene. Karl Zittel, *Handbuch der Palaeontologie* (Munich, 1891–93), fig. 374.

many of the limb bones of the large herd animals were fractured; he speculated that a catastrophic forest fire stampeded the animals over a precipice.[27]

It was the Pikermi fossils that allowed the nineteenth-century scientists to advance the bold theory that Greece had been the crossroads in the mass migrations of many animal groups, especially proboscideans, between Asia, Europe, and Africa in the Miocene epoch, 23 to 5 million years ago. Some of the extinct species resembled living elephants and other animals of Africa and Asia, yielding important knowledge about origins and stimulating a wider search for transitional mammals. Fossil remains like those at Pikermi also exist in Europe, northern Greece (Euboea, Boeotia, Thessaly, Macedonia), on Samos and several other Aegean islands, in western Turkey, North Africa, and the Siwalik Hills of India and Pakistan. The largest Pikermi-type fossil skeletons account for many of the bones of giants, monsters, and dragons described by ancient authors in the following chapter.[28]

Fossils of Samos

While others were digging at Pikermi, paleontologist Charles Forsyth Major was inspired by Plutarch's description of the monstrous Neades and Dionysus's war elephants to look for large vertebrate fossils in Samos. After learning from the village doctor of the bone beds near Mytilini, Major was the first to identify the skull and femur of the *Samotherium*, the great Miocene giraffe, in 1885 (fig. 2.20). He amassed a rich collection of more than 2,000 fossil specimens for Swiss and British museums. From what proved to be some of the most prolific fossil beds in all Europe, paleontologists are still unearthing the bones of samotheres, mastodons, deinotheres, rhinoceroses, chalicotheres, early horses, large hyenas, ostriches, and other Pikermi-type species from deposits in Samos.[29]

The place where Plutarch said Dionysus's elephants defeated the Amazons, the Blood-Soaked Field, has never been securely identified by ancient historians. I described this ancient topographical puzzle to paleontologist Nikos Solounias, who discovered his first

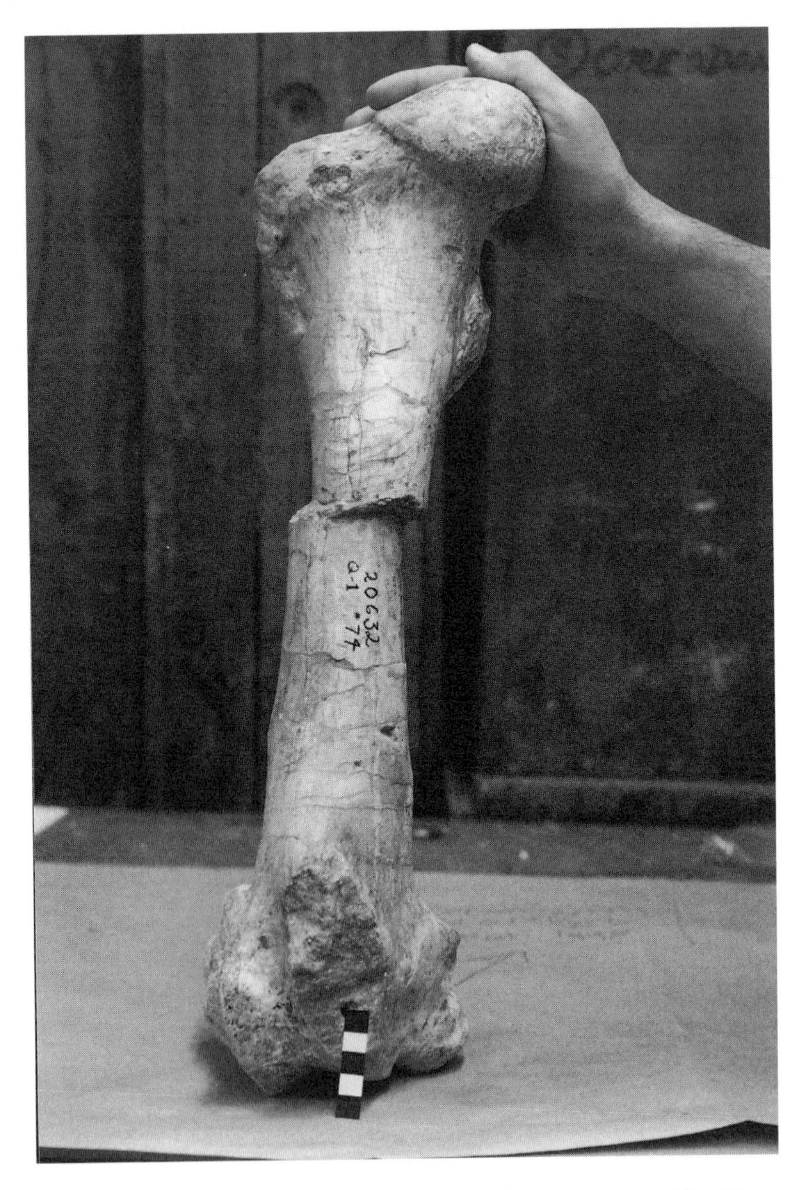

2.20. *Samotherium* femur, about 20 inches long, excavated by Barnum Brown, Samos, Greece, 1923–24. Note shearing by seismic activity. Photo courtesy of Nikos Solounias.

fossil (a *Samotherium* femur) as a boy near Mytilini and has excavated extensively in Samos. Solounias immediately recalled a landmark that matched Plutarch's description. The fossils of Samos are embedded in white, not red, sediments. But not far from Mytilini, between the two most abundant fossil exposures on the island, lies a flat plateau of unusual red soil. This unique red surface contrasts starkly with the surrounding hills of white sediment containing masses of chalk-white bones. The place corresponds to the ancient image of a blood-soaked battlefield with the bleached bones of war elephants and mighty Amazons piled around it.

The other site where Plutarch said big bones were displayed was Phloion. The location of this place puzzles classicists, who have tried to link the name to religious cults. Knowing that *phloion* means "crust of the earth," however, Solounias searched for a geological formation that would match Plutarch's description of tremendous beasts destroyed by collapsing earth. He located a dense concentration of Miocene fossils trapped under a massive shelf of limestone thrust up by earthquakes. The ancient Greeks logically explained the great block of faulted limestone overlying the immense bones of mastodons and samotheres as the result of a terrible earthquake in the deep past.[30]

Who first discovered the prehistoric remains of Samos? Since the most fertile soil occurs in the Neogene sediments, ancient farmers would have plowed up big bones in their fields (fig. 2.21). The search for valuable minerals probably also led to fossil discoveries. In chapter 1 we saw that Scythian nomads combing the landscape for gold encountered dinosaur remains, Siberian natives collecting gold came upon Ice Age mammoths, and Austrian quarrymen discovered the rhinoceros skull known as the Klagenfurt dragon. In the New World, we know that Native Americans collecting salt encountered dinosaur and mammoth bones. In classical antiquity, too, prospectors would have found fossils on Samos. In his fourth-century B.C. treatise *On Stones*, the natural philosopher Theophrastus told how miners dug deep shafts through different layers of rock to extract "Samian earth," a dense white kaolin. This rare earth was a precious commodity, mined for medicinal and other uses. The kaolin deposits were located by prospectors who

2.21. Fossil Miocene rhinoceros jawbone on Samos, Greece. Photo © Kevin Fleming.

"crawled over the ground searching for veins among the rocks," according to Pliny. Nikos Solounias confirms that kaolin does occur near the fossil beds around Mytilini, so it seems likely that ancient mineral collectors were among the first to call attention to the extraordinary bones weathering out of the soil.[31]

Like the Pikermi fossils dispatched to faraway museums, the bones of the ancient Neades have been dispersed around the globe. After Major left Samos, German excavators took away thousands of remains between 1890 and 1920. A German exporter of Samian wines, Karl Acker, paid farmers for tons of fossils, which he packed into wine crates and sold to numerous museums in Germany. Fossils became an essential cash crop for many Samos farmers, recalling the traditional harvest of dragon bones in China (chapter 1). In 1924, the great fossil hunter Barnum Brown shipped what is still the world's largest trove of Samos fossils (some 5,000 specimens) to the American Museum of Natural History (AMNH) in New York. Between 1850 and 1924, more than *30,000* fossil bones had been taken from Samos by foreigners to enrich museums and private collections.[32]

Whose Bones?

Notice a pattern here? After local people discover spectacular skeletons, outsiders often arrive to remove the fossils from their home soil for "safekeeping" and "proper" interpretation. This is a reccurrent theme in paleontology, from earliest antiquity to today's legal disputes over *T. rex* ownership and the astronomical prices paid for dinosaur and other fossil remains. Two related issues are at stake: Who assigns meaning to remarkable bones? Who should own them?

Wrangling over possession of fossils crops up whenever sensational bones crop up. Beginning in the sixteenth century, crates of mammoth remains were shipped from the Americas to Europe. As we saw in chapter 1, the Mongolians were not pleased with Roy Chapman Andrews's profitable exploitation of their dinosaur eggs in New York. More recently, in 1996, the Dutch formally requested the repatriation of their fossil *Mosasaurus* skull from France. The skull of the gigantic Mesozoic marine reptile had been discovered in 1770 by quarrymen in a chalk mine near Maastricht. Then, in 1794, Napoleon seized the fossil and took it to Paris, where it became the key to Georges Cuvier's theory of extinction. The Dutch now claim that the great skull is intrinsic to their national identity. But the Paris Museum of Natural History intends to keep the relic that "changed the history of paleontology." In earliest antiquity, too, heavy petrified bones were transported great distances for political gain and to enrich museums, and intense feelings arose over the possession of and the authority to interpret such marvels. Plutarch and Pausanias preserved disputes over the meaning of huge bones in Samos and Asia Minor, and the next chapter shows how Greek and Roman authorities interpreted and appropriated valuable fossils found by ordinary folk.[33]

By the time Barnum Brown sailed to Samos in 1921, the Greek government had passed laws forbidding the foreign plunder of its antiquities, including fossils. Brown found it curious that even though ancient writers like Plutarch "proved their knowledge of the presence of fossils" on Samos, not "a single trace" of that ancient knowledge appeared in modern writings about Samos.

2.22. Refugees excavating for Barnum Brown, near Mytilini, Samos, Greece, 1923–24. From Brown 1927. Courtesy of Department of Library Services, American Museum of Natural History.

Brown blamed the Greek government's restriction on exports "of antiquities, which has been warped to include fossils" for keeping the great Aegean bone beds relatively unknown to the rest of the world. Over the protests of the fossil merchant Karl Acker, Brown hired destitute Greek refugees (expelled from Turkey in 1922) to quarry fossils near Mytilini. Men loaded the unwieldy bones of mastodons and giant giraffes on mules or carried them slung between poles down switchback trails to the harbor (figs. 2.22, 2.23). Brown then requested extraordinary permission from the Greeks to ship everything he dug up back to the AMNH. Professor Theodore Skoufos, the paleontologist at Athens University, hoped to retain the fossils for Greece, but in the end the Americans pressured the Greek government to let thousands of bones, considered priceless antiquities, leave Samos in 1924.[34]

2.23. Barnum Brown (right) excavating bone bed near Mytilini, Samos, Greece, 1923–24. Note large limb bone in left foreground. From Brown 1927. Courtesy of Department of Library Services, American Museum of Natural History.

Brown wanted to excavate the newly discovered bone beds of Megalopolis in Arcadia (central Peloponnese) next. But that site was already being dug by his nemesis, Professor Skoufos. Megalopolis, an important urban center in the time of Aristotle, is another crucial site in the lost history of ancient paleontology. Pausanias, Herodotus, and others tell us that enormous bones were collected there and revered as the relics of giants and heroes of myth, but today the fossil riches of Megalopolis are even less known than those of Pikermi and Samos. In the 1960s, the translator of Pausanias, Peter Levi, tried to track down the Megalopolis fossils excavated by Skoufos. Levi finally found the big bones languishing in the "old petrological museum close to the University of Athens, a ruinous building that can be visited only by personal negotiation."[35]

2.24. Professor Theodore Skoufos excavating a fossil skull and tusk (*M. meridionalis*?) at Megalopolis, Peloponnese, Greece, 1902.

Megalopolis Megafauna

Like the Pikermi fossils, the fossils of Megalopolis came to the attention of scientists because of a chance find. In 1902, a woodcutter searching for a lost ax discovered tusks of outrageous dimensions in a steep ravine near Megalopolis. Word of the find reached Professor Skoufos in Athens, who began extensive excavations. In a few months he transported five tons of Pleistocene fossils from Megalopolis to Athens University (fig. 2.24).

Skoufos was the first trained geologist to study the fossil exposures around the ruins of ancient Megalopolis, but of course the people of Arcadia had known about the prodigious skeletons since earliest times. The bones and tusks of prehistoric elephants and mammoths turned up under farmers' plows and well-diggers' shovels, all along the Alpheios River and other valleys cutting through Pleistocene sediments. Ordinary people in Arcadia still encounter big fossils, which are displayed in the museums of Dimitsana, Megalopolis, Olympia, and other towns.[36]

In 1994, for example, road-builders working northwest of Olympia unearthed two huge tusks, each about 10 feet long (3 m). Since they are straight, they probably belonged to the 13-foot-tall (4-m) Pleistocene "ancient elephant" *Palaeoloxodon* (or *Elephas*) *antiquus*, whose remains are common in Arcadia, Italy, and Eurasia. Those great tusks, now stored in the museum at Olympia, recapitulate similar exciting discoveries in antiquity: Pausanias tells us that an enormous shoulder blade was displayed in Olympia in the time of the Trojan War, and that a pair of gigantic tusks was kept in the temple at Tegea, southeast of Olympia. In 1997, some big bones came to light in a lignite mine operated by the Greek electric company near Megalopolis. In antiquity, the spontaneously burning lignite soils of Arcadia were identified as still-smouldering battlegrounds where Zeus's cosmic lightning bolts had destroyed mythical giants and monsters during the Gigantomachy. Now those same combustible lignite deposits provide modern Greeks with electricity. The big bones found in the lignite mine in 1997 are exhibited in the museum of antiquities at Megalopolis—the same city where Pausanias saw the enormous skeleton of a fallen giant displayed in a temple nearly two thousand years ago.[37]

Giant Bones around the Greco-Roman World

The most impressive bones embedded in the lands known to the Greeks and Romans belong to numerous species of proboscideans of the late Tertiary and Quaternary: the gigantic Miocene-Pliocene mastodons and early elephants that evolved into the great mammoths and elephants of the Pleistocene and Holocene. Imagine the excitement of unearthing an ivory tusk 10 feet long, or plowing up a petrified thigh bone nearly as tall as yourself! The femurs of prehistoric elephants usually measure between 3 and 5 feet in length (1–1.5 m). The longest pair of tusks ever found in Greece was discovered in Macedonia in 1997 by Evangelia Tsoukala, paleontologist at Aristotle University, Thessaloniki. Those tusks, over

2.25. The longest ivory tusks ever found in Greece (about 14 feet in length). Mastodon *Zygolophodon (M.) borsoni*, excavated in Macedonia, 1998. Note limb bone encased in plaster at bottom left. Photo courtesy of Evangelia Tsoukala, Aristotle University, Thessaloniki.

14 feet long, belonged to the mastodon *Zygolophodon* (*Mammut*) *borsoni*, whose remains occur in Italy, Greece, the Aegean, and western Turkey (fig. 2.25).[38]

The ancestral mammoth *Mammuthus meridionalis* also ranged from southern Europe to Asia during the Pliocene-Pleistocene. It was the same height as the Eurasian "ancient elephant" *P. antiquus* (about 13 feet at the shoulder) but its tusks were slightly curved and about 8 feet (2.5 m) long. The woolly mammoth, *Mammuthus primigenius*, with its strongly curved tusks, is usually associated with northern Europe and Siberia, but during the last Ice Age it roamed southern Europe, leaving huge remains as far south as Megalopolis (figs. 2.10, 2.26). Woolly mammoths stood 9 to 11 feet high (about 3 m). The colossal steppe mammoth of Eurasia, *Mammuthus trogontherii*, towered over 14 feet (4.3 m) at the shoulder. The remains of numerous mastodon species, such as *Gomphotherium angustidens*, and the immense *Anancus arvernensis* with straight tusks nearly 10 feet long, also exist in the old Greco-Roman world. In North Africa, the largest prehistoric land species are mastodons such as *Anancus osiris* and the early mammoth *Mammuthus africanavus*.[39]

Other large, unfamiliar animals besides elephants left perplexing skeletons for the ancients to find. The remains of giant giraffes of the Neogene turn up in Greece, Asia Minor, Egypt, and the foothills of the Himalayas, and numerous rhinoceros species flourished across Europe and Asia in the Tertiary and Quaternary periods. Three species of rhino, including the southernmost examples of the great Ice Age woolly rhinoceros, *Coelodonta antiquitatis*, inhabited the Peloponnese. The Black Sea area has the remains of the colossal rhinoceros *Elasmotherium*. Recent excavations in northern Turkey have turned up a new species of embrithopod, a massive rhinoceros of the Oligocene similar to the bizarre *Arsinoitherium* of Egypt. The remains of large and giant hippopotamuses (*H. amphibius* and *H. major*) occur across southern Europe, the Levant, and North Africa. The skeletons of giant cave bears (*Ursus spelaeus*) and very large hyenas, lions, and tigers also exist in southern Europe, Greece, Asia Minor, and the Black Sea area.

2.26. *Mammuthus primigenius.* Photo by author.

Another common Eurasian fossil of unexpected size is the "progenitor ox" (*Bos primigenius*). These giant wild cattle (or aurochs) stood 6 feet (1.8 m) at the shoulder and weighed about a ton, with horns 3 feet long (1 m). Once widespread throughout Europe and the Levant, they disappeared in central Greece and Italy by about 1850 B.C. *Bos primigenius* (and woolly rhinos) are familiar to anyone who has admired the Lascaux cave paintings in France, and *Bos* also appears in the famous Minoan frescoes of Knossos, Crete, showing men and women performing acrobatic feats with king-size bulls. I think that the huge remains of Geryon's fabulous cattle plowed up by farmers in Lydia must have belonged to *Bos* (see Pausanias's account, above). According to the natural historian Aelian, oversize wild cattle found living in northwest Greece and Albania were believed to have descended from Geryon's original herd.[40]

In the stories of extraordinary bones in the next chapter, the chronological and geographic scope is astonishing: from the time of the legendary Trojan War (ca. 1250 B.C.) to the end of the Roman Empire (fifth century A.D.), Greeks and Romans were the first to record descriptions of significant prehistoric fossil remains in Greece, Italy, North Africa, Egypt, Asia Minor, and the Himalayas. With the modern paleontological knowledge surveyed in this chapter, we can read those chronicles in a brand-new light. No longer isolated curiosities, or tales of superstition and fantasy, these accounts emerge from the oblivion of footnotes to stand as a coherent body of evidence for fossil discovery and paleontological inquiry in classical antiquity.

CHAPTER

3

Ancient Discoveries of Giant Bones

The Giant Shoulder Blade of Pelops: A Fossil Odyssey

THE GREAT WAR against Troy had been dragging on for nearly ten years. The battle-weary Greeks captured a Trojan seer and forced him to reveal secret oracles. The seer predicted that his city would never fall unless the Greeks brought a bone of the great hero Pelops to Troy as a talisman. The Greeks immediately dispatched a ship to fetch Pelops's enormous shoulder blade from Olympia.

As Pausanias relates this story (already centuries old in his day, about A.D. 150), some bones of heroic size were acclaimed as the remains of the mythic hero Pelops sometime before the Trojan War (ca. 1250 B.C.). The big bones were kept in a bronze chest at the Temple of Artemis at Olympia. Pelops's shoulder blade—reputed to have magical powers—was apparently displayed in its own shrine. In myth, Pelops was Heracles' great-grandfather and a founder of the Olympic Games. As a youth Pelops had been chopped up and served to the gods as a grotesque sacrifice. But as soon as the gods realized what they were eating, they restored Pelops to life. His shoulder, however, had already been been eaten, so the gods replaced it with one of ivory. And it was this ivory shoulder blade that was enshrined at Olympia. Scholars have long tried to explain the mystery of Pelops's ivory shoulder, comparing the sheen of ivory to radiant skin and so on. But if we take into account the great size of Pelops's bone and its place of discovery, a more plausible explanation emerges.[1]

The story of the huge shoulder bone made of ivory and stored at Olympia suggests that prehistoric remains from Arcadia contributed to the myth of the great eponymous hero of the Peloponnese. Olympia is on the Alpheios River: the surrounding valleys contain dense concentrations of the bones of large Pleistocene mammals, including mammoths. Shoulder blades and thigh bones are large, durable, and easily recognized mammal bones—and they convey an immediate sense of a creature's size. These two bones each had a special meaning in ancient Greek rituals, according to Walter Burkert, a scholar of Greek religion. The shoulder blade in particular played a prominent role in sacrifices and in foretelling the future. An impressive mammoth scapula would certainly stand out in a fossil bone assemblage, and its dimensions would match the ancient search image for mythical heroes who towered over present-day men.

Very old bone strongly resembles ivory, especially when polished, says Kenneth Lapatin, an authority on ivory in antiquity. If people of early antiquity found a semifossilized shoulder blade that looked human but dwarfed a man's, it's easy to imagine that they might burnish it to bring out an ivorine quality and might revere the bone as that of their mighty hero Pelops. Once polished and displayed, the bone would inevitably accrue stories to explain its special appearance and origins. And the stories would in turn make the relic that much more valuable, worthy of special treatment. Burkert's and Lapatin's findings thus help us to explain how a huge "ivory" shoulder blade came to be incorporated into the myth of Pelops, and why such a relic would have deserved its own shrine.[2]

Pausanias claims that the ivory shoulder bone was actually shipped to Troy during the war. The relic would have been bundled in straw for its journey by mule from Olympia to the port of Cyllene on the coast. There it was loaded onto a boat bound for Troy, either lashed to the deck with ropes or placed below as ballast. The big flat bone was unwieldy and heavy. If the shoulder blade came from the ancestral mammoth *M. meridionalis*, the woolly mammoth *M. primigenius*, or the ancient elephant *P. antiquus*, it could measure 3–4 feet long and 2–3 feet across (figs. 3.1, 3.2). The mineralized scapula of a large prehistoric proboscidean weighs 66–110 pounds (30–50 kg).

After weeks at sea, Pelops's bone arrived safely in Troy, and, true to the Trojan seer's prediction, it witnessed the sack of the city by the Greek heroes. After the victory, the Greeks loaded the relic onto a ship for the return home. But a storm wrecked the ship off the treacherous coast of Euboea. The precious cargo was lost at sea.[3]

A great many years passed. One day a young fisherman named Damarmenos of Eretria hauled in his net to find that he had caught a bone of staggering size. Damarmenos was fishing the shallow seas around Euboea, which we now know to be sunken Neogene valleys that were once part of the Aegean land bridge populated by the extinct animals described in chapter 2. The bone he netted may have come from a large Pikermi-type mammal, perhaps a mastodon, rhinoceros, chalicothere, or deinothere (fig. 3.3). For modern instances of large prehistoric land mammal remains netted at sea, we have only to consider the thousands of mammoth bones routinely hauled in by fishermen in English Channel and the North Sea.[4]

Convinced he had found something important, but worried that it could get him in trouble, the fisherman buried the great bone in the sand and told no one. Over the years the secret nagged at him. Finally Damarmenos traveled to Delphi to ask the oracle of Apollo whose bone it was, and what he should do with it. By coincidence, ambassadors from Elis (the territory within which Olympia lay) were in town consulting the oracle about how to cure a terrible plague.

After hearing the two inquiries, the oracle declared that only the shipwrecked shoulder blade of Pelops could banish the plague in Elis, and ordered the fisherman to restore the bone to its home soil. The Eleans arranged for Damarmenos and his family to transport the long-lost bone by sea from Euboea around the Peloponnese

3.1. Scapula of *Mammuthus primigenius.* A large fossil elephant shoulder blade from the region of Olympia, southern Greece, may have been enshrined as the shoulder bone of the giant hero Pelops in antiquity. Photo by author.

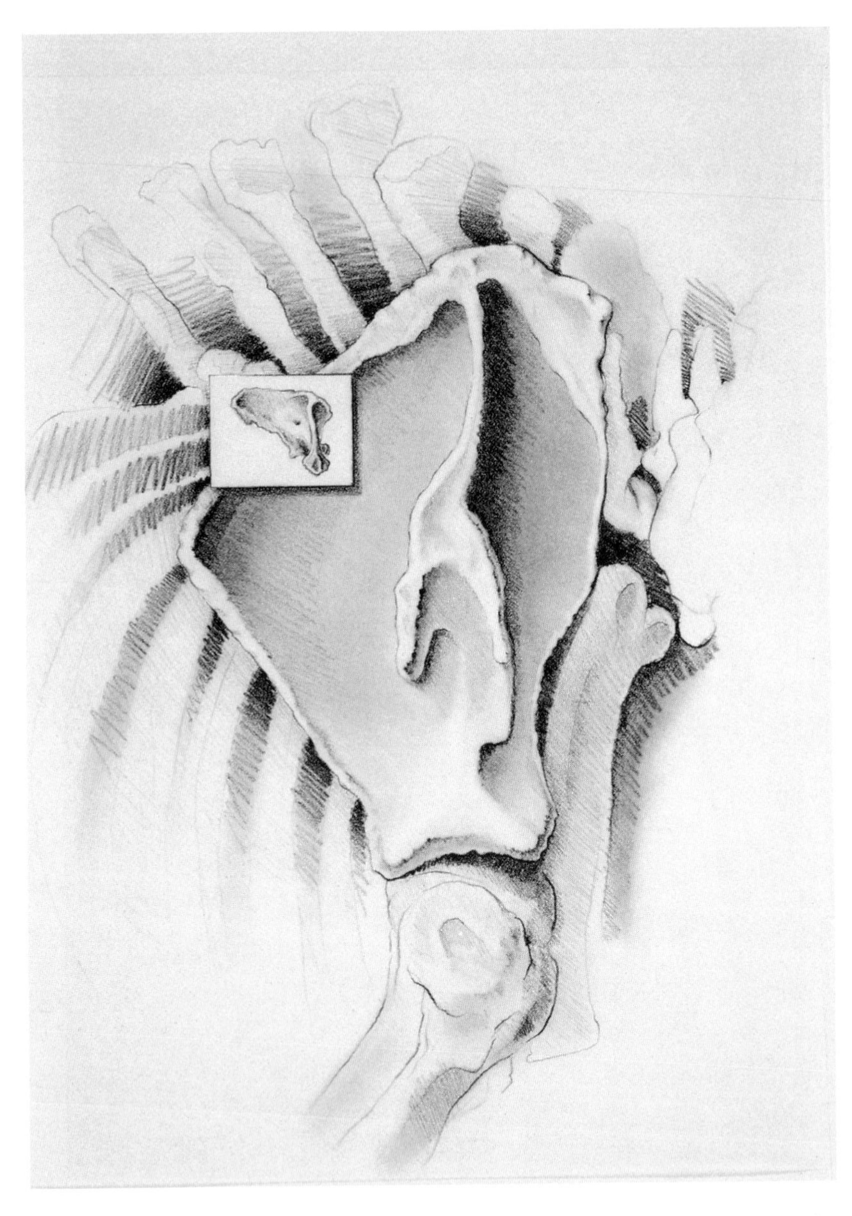

3.2. Human shoulder blade (in box) compared to mammoth scapula. Drawing by Kris Ellingsen, © 1999.

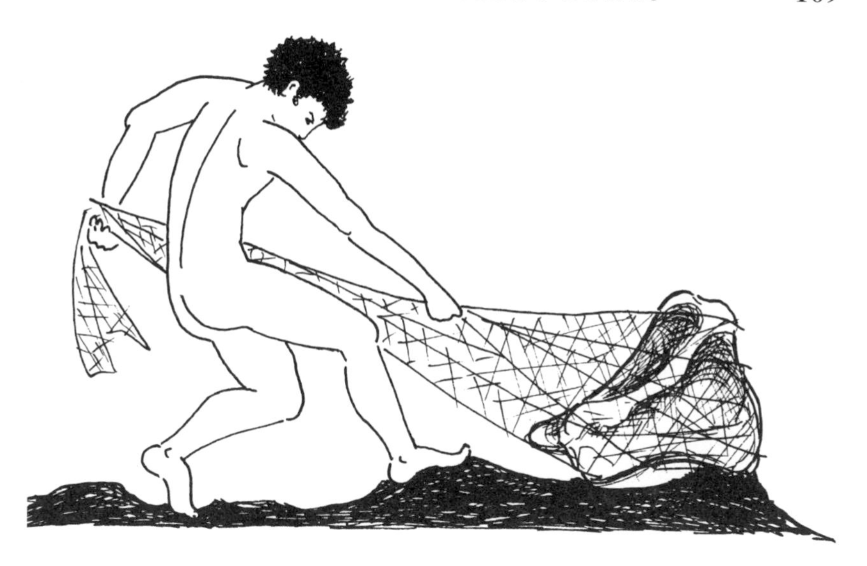

3.3. If the huge bone that the fisherman Damarmenos netted off Euboea belonged to a Neogene mastodon, this sketch indicates its approximate size. Drawing by author.

to Cyllene, and then overland to Olympia. Thereafter, the fisherman and his descendants became the guardians of the Pelops bone in the Pelopion, Pelops's shrine at Olympia. Archaeological evidence dates the ruins of the Pelopion to the seventh century B.C., a period when the bones of Hellenic heroes and ancestors of the glorious past were much-sought-after relics. We might guess that the bone's miraculous rediscovery occurred during the emerging cult of ancient heroes' bones.[5]

Pausanias was eager to see that venerable shoulder blade when he visited Olympia in the second century A.D. He located the ruins of the old Temple of Artemis, all overgrown with vines, and found the Pelops shrine still standing. But the ancient relic had long since crumbled to dust. "In my opinion," remarked Pausanias, the bone was "worn away" by its arduous travels, "centuries on the sea floor, and other processes of time." (Indeed, if we assume that the original bone was found in fossil deposits near Olympia, and Damarmenos's bone came from a submerged Neogene valley off Euboea, then a Pleistocene bone lost at sea was replaced by a Miocene bone *millions* of years older.) But even after Pausanias reported that the relic had crumbled away, the great ivory bone was not forgotten. The early Christian father Clement of Alexandria, in his *Exhortation to the Greeks* (ca. A.D. 190), railed against the pagan worship of Pelops's shoulder blade, even as he recounted its fantastic history.[6]

The Pelops bone episode related by Pausanias is the most ancient instance of prehistoric fossils identified as the remains of an extinct race of giant heroes. The following narratives, from the fifth century B.C. to the fifth century A.D., record other early paleontological events, as the Greeks and Romans strove to locate, measure, compare, explain, and visualize the gigantic creatures whose bones they encountered all around their world. Timeless themes that mark the long history of human encounters with extraordinary fossils are evident here. The big bones were believed to be vestiges of the glorious past described in myths, and they brought prestige to their possessors. The meaning of the fossils was debated, and public display was an important aspect of the bones' power. Religious oracles authenticated the relics found by ordinary folk, and political authorities tried to exploit them.

Since individual authors often compared remarkable remains they saw in their own lifetimes with giant bones found in the past in far-flung locales, the following pages take us all around the ancient world. By the Roman era, the world was expanding rapidly, and the geographical distibution of gigantic remains widened, ranging from the Atlantic coast of Morocco to the English Channel, and as far east as the foothills of the Himalayas. (See maps 1.1, 2.1, 2.2, 3.1, 3.2, 3.3, the Historical Time Line, and appendix 1.)

The Ancient Bone Rush

In about 560 B.C., the oracle at Delphi told the Spartans that they needed to find the bones of the hero Orestes before they could defeat their regional rival, Arcadian Tegea. Herodotus (ca. 430 B.C.) described how the acquisition of those heroic remains led to Sparta's military dominance in the Peloponnese.

Frustrated by their failure to defeat Tegea in battle, the Spartan leaders sought advice at Delphi. "Bring Orestes to your city" was the oracle's reply. Unable to find his tomb, the Spartans returned to Delphi and pestered the oracle for a hint. They received a cryptic verse about a forge. Still no big bones turned up. But then a retired Spartan cavalryman, Lichas, happened to be in Tegean territory during a lull in hostilities. He struck up a conversation with a smith working at his forge, who mentioned an astounding find in his yard. "As I was digging my well, I came on a huge coffin—7 cubits long! [10 feet; 3 m]. I couldn't quite believe that men in the past were bigger than they are today, so I opened it—and the skeleton was as big as the coffin! I measured it, and then I shoveled the earth back."

Pretending to be an exile from Sparta, Lichas leased a room from the smith. Then he secretly dug up the grave and absconded with the big bones. Sparta trumpeted the recovery of "Orestes' bones" and reburied them in the city with great honors. Once they possessed this powerful talisman, says Herodotus, Spartan rule in the Peloponnese soon became absolute.

Tegea lies in a prehistoric lake basin that contains the remains of mammoths and other Ice Age mammals like those found around Megalopolis and Olympia. With this in mind, classical scholar George Huxley suggests the following sequence of events: In the eighth or seventh century B.C., when the cult of hero relics began, "large bones of a Pleistocene date were discovered" and "given a respectful burial in a 7-cubit coffin fit for a hero." A century or so later, the blacksmith "found the reburied bones when digging a well [and] news of the discovery reached Delphi. Lichas found the bones and took them to Sparta."[7]

The Tegean Affair was resolved by the Orestes Bones Policy, as this ancient paleontological discovery is known to historians. It was a brilliant propaganda move that set in motion the long chain of events that eventually played out in the Peloponnesian War. The policy was also the most prominent incident in a long-lasting Panhellenic bone rush. City-states all around the Greek world scrambled to recover the huge remains of heroes in the seventh, sixth, and fifth centuries B.C. Chance fossil finds now spurred deliberate bone hunting. Every city sought the "peculiar glamour"—the religious anointment and political power—conferred by heroes' remains. The impressive bones were a vital physical link to the glorious past.[8]

Athens was caught up in the fever for king-size skeletons, too. The Delphic oracle, which enhanced its prestige by gaining a reputation as the central source of information about where to find big bones and how to identify them, advised the Athenians to recover the bones of their own local hero Theseus from the island of Skyros. According to Athenian legend, he had been murdered there—pushed off a high cliff in the northeast of the island sometime in the ninth century B.C. But the residents of Skyros denied the murder and refused to allow a search for his remains.

In 476 B.C., the Athenian general Kimon captured the island and made it his personal mission to hunt down the sacred bones of Theseus. Noticing an eagle tearing at a mound, he ordered his men to dig there. They unearthed some oversize bones lying beside a bronze-pointed spear and sword (the sort used in the Bronze Age). Kimon loaded the bones and weapons onto his trireme and sailed back to Athens. Theseus's relics were welcomed with magnificent processions and interred in the heart of the city, and Kimon reaped many political points. Skyros has rich early Greek settlements and tombs dating to 1000–700 B.C., and the fertile northeast part of the island does have Miocene sediments where large Pikermi-type fossils might be found. It seems likely that Kimon rediscovered some large prehistoric bones that had been given a hero's burial centuries earlier, just as George Huxley suggested for Orestes' bones.[9]

Many Delphic oracles interpreting chance finds and ordering searches for heroes' bones have survived in ancient texts. Pausanias alone records more than two dozen cases. In the mania for remains, Sparta also obtained the bones of Orestes' son Tisamenus from Helike on the Gulf of Corinth. The people of Olympia installed the bones of Pelops's wife Hippodamia in a shrine near a mound said to contain the bones of her father, the giant son begotten by the god Ares upon the daughter of the giant Atlas. Nearby, on the north bank of the Alpheios River, Saurus's ("Lizard's") Ridge was named after a giant killed by the mythical hero Heracles.

To cure a plague raging in Boeotian Orchomenos in the fifth century B.C., the oracle advised that a crow would reveal the lost remains of the native poet Hesiod. Sure enough, the people told Pausanias, a raven was seen scratching at a hollow mound where some large bones spilled out. The poet's relics were reverently reburied in the agora. The city of Mantinea retrieved the bones of the eponymous Arcadian hero Arcas near Tegea in about 422 B.C. The Messenians imported the bones of their hero Aristomenes from faraway Rhodes, a fact that puzzled Pausanias since the Spartans had supposedly killed him in the Peloponnese. (The island of Rhodes does have large prehistoric remains, but so does Messenia.) Pausanias also ridiculed a small bronze urn in Argos alleged to hold the bones of Pelops's giant father Tantalus. Pausanias said he saw a more capacious tomb more suitable for Tantalus's remains in Sipylos (Asia Minor).

Pausanias's fascination with celebrity bones became something of an obsession—he admits that he "made a nuisance of himself" in Athens trying to find out how Oedipus's bones were spirited away from Thebes, a city that once boasted several heroic shrines. The skeleton of the hero Melanippos was transferred from Thebes to Sikyon in about 600 B.C., and a Theban delegation sailed to Troy to recover the bones of the great Trojan hero Hector from the old battlefield, after the oracle promised that his relics would make the city rich. When Thebes fell to the Macedonians in 338 B.C., Philip II (father of Alexander the Great) looted the bones of the mythical musician Linus. (Soon after, however, a guilty dream prompted Philip to send the bones back to Thebes.)[10]

Despite Pausanias's diligence, his work shows only the tip of the iceberg of the great archaic-classical Greek bone rush. Many other less prominent "heroes" were no doubt discovered and enshrined, either with the help of oracles or without. This vigorous early traffic in celebrity relics helps explain how the term "heroes' bones" came to mean any large prehistoric skeleton that came to light in later Roman times. To pass muster as supersize heroes of the past, the remains almost certainly belonged to very large prehistoric mammals. If some nonhuman features were detected in fossil assemblages, they could be explained by the mythological paradigm. Everyone knew that giants and heroes of myth were not merely

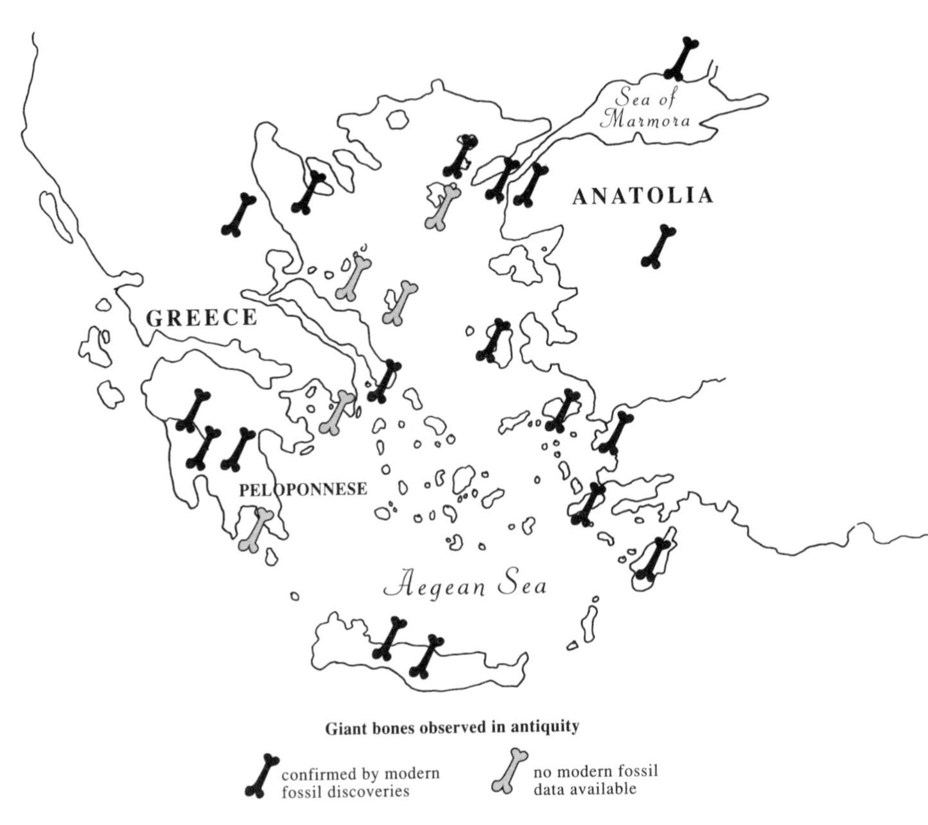

Map 3.1. The Aegean world. Giant bones reported by ancient sources compared with modern vertebrate fossil discoveries. Map by Michele Mayor Angel.

bigger and stronger than ordinary humans, but they could also have grotesque anatomical features, such as multiple heads or animal parts (see chapter 5).

Heroes' Bones of Troy

The heroes of the Trojan War were venerated in the Troad along the banks of the Hellespont (Dardanelles, Turkey). Some years before Pausanias began his travels, a buzz of excitement surrounded the discovery of the bones of the great Greek champion Ajax of

Salamis. According to Homeric myth, Ajax's grave was on the headland of Rhoeteum, where the Greek ships had landed to attack Troy. A local man explained to Pausanias how the sea had suddenly washed out the beach there and revealed some enormous bones. "The man told me to judge the size of Ajax like this," says Pausanias. "The kneecaps, which doctors call the millstones [patellas], were exactly the size of a discus for the boys' pentathlon."[11]

This report is arresting for several reasons. First, Pausanias uses a technical anatomical term for kneecap. If Pausanias was a doctor, as some believe, his vocation would explain his use of medical terms, his fascination with the anatomy of extraordinary skeletons, and his interest in visiting sanctuaries of Asklepios, the god of medical arts, where gigantic remains were commonly exhibited. In the sanctuary of Asklepios at Asopos (southern Peloponnese), for example, Pausanias examined bones that were "enormous but apparently human." At a sanctuary of Asklepios at Megalopolis, another collection of immense bones seemed "much too vast for a human being," but the curators told Pausanias that they belonged to one of the giants of early myth. (If the giant remains in these temples were mammoth bones, the skulls must have been missing, since Pausanias tells us he examined an elephant skull and tusks in a temple in Italy.)[12]

The device of using an analogy with a familiar object to convey the enormity of prehistoric bones is an interesting feature in big bone narratives. Pausanias's Ajax passage excites archaeologists because it is proof that boys' discuses were smaller than those thrown by men. Numerous discuses used by adult athletes have been excavated—they range between 6.5 and 9 inches (16.5–23 cm) in diameter. So, just how big were Ajax's kneecaps? Probably about 5 or 6 inches (14 cm) across, a little larger than a compact disc. Bones big enough to be worthy of the mighty Ajax would most likely belong to a Miocene mastodon or rhinoceros, whose patellas can measure nearly 5 inches across. These and other large mammal remains are abundant in the Neogene-Pleistocene deposits of Troy's eroding coast. The salient point here is that Pausanias's local informant resisted the temptation to impress him with a fantastic figure. Instead of comparing the big patella to, say, a shield or even a full-size discus, the eyewitness strove for precision over hyperbole. Reports like this increase the likelihood that actual remains of prehistoric animals were being measured more or less accurately and discussed rationally.[13]

On Heroes

Across the narrow Dardanelles from Troy, the ancient Thracian Chersonese (the Gallipoli Peninsula) was the setting of a unique dialogue. *On Heroes*, written in about A.D. 218 by the sophist Philostratus of Lemnos, is usually considered a literary romance devoid of historical merit. But the work is peppered with realistic accounts of large fossil exposures. Of the fifteen giant bone finds it mentions, thirteen occur in places where prehistoric deposits are now known to paleontologists, and the other two have geology compatible with fossil beds. Philostratus wrote for a sophisticated audience steeped in mythohistory and interested in curious current events. He and his readers seem to be aware of a long-standing philosophical silence regarding giant bones (a silence considered in chapter 5). His readers could find a practical opinion on the matter in Philostratus's biography of Apollonius of Tyana, the philosophically educated traveling sage who concluded that giant creatures really did exist in the past because their remains are revealed all around the world.[14]

In *On Heroes*, Philostratus self-consciously mediates a gap between philosophy and popular knowledge about heroes' bones by casting his fictional paleontological investigator as a rustic philosopher. The dialogue takes place at Elaeus, at the mouth of the Hellespont, as an old grape-farmer entertains a seafaring merchant from Phoenicia who is awaiting good sailing omens. The farmer is a student of philosophy and claims to know all about the heroes of the Trojan War. Intrigued, the merchant wonders if there is any empirical proof for the general belief that ancient heroes averaged 10.5 cubits in height (15 feet; 4.5 m).

The farmer counts off the most famous giant bones of the past, such as Orestes' 7-cubit skeleton abducted from Tegea by the Spartans; a huge skeleton found by a shepherd after an earthquake in Lydia (described by Plato; see chapter 5); and the more recent discovery of the giant Aryades on the banks of the Orontes (Pausanias's version of this event is mentioned in chapter 2). The farmer's own grandfather had told him how the Roman emperor Hadrian came to pay homage to Ajax's relics after the skeleton with memorably big kneecaps was washed out by the sea just across the strait. Hadrian "embraced and kissed the bones and laid them out," apparently in the configuration of a man about 15 feet tall. Then the emperor built a fine tomb for Ajax's bones at Troy. (There is nothing fantastic about this account: Hadrian is known to have toured Asia Minor and made restorations at Troy in A.D. 124 and 129.)[15]

"But," the merchant asks—posing the question that every scientist, anthropologist, and folklorist asks a local informant—"have *you* personally seen any heroes' bones?" "Less than fifty years ago," recalls the old farmer, "I myself sailed across the strait to Sigeum" to view a vast skeleton that eroded out of a rocky cave on the cape. (Sigeum, near Rhoeteum, was the Greek hero Achilles' traditional burial place.) The farmer explains that this event (in about A.D. 170) was a public sensation. In the two months that the remains were visible, before they crashed into the sea, throngs of spectators sailed to Sigeum from the Hellespont, from the Anatolian coast as far south as Izmir, and from the Aegean islands to get a close look. The upper body was hidden in a cave, but the bones extended outward some 33 feet (10 m) on the cape. They seemed human. Everyone had a different explanation, until finally an oracle hinted that the remains belonged to the great warrior Achilles, a taxonomy consistent with the geography of Homer's *Iliad*. The dimensions would be absurd if we imagine them as describing the remains of a single animal. But remember the conditions in which Aegean fossil remains are found: the grape-farmer has given us a realistic account of a large fossil bone assemblage eroding out of the collapsing bluffs of the Sigeum promontory.[16]

"Four years ago," says the farmer, "my friend Hymnaios was digging vines on Alonnisos when the earth rang under his shovel." He and his son cleared away the dirt, and "there lay a skeleton nearly 18 feet [5.5 m] long." They consulted an oracle, which declared it was one of the giants slain in the Gigantomachy, the mythical war between the gods and the giants. They carefully reburied the bones. The island of Alonnisos (ancient Ikos) is a tectonically uplifted block sheared off from Euboea. Alonnisos has Miocene sediments, but no paleontologists have explored there. It is a reasonable assumption that the man and his son uncovered a Pikermi-type mammal, perhaps a mastodon or even a *Deinotherium*, which towered 15 feet at the shoulder.[17]

But there were even more recent finds: "Last year" the grape-farmer and some friends sailed to Lemnos to view another giant skeleton found by Menecrates of Steira after an earthquake. "We saw that the bones were completely shaken out of their proper position. The backbone was in pieces and the ribs were wrenched away from the vertebrae." But "as I examined the entire skeleton and the individual bones, I got an impression of terrifying size, impossible to describe." The creature's skull "held more than two Cretan wine amphoras!" Classicists who take this statement as evidence for some sort of funeral rite miss the point. I think the ancient investigators were trying to determine cranial capacity, not making ritual libations of wine.

Philostratus conveys the magnitude of the skull by reference to an everyday measurement of volume, an amphora. The grape-farmer, as a wine-maker, would naturally think in terms of a terracotta wine jug to measure volume. We know that the standard Greek amphora held about 39 liters and the Roman amphora about 26 liters. But this passage is the only surviving ancient reference to a Cretan amphora as a measure. Actual Cretan amphoras have only recently been discovered—by French archaeologists who excavated and measured a series of amphoras in Crete. They determined that the jars held about 20–24 liters. To the French archaeologists, this unique passage in *On Heroes* is proof that the Cretan amphora was a metrological standard like the Greek and Roman jugs. For us, it shows the grape-farmer's intention to convey a precise measurement, a volume slightly less than that of the Roman amphora.

The grape-farmer and his friends on Lemnos estimated the skull capacity as the equivalent of 40–48 liters. I contacted prehistoric proboscidean experts Adrian Lister and William Sanders to learn whether this volume matches that of an early elephant skull. Sanders notes that if the pneumatic bone was broken away from the cranium, viewers would have "an impression of a huge endocranial capacity." According to figures provided by Lister, someone looking at a adult male elephant skull might guess the volume at 70–100 liters; for a 10-year-old elephant, 40–50 liters. The skull of a large extinct mastodon might appear to hold roughly 20–70 liters. The grape-farmer's estimate is thus not at all out of line for a big Miocene mammal skull, similar to the types found on Samos and the mainland. Lemnos was a peninsula of Asia Minor until 20,000 years ago; the island also has sediments of the Eocene and Oligocene epochs (55–23 million years ago). Modern paleontological studies are lacking for Lemnos, but the account by Philostratus, a native of Lemnos, suggests that exploration there might be worthwhile.[18]

Finally, the farmer's account comes up to the present: "This year," he says, an enormous body appeared at Naulochus on the small island of Imbros (Imroz, Turkey), when part of the southwestern promontory fell into the sea. "The broken-off chunk of earth carried the giant with it. If you don't believe me, we can sail there now—that skeleton is still visible and it's a short trip." And believe him we should. Some 1,800 years after the farmer's challenge, Turkish paleontologist Sevket Sen told me that a French tourist found a large bone on Imroz in the summer of 1997. Sen identified it as the femur of a Miocene mastodon, common remains in the southwestern part of that island.[19]

For further proof of extinct colossi, the farmer advises the merchant to sail to Kos, where the daughters of giants were buried, or visit Phrygia to see the giant Hyllos or Thessaly to see the bones of fallen giants. Near Mount Vesuvius the Italians display the remains of giants killed by the gods at the Burning Fields (Phlegra), and in Olympia one could marvel at the monstrous bones of Geryon. Legions of giants are buried in the soil of Pallene (the Kassandra Peninsula, Chalkidiki, northern Greece), where "thunderstorms and earthquakes expose their colossal bones on the surface of the ground." Not surprisingly, each of these locales contains fossil beds.

The farmer ends with a gory description of a battle in the time of the Trojan War, in which Achilles' carnivorous war horses attacked the mighty Amazons and scattered their limbs on the tiny island of Leuke (in the Black Sea across from the mouth of the Danube). Leuke (known as Zmeinyi Island, "Island of Serpents"), was a secret Soviet military base, now controlled by Ukraine. It's not open to paleontological research but probably contains deposits similar to the mainland's Pleistocene elephant, rhino, and horse fossils. The bloody battle between Amazons and war horses recalls Plutarch's earlier story of a battle between Amazons and the war elephants to explain the huge bones found on Samos (chapter 2).[20]

The Measure of Giants

On Heroes places current finds by ordinary people within the context of historical discoveries of what were presumed to be the remains of extinct races of heroes and giants around the Mediterranean. The conditions of exposure are naturalistic and measurements are recorded. The emphasis on measuring, even if exaggerated, in this and other giant bone narratives is "surprising," comments Geoffrey Lloyd, historian of ancient science. "Measurements are rare" in the ancient sources, "and they are absent in many contexts where one might expect them."

But we have seen that measurements, along with taphonomic details, were part of the tradition of reporting extraordinary remains. The use of exact (even if spurious) measurements shows a desire to express the wonder of the physical evidence and to record finds with precision for posterity. Keep in mind that large mammalian limb bones were probably arranged vertically to match the bipedal giant search image, which would considerably extend a four-legged animal's "height" (figs. 3.4, 3.5). A more recent incident in a village in Moldavia (ancient Dacia, northwest of the Black Sea) lets us imagine what it might have been like when people came upon enormous remains in antiquity. In 1843, some peasants plowed up a great number of big bones in a field. A shepherd tied the bones together to make an upright giant. People "flocked from all over, singing and dancing" around the figure and praising the shepherd for his fine restoration of a saint—although the military governor maintained that it was a giant Roman soldier with "unusually large cheek teeth." Soon religious officials seized the skeleton and hacked it to pieces. Then priests buried the bones with a formal ceremony and sowed crops over the spot so it could never be found. But an old woman secretly hid away the huge jawbone, which allowed later paleontologists to identify the giant as an extinct rhinoceros (figs. 3.6, 3.7).[21]

Some extreme dimensions recorded by ancient authors may represent attempts to convey the magnitude of jumbled masses of bones of several species. People may have exaggerated spectacular finds, especially as stories circulated by word of mouth and in retranscribed written copies over centuries. Some finds were conflated or statistics got inflated. A few reports of outlandish sizes have led scholars to doubt the veracity of all ancient reports of "giant bones." Yet several oft-repeated dimensions remained constant over centuries, such as the 7-cubit Orestes (the earliest fossil measurement ever recorded, by Herodotus in the fifth century B.C.). And we saw that some informants avoided fabulous figures by employing self-consciously moderate comparisons (such as a boy's discus or a Cretan amphora). Table 3.1 shows the range of sizes, from realistic to outrageous, preserved by the ancient authors. The range reflects the ongoing tension in popular science between the impulse to exaggerate strange phenomena for rhetorical effect and the desire to maintain accuracy in the search for meaning.[22]

A Giant Skeleton in Morocco

These factors came into play in the reports of the giant skeleton encountered by the Roman commander Quintus Sertorius in Tingis (Tangier, Morocco) in about 81 B.C. The people of Tingis were proud of a great mound that they said contained the remains of the giant Antaeus, the founder of their city. Antaeus was a leg-

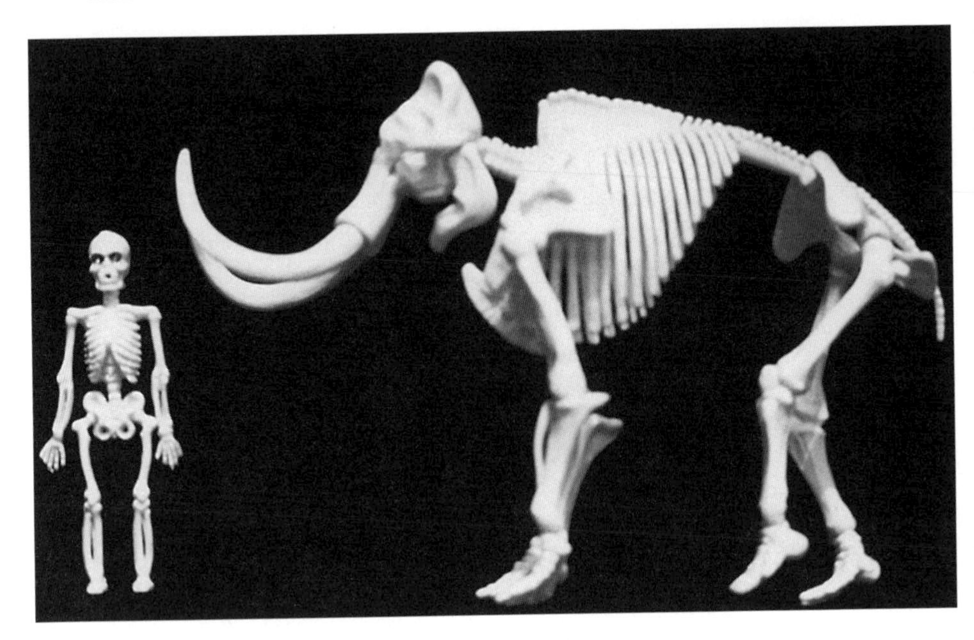

3.4. Models of mammoth and human skeletons. Photo by author.

endary ogre of North Africa who was feared for his lethal wrestling contests until Heracles slew him. Locals believed that Heracles then took up with Antaeus's giant widow Tinga and fathered Sophax, their first king. Skeptical, Sertorius ordered his soldiers to dig up the mound. So dumbfounded was he by the skeleton—supposedly 60 cubits long (85 feet; 26 m)—that he personally affirmed the Tingis legend and reburied the giant Antaeus with great honors.

"Unless the mound contained the bones of some antediluvian animal, Sertorius of course did not find a sixty-cubit skeleton." Sarcastically dismissing that possibility without troubling to check Moroccan paleontology, a recent commentator asserts that Plutarch's narrative simply shows how Sertorius cleverly manipulated native beliefs. But for our purposes, the incident is an important event in paleontological history: this appears to be the earliest recorded investigation of the significant Neogene fossils of Morocco. The local story of the giant Antaeus was consistent with the traditional geographical distribution of the extinct race of giants featured in Greek myths, and the bones' location corresponds to ac-

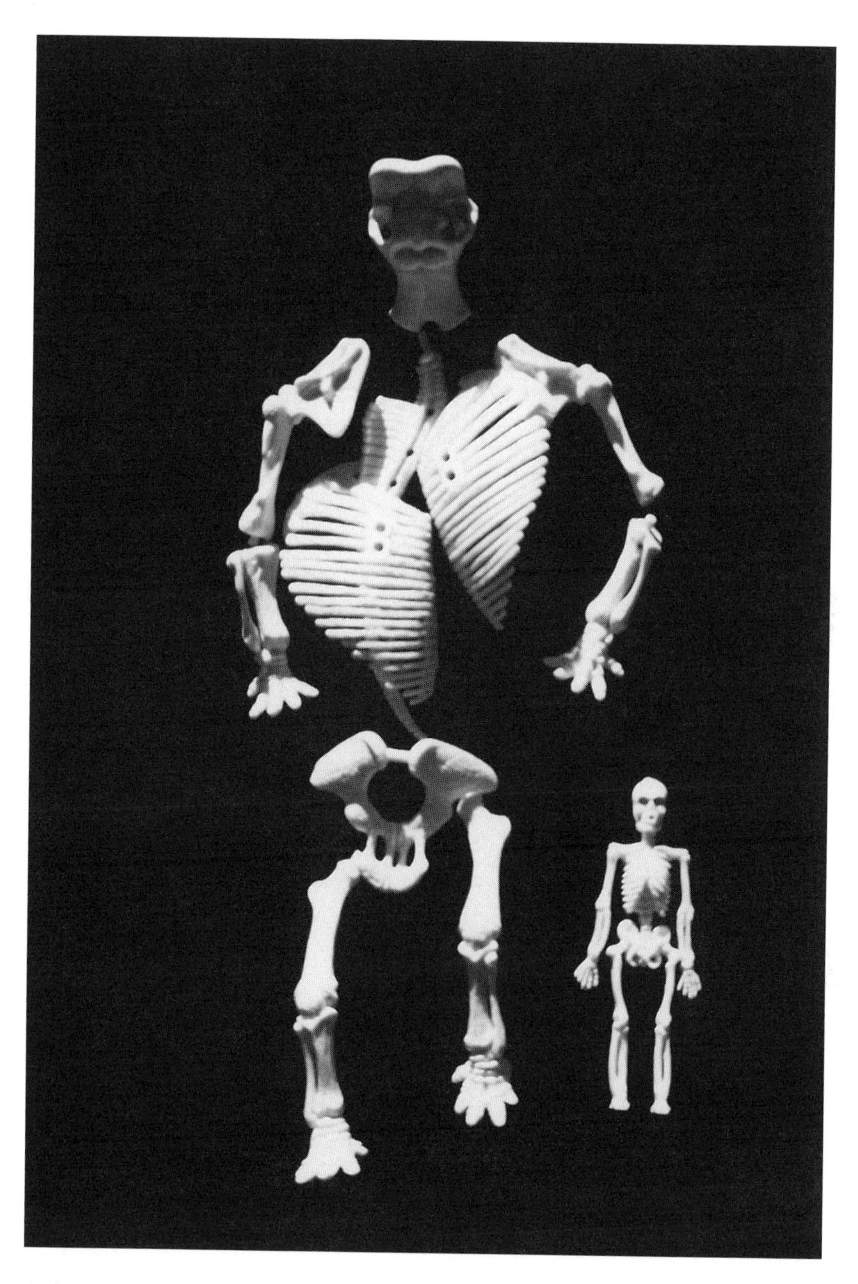

3.5. Bipedal mammoth monster and human skeleton. If we rearrange a mammoth skeleton into a two-legged giant and place it next to a human skeleton, we can begin to sense how the ancient Greeks interpreted many of the giant bones they observed around the Mediterranean. Photo by author.

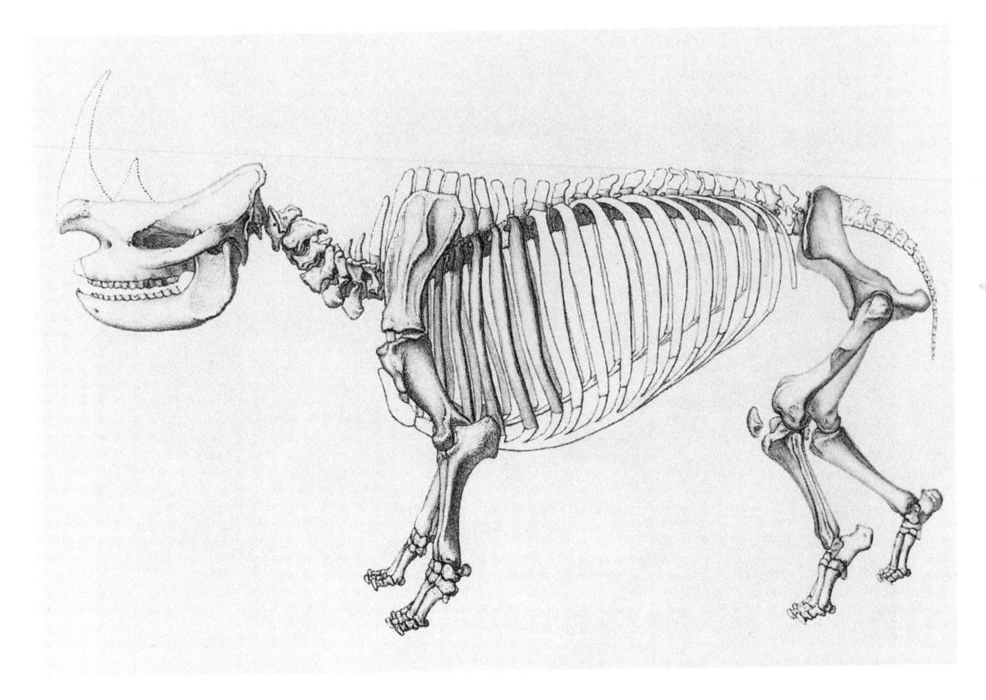

3.6. Fossils of numerous rhinoceros species are common in the Balkans and Eurasia. The Moldavian peasants restored a skeleton similar to this one in the configuration of a bipedal giant. This example (*Rhinoceros pachygnathus*) is a Neogene species from Pikermi, Greece. It stood about 5 feet at the shoulder; the femur was about 20 inches long. Gaudry 1862–67, plate 31.

tual fossil beds. The historians Gabinius (now known only from fragments) and Strabo say that the giant was buried in Lixus (Larache) about forty miles south of Tingis on the Atlantic coast. Pliny also placed Antaeus in Lixus, but he refused to write down any of the region's "fantastic" legends, complaining that the native words "are absolutely unpronounceable."

What kind of skeleton was identified as the giant Antaeus? As Cuvier remarked in his discussion of this event, the ancients "often exaggerated giant skeletons by eight or ten times the size of the largest fossil elephant." Applying Cuvier's formula, we might estimate the skeleton to have been 9 feet long. Rich bone beds around ancient Tingis and Lixus contain the remains of Neogene

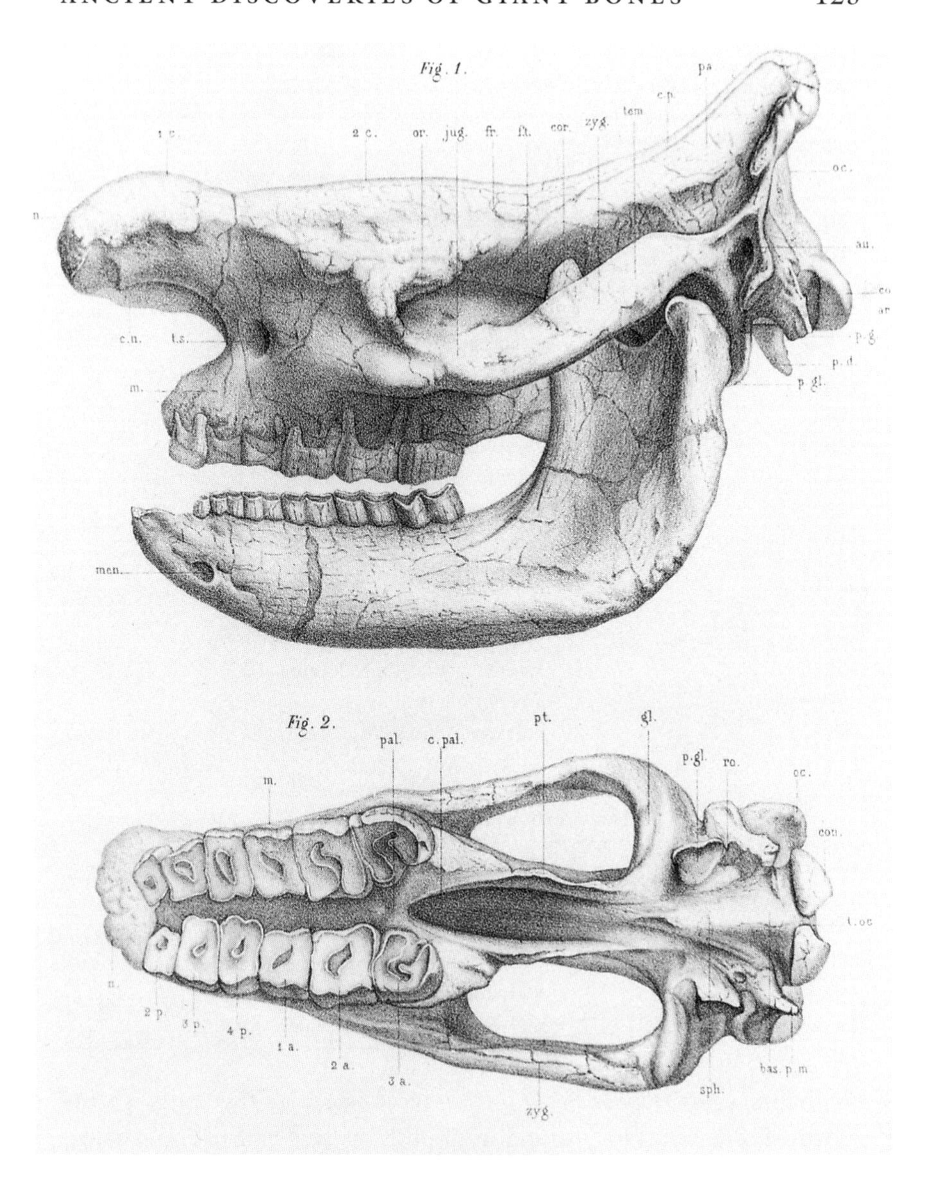

3.7. Rhinoceros skull from Pikermi, Greece, about 26 inches long. A similar skull was taken for that of a giant human in Moldavia in 1843. Gaudry 1862–67, plate 27.

Table 3.1
RANGE OF REPORTED DIMENSIONS OF GIANT SKELETONS

Approximate "height" (in feet)	*Giant bones (source)*
10	Orestes (Herodotus, Philostratus, Pliny, Solinus)
10+	Augustus's giants (Pliny)
14	Protesilaus (Philostratus)
15	Asterios (Pausanias)
15+	Ajax (Philostratus)
15+	Orontes or Aryades (Pausanias)
18	Alonissos giant (Philostratus)
33	Achilles (Philostratus)
34	Taman Peninsula skeleton (Phlegon)
34	Carthage giants (Phlegon)
40	Scaurus's sea monster (Pliny)
45	Aryades (Philostratus)
47	Cretan giant (Solinus)
53	giants in Thessaly (Philostratus)
68	Cretan giant (Philodemus)
69	Cretan giant (Pliny)
85	Antaeus (Plutarch, Strabo)
140	Makroseiris (Phlegon)
	Individual bones, teeth, and tusks
40–48 liters	skull on Lemnos (Philostratus)
5 inch diameter	patella of Ajax (Pausanias)
1 foot long	tooth of Pontic hero (Phlegon)
100 times human	molar at Utica (Augustine)
3 feet long	Calydonian Boar tusk (Pausanias)
27 inch circumference	Calydonian Boar tusk (Procopius)

elephants (*Tetralophodon longirostris*, *Anancus*), the early mammoth *M. africanavus*, and giant giraffids, as well as Eocene whales, which could measure up to 70 feet long. Any of those skeletons would stagger the ancient imagination.[23]

Cretan Giants

Around the time that Sertorius was in Morocco, Roman generals in Crete encountered another impressive skeleton. During the war

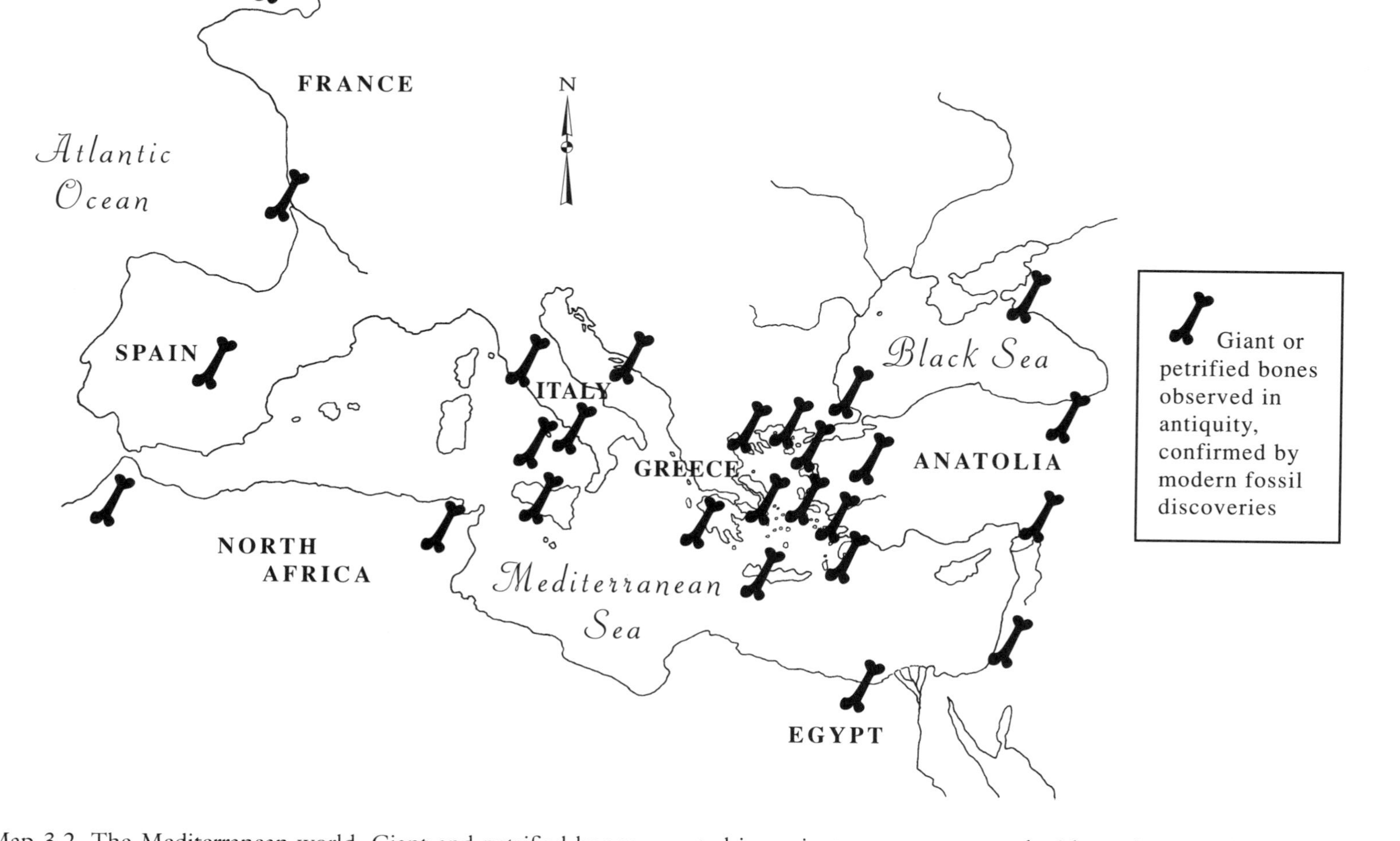

Map 3.2. The Mediterranean world. Giant and petrified bones reported in ancient sources compared with modern vertebrate fossil discoveries. Map by Michele Mayor Angel.

against the Cretan pirates (ca. 106–66 B.C.), a tremendous thunderstorm caused rivers to flood. After the high water receded, says the historian Solinus, the riverbanks collapsed, laying bare a skeleton said to measure 33 cubits (47 feet; 14 m). Word reached the Roman commanders Metellus Creticus and Lucius Flaccus, who took a break from pirate bashing to investigate this marvel for themselves. We can imagine them elbowing through throngs of villagers gaping at the giant bones; perhaps the generals ordered a ceremonious reburial of the relics. Modern scholars dismiss Solinus (ca. A.D. 200) as a mere copier of Pliny, yet only Solinus described this particular paleontological event. Pliny told of a larger Cretan giant revealed by an earthquake, and Philodemos of Gadara (ca. 110–40 B.C.) reported yet another skeleton of 48 cubits in Crete.[24]

In Crete, deep layers of Pleistocene and Holocene (the last 10,000 years) mammal fossils lie in coastal caves and other sites. Early European travelers described candlelight tours of these mysterious bone grottoes. Monks went "to and fro with their tapers" illuminating what they said were giants "from ages ago." The monks identified the femurs and shoulder blades by touching the corresponding bones on their own bodies. Inspired by these early accounts, paleontologist Dorothea M. A. Bate of the British Museum (Natural History) decided to explore Crete in 1903, disguised as a man. Villagers guided her to rewarding discoveries along the rugged coast: she identified seven extinct mammal species including both large and small Pleistocene elephants.

Bate's finds of large elephant bones are controversial. Some paleontologists doubt the presence of large proboscidean remains in Crete and acknowledge only dwarf species. But the reports of huge skeletons in Crete by Solinus, Philodemos, and Pliny—even if exaggerated—seem to confirm Bate's finds and indicate that larger-than-expected prehistoric remains may well exist in Crete.[25]

The Giant Bones of Pallene

After his description of the Cretan giant, Solinus referred to Pliny's "manner of measuring a man," a formula derived from anatomical clues, such as arm span to determine height. Then he gives a detailed original description of how big bones erode out of the soil of Pallene (Kassandra Peninsula, Chalkidiki). "Before there were any humans on Pallene, the story goes that a battle was fought between the gods and the giants." Traces of the giants' destruction "continue to be seen to this day, whenever torrents swell with rain and excessive water breaks their banks and floods the fields. They say that even now in gullies and ravines the people discover bones of immeasurable enormity, like men's carcasses but far bigger." Note that Solinus places the giants in the prehuman past. Philostratus (above) also remarked that thunderstorms and earthquakes revealed giants' bones in Pallene, and Pausanias and many other authors knew the peninsula as the giants' birthplace and one of the major battlefields of the Gigantomachy.

But it was not until 1994 that paleontologists learned about the giant bones of Pallene, after a villager found some huge teeth while building a road. The traditional homeland of the mythical giants turns out to be the old stomping grounds of the stupendous *Deinotherium giganteum* and Miocene mastodons. Evangelia Tsoukala (Aristotle University, Thessaloniki) is currently excavating bone beds on the Kassandra Peninsula containing abundant mastodons; elsewhere on the peninsula there are more deposits of mastodon, cave bear, and rhinoceros fossils.[26]

Monster and Dragon Remains

Not all prehistoric skeletons fit the ancient image of humanoid giants or heroes. We saw that the enormous bones of Miocene mastodons and giraffes on Samos were recognized as primeval *animals* of the prehuman past as early as the fifth century B.C. (chapter 2). Other descriptions of monsters and dragons may have been influenced by observations of unusual, large vertebrate fossils.

The Dragons of India

Dragons of enormous size and variety infested northern India, according to Apollonius of Tyana, who traveled from Asia Minor to the southern foothills of the Himalayas in the first century A.D. The countryside was full of dragons and "no mountain ridge was without one," relates Philostratus, who compiled a biography of Apollonius based on the sage's lost letters and manuscripts. The local people spun fantastic tales about hunting these dragons, using magic to lure them out of the earth in order to pry out the gems embedded in their skulls.

Trophies of these dragon quests were displayed in Paraka, an important city at the foot of a high mountain: "In the center of that city are enshrined a great many skulls of dragons." Ancient Paraka has never been identified. Could *Paraka* be *Parasha*, an ancient name for Peshawar? *K* and *sh* are linguistically related in Indo-European languages, and by the first century A.D. Peshawar was an important center at the east end of the Khyber Pass, north of Taxila (Pakistan). And it's interesting to note that a famous Buddhist holy place north of Taxila was known as "the shrine of the thousand heads," according to the writings of Chinese pilgrims in A.D. 500–640.[27]

Apollonius and his companions traveled through the vale of Peshawar to Taxila and proceeded southeast on the Royal Highway to the Ganges plain. The route skirted the Siwalik Hills, a range of foothills paralleling the Himalayas for over a thousand miles from Kashmir to Nepal. The Siwalik range is a remarkable geological formation of vast and rich vertebrate fossil beds. The "extensive remains of late Tertiary creatures are found all along the barren foothills of the Himalayas, and deposits certainly occur around Taxila," observes paleontologist Eric Buffetaut. On the slopes, in eroding cliffs, and along marshy streambeds, from Kashmir to the banks of the Ganges, the local people would have observed a host of strange skeletons, including enormous crocodiles of great bulk (20 feet long; 6 m); tortoises of incredible dimensions (like those at Pikermi, the size of a small car); prehistoric elephants such as shovel-tusked gomphotheres, stegodons, and *Elephas hysudricus* with its bulging brow; rhinoceroses; bizarre chalicotheres and anthracotheres; the large giraffe *Giraffokeryx*; and the colossal *Sivatherium* (named after the Hindu god Siva), a mooselike giraffe as big as an elephant and carrying massive antlers.

I believe that the array of dragon heads exhibited at Paraka included the skulls of some of these strange creatures (figs. 3.8, 3.9, 3.10, 3.11).

In the wealth of Apollonius's description of the dragons, several details catch the eye of a paleontologist. The dragons of the high ridges were said to be larger than dragons of the marshes, which had sharp, twisted tusks. When marsh dragons fought elephants, both perished, and their entwined bodies were a great find for the dragon-hunters. The dragons of the ridges had long necks, very prominent brows, and deeply sunken eye sockets, giving them a frightening aspect. Impressive "crests" grew from their heads, which were of moderate size on the young, but grew to huge proportions on the adults. Men and boys hunted these creatures for the gems stored inside their skulls: the gems were iridescent, "flashing out every hue." People said the dragons made a great clashing noise and shook the earth when they burrowed in the ground. The imagery recalls the monstrous Neades of Samos and their association with earthquakes—and indeed, the Siwaliks are prone to severe quakes.

The lowland dragons with distorted tusks, found along with the remains of familiar elephants, could have been inspired by fossil assemblages of mixed proboscidean species, some similar to living elephants but others with oddly formed jaws and tusks. *E. hysudricus* skulls are distinguished by overhanging brows. Glowering browridges over deep-sunk eyes and long necks would also fit the appearance of the great giraffid skulls of the *Giraffokeryx* and *Sivatherium giganteum* (whose skull is nearly 3 feet or 1 m long). As for the crests, both giraffe skulls have two pairs of prominent bony excrescences behind and over the eye sockets. The *Sivatherium*'s palmated antlers are truly massive, while the longer-necked *Giraffokeryx*'s four ossicones or horns project back laterally from its long, streamlined skull (about 20 inches, or 50 cm, long). These structures would indeed be smaller on immature specimens.

And what about the gems in the dragons' skulls? That clue points to a taphonomic feature of certain fossils. Remember the sparkling "diamonds" (calcite crystals) that encrusted the Pikermi skull stolen by the Bavarian soldier? And the selenite crystals that

3.8. Fossil skull of a giant crocodile (gavial, 12 million years old) embedded in the ground, near Taxila, Siwalik Hills, Pakistan. *Leptorhynchus crassidens* reached 17 to 25 feet in length. Photo © John Barry, Peabody Museum, Harvard University.

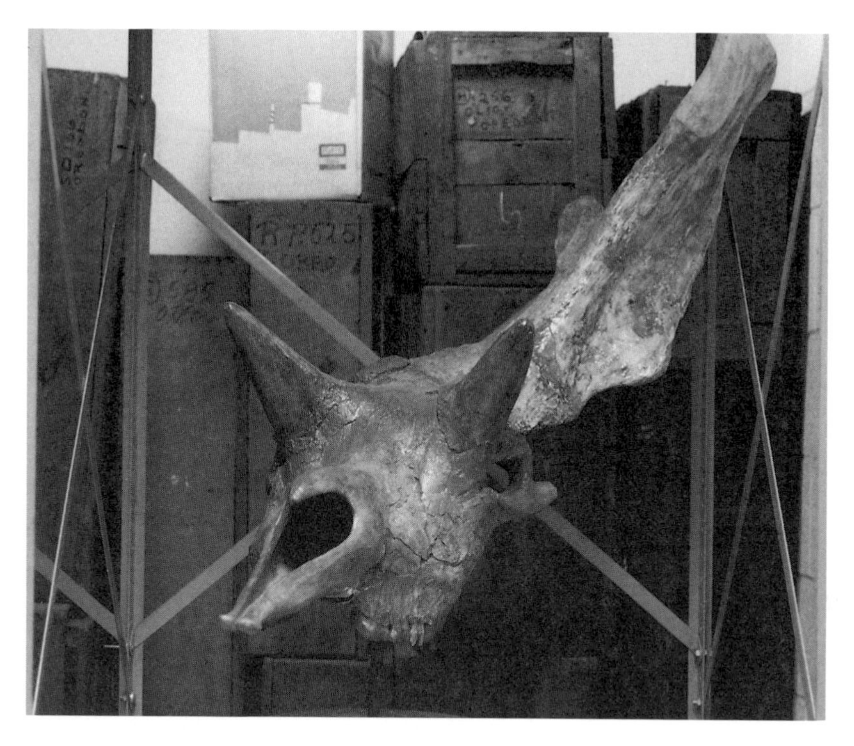

3.9. Massive *Sivatherium* skull, from the Siwalik Hills, about 3 feet long. One antler is missing. Photo courtesy of Nikos Solounias.

replaced the bone marrow of petrified animals in Spain described by Pliny? I think the Indian lore about special gems prised out of dragon skulls alludes to the crystals that can form on mineralized bones. The detailed observations of the first modern investigator of the Siwalik fossils confirm my theory: large, glittering calcite crystals and tubular selenite crystals are common in the Siwalik fossils.

In 1834–42, paleontologist Hugh Falconer was the first scientist to excavate the abundant deposits of the Siwalik Hills, after reading old Persian legends and travelers' accounts of enormous bones in the region. A local rajah showed Falconer some remains of giant beings, known as *Rakshas* ("Titans"), destroyed by a mythical hero of Indian epic. Natives of the Siwaliks had long gathered these *Bijli ki har*, "lightning bones," for their magical

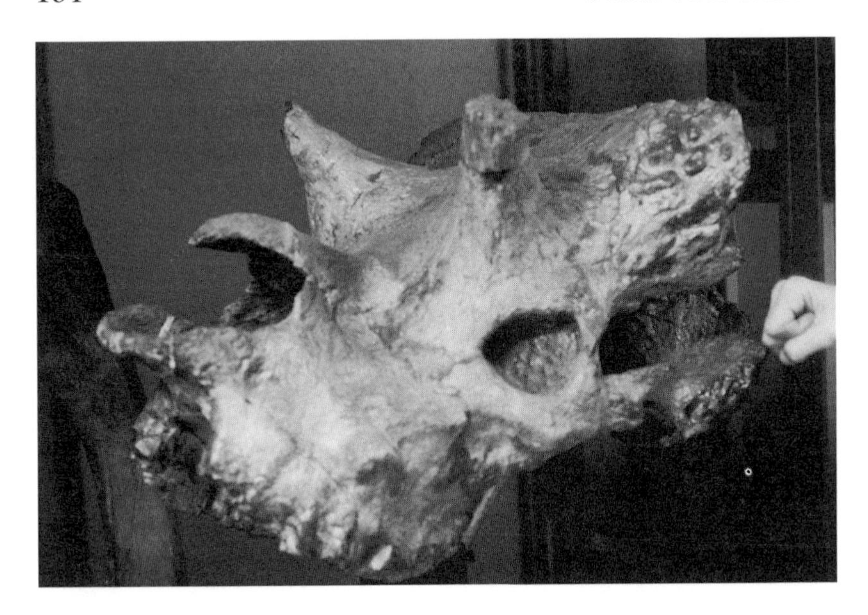

3.10. Part of a *Sivatherium* skull (antlers missing). Siwalik Hills. Photo courtesy of Nikos Solounias.

3.11. Dragonlike *Giraffokeryx* skull, from the Siwalik Hills. Photo courtesy of Nikos Solounias.

powers. On his first visit to the Siwalik beds, Falconer collected more than 300 big bones in six hours. He was struck by the colorful minerals and "ornamental" quality of the fossils: the bone-marrow crevices of the black bones were filled with sparkling crystals. In 1836, Falconer amassed more than 250 proboscidean and giraffid heads (many of them complete). In 1848 he shipped over five tons of fossils to London museums. Since his discoveries, many paleontologists continue to excavate the rich deposits of Siwalik remains. They too have noticed large calcite crystals on the fossils once taken for dragon bones.[28]

The folklore about dragons in northern India gathered by Apollonius and preserved by Philostratus was criticized in antiquity as nonsense, and modern scholars dismiss the dragon passages as sensational travelers' tales. But if we consider the fossiliferous topography of Apollonius's route and read the Indian tales in the same light as we read the Scythian tales of griffins or Chinese accounts of dragon bones, there seems to be little doubt that the ancient dragon lore of northern India was influenced by observations of the world-famous Siwalik fossil beds.

Flying Reptiles of Egypt

When Herodotus traveled to Egypt, he heard tales of flying reptiles or dragons with membraned wings. Seeking proof, he made a special trip to the vicinity of Buto, whose location is unknown. In a narrow pass opening to the desert, Egyptian guides showed him "bones and spines in incalculable numbers, piled in heaps, some big and some small." This passage is one of the most cryptic in Herodotus. Classicists and cryptozoologists have long puzzled over what he saw and where. A horde of dead locusts? "Parachuting" lizards? Pterosaur fossils? All these possibilities have been proposed.

Given that Herodotus was shown skeletons and not living specimens, it's worth considering what kind of fossils exist in Egypt. Could the tales of winged reptiles have been based on attempts to restore the remains of pterosaurs or even spinosaurid dinosaurs?

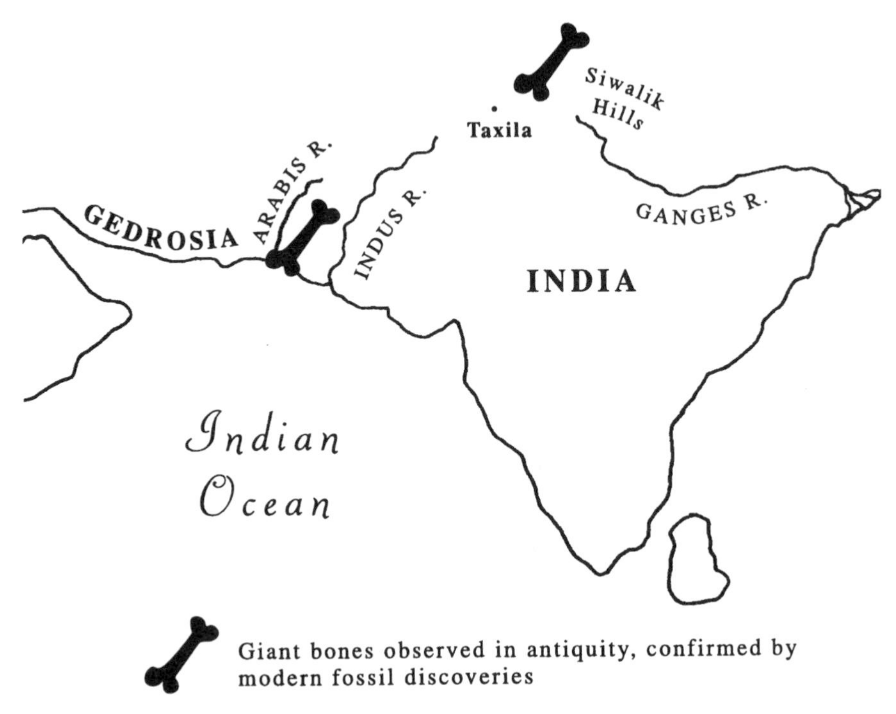

Map 3.3. South Asia. Giant bones reported in Greco-Roman sources compared with modern vertebrate fossil discoveries. Map by Michele Mayor Angel.

Spinosaurs were large Cretaceous reptiles with a dorsal array of membraned spines. Buto is thought to be east of the Nile, while spinosaurs are known only in western Egypt, but this would match the later Roman writer Cicero's statement that winged reptiles occurred in the western desert. Whatever they were, the myriad spines and bones that Herodotus saw may have been fossil remains collected and placed at a shrine by the Egyptians. His description recalls the tons of fossilized bones that archaeologists found heaped at two ancient shrines along the Nile (chapter 5).[29]

Aegean Monsters

A terrible forest fire swept through a glen below Mount Pelinnaeus on the island of Chios, destroying a thick forest sometime in the

Roman era. Poking through the bare, scorched rocks afterwards, someone, perhaps a shepherd, came upon a monstrous skeleton. Farmers gathered to marvel at the backbone and skull of terrifying size. We can imagine a holiday atmosphere lasting for days after the conflagration, as families flocked to see the spectacle. The Roman-era naturalist Aelian tells us that the people identified the beast's remains as those of a legendary dragon that had once terrorized their island, a dreadful monster described in Chian histories consulted by Aelian but now lost. "From these gigantic bones," says Aelian, "the villagers were able to observe how immense and awful the monster was when it was alive."

Chios is a geological extension of Asia Minor; the island does contain large Miocene skeletons, including *Deinotherium* and mastodons. A forest fire may well have revealed a fossil assemblage. Indeed, paleontologist Nikos Solounias says that he was able to discover the fossil bed known in antiquity as Phloion ("Earth's Crust") on Samos after a forest fire had raged through the landscape, exposing rock formations and fossils.

Fossil remains may explain the marvel that ancient quarrymen found inside a slab of rock on Chios. The workers described an impression within the stone resembling "the head of Pan." The island's pink marble was in great demand in the late first century B.C., and news of the curious find in the quarry circulated back in Rome, according to Cicero. Archaeologists have located the ancient marble quarries of Chios in a block of Triassic limestone thrust over surrounding Neogene sediments, which contain fossils. It's interesting that Pliny reported a very similar discovery in a marble quarry on the island of Paros in the next century. "When stone-breakers split a rock vein with wedges, a naturally imprinted image of Silenus was inside."

In the popular imagination, Pan and Silenus resembled Satyrs, generic contemporaries of the primeval giants (see chapter 5). Satyrs had large shaggy heads with grotesque semihuman features, equine or goat ears and horns, tails, human or animal thighs, and hooves. We know that the natural philosopher Xenophanes had described fossil fish impressions in deep mine shafts in Paros as early as the sixth century B.C. (chapter 5), and archaeologists have located the very deep tunnels where Parian marble was quarried in antiquity. Paros does have Pliocene limestones and Miocene sediments, but no modern paleontological studies have been published, so it's difficult to guess what species of fossils the quarrymen might have seen.[30]

Monster Remains on Display

The Romans loved sensational displays of natural curiosities. But the skeleton of the monster of Joppa surpassed all expectations for Marcus Aemilius Scaurus's extravagant victory celebrations in 58 B.C. According to Greek myth, the hero Perseus rescued the maiden Andromeda, who was chained to a rock and exposed to the monster at Joppa (Jaffa–Tel Aviv). The rescue was a favorite scene in Roman art. People said that the sea around Joppa was still stained red with gore from the monster (Perseus petrified it with the Gorgon's head and smashed it with rocks). Traces of Andromeda's chains were pointed out to travelers on the promontory at Joppa, according to the Jewish historian Josephus. And now Scaurus had obtained the monster's skeleton. His men loaded the thing onto a huge grain transport vessel sailing counterclockwise from Alexandria by way of Joppa to Rome. After a two-month voyage, the assemblage was unloaded at Ostia, hauled to Rome, and reassembled as the centerpiece of Scaurus's "marvels from Judaea." Pliny tells us that the backbone was 40 feet long (12 m) and 1.5 feet thick, with ribs taller than an Indian elephant.

What kind of skeleton was billed as Andromeda's dragon? Most scholars assume that Scaurus shipped the bleached bones of a whale from Palestine. The huge proportions seem to match a cetacean, rather than a prehistoric land mammal. Jewish lore located Jonah's whale at Joppa, and sperm whales (up to 60 feet long; 18 m) are known to beach on those shores. The immense "dragon" at Macras (near Beirut) reported by Posidonius a few years earlier (ca. 90 B.C.) also sounds like a whale carcass. Nearly a *plethrum* long (100 feet, 30 m), the carcass was so thick that men on horseback couldn't see over it, and the jaws were wide enough to swallow a man on a horse. Pliny and other Roman writers described whale strandings in the Mediterranean and Arabian Seas, as well as the Atlantic. In about 54 B.C., a spectacular naval battle was staged with a live whale that was trapped in Ostia harbor. And the emperor Septimius Severus (A.D. 193–211) constructed a lifelike restoration of the skeleton of a beached whale in the amphitheater in Rome (fifty live bears were led inside the belly). But if whales were familiar wonders, would Scaurus bother to transport a cetacean skeleton to Rome?

There is evidence that some monsters of myth were fabricated as hoaxes in Roman times, so we have to wonder whether Scaurus's "sea monster" was constructed out of whale and other bones and materials. The problem recalls a modern sea monster hoax that relied on whale bones. In 1845, a fossil collector named Albert Koch exhibited a 114-foot (35-m) skeleton in New York City as the extinct ancestor of the classical sea serpent like the one Perseus killed. Paleontologists exposed Koch's monster as a fake concocted from a number of Eocene *Zeuglodon* whale vertebrae wired together with other bones. (A few years earlier, Koch had wowed Londoners with another monster made out of mammoth and mastodon bones.) A similar composite monster of real and fossil bones could have been constructed in Roman times.[31]

The magical petrifaction of Andromeda's monster in the Greek myth raises the possibility that large mineralized skeletons of extinct mammals might have influenced the Joppa lore, so it's worth mentioning that the impressive remains of extinct proboscideans exist in Israel. Josephus alluded to ancient Hebrew traditions about enormous bones dug up around what is now Hebron, Israel. In *Jewish Antiquities* (written for a Roman audience in the first century A.D.) he stated that the early Israelites had wiped out "a race of giants, who had bodies so large and countenances so entirely different from humans, that they were amazing to the sight and terrible to the hearing. These bones are still shown to this very day, unlike any credible relations of other men."[32]

Bone Relics as Popular Science

Whether whale, fossil, or fake, the Joppa monster shows the great enthusiasm for curiosities in the Roman Empire. No expense was spared to obtain up-to-the-minute natural marvels and old relics.

Emperors and entrepreneurs began to plunder the collections that had accumulated over centuries in temples. Those sacred museum inventories included foreign armor and antique weapons, pickled mermen, celebrities' jewelry, the egg that Leda laid after her affair with Zeus in the form of a swan (probably an ostrich egg), great ivory tusks, the Golden Fleece, strands of Medusa's hair, gigantic snakeskins, and myriad other miscellaneous mementos, including the bones and teeth of heroes, giants, and monsters. People brought curious objects to sanctuaries, not just as private offerings to gods, but in the spirit of creating communal museums where men, women, and children could contemplate natural wonders and try to puzzle out their meaning.

Many cultures enshrine special objects with a mythic aura. Sri Lanka has the Temple of Buddha's Tooth, and the Smithsonian displays Judy Garland's red shoes from *The Wizard of Oz*. Think of the mystique surrounding George Washington's false teeth, John Dillinger's penis, Einstein's brain, a fragment of John F. Kennedy's skull. When it comes to relics, the lines between myth and history, and piety and spectacle, are thin. But the general assumption that in antiquity people gazed in awe at relics without attempting to make sense of them is doubtful. That view is more appropriate to seventeenth-century European cabinets of curiosities, which were intended to evoke a kind of decontextualized wonder at the random juxtaposition of unlabeled exotic items. In contrast, in antiquity the mythohistorical relics displayed by temple curators were *not* random objects stripped of historical meaning but physical links to a culture's shared legendary past. In the case of giant bones, we have already seen that their meaning was part of lively public discourse about the mythical past.

It was in encounters with items of remotest antiquity that the Greeks, Romans, and other ancient people were able to experience "a sense of the immense distance between present and past," observes ancient historian Sally Humphreys. Relics evoked "a convergence of nostalgia, local history, erudition . . . and ritual" as people tried to reconcile myth with contemporary life. Humphreys argues that it was through relics that ancient people developed a "historical consciousness" of time, a consciousness that infused "popular as well as elite culture, oral tradition as well as written texts." We have seen how the physical vestiges of extinct giants gave individuals of all walks of life a sense of the earth's awesome chronology. People compared local bone lore with archaic myths, and went out of their way to view relics and marvels, either in situ or in sanctuaries. As a Roman poet of the first century A.D. remarked, it is by braving "the dangers of land and sea, greedily seeking the tales of old lore passed from folk to folk" and gazing at old relics in temple treasuries, that "we relive the ancient times."

The ancient passion for relics, notes Humphreys, should help historians recognize that "world-views once dismissed as irrational" played "a central role in organizing research" in historiography and natural science. Stressing that modern standards of evidence, truth, and progress should not be imposed on ancient history and legend, Humphreys asks, "Why does paradox still seem problematic as an instrument for discovery?" Similar insights are expressed by historian of science Thomas Kuhn and philosopher Philip Fisher. Kuhn maintains that "science" does not flow from a logical progression of objective truths but is strongly influenced by nonrational concepts that attempt to embrace accumulating anomalies. Fisher studies the experience of wonder as the engine of scientific curiosity and creativity since antiquity (chapter 5). The tension between philosophical standards of truth and the folk knowledge expressed in myths was also felt by thoughtful historians in antiquity. "I am not unaware of the difficulties besetting those who undertake to account for ancient myths," wrote Diodorus of Sicily in about 30 B.C. The very magnitude of time encompassed in folk memories "makes legends seem incredible," and "some readers set up an unfair standard and require of the myths the same exactness as in the events of our own time."[33]

The proliferating lists of big bone finds in ancient literature are strong evidence that in antiquity a culture of popular knowledge, or popular science, about the earth's history coexisted with established "academic science," which ignored the evidence of giant bones (see chapter 5). Paradoxical physical evidence that philosophers and scientists are unable to explain encourages the development of alternative natural knowledge and beliefs in any era.

Anomalies are magnets that either attract or repel curiosity. Are irregular phenomena, such as oversize skeletons, exceptions that prove the rule, as the ancient natural philosophers seemed to assume, or are anomalies evidence of nature's secrets, as many others suspect?

Imperial Fossil Inventories

With widening communication and travel in the Roman Empire, examples of giant bones multiplied. Paradoxographers and other compilers of natural wonders began to compare the new paleontological finds with historical ones. Meanwhile, as noted above, Roman rulers began to grab up unusual remains from sacred sites around the empire, just as Napoleon would later seize the famous *Mosasaur* dinosaur fossil from Holland as a prize for Paris in 1794. In 200 B.C., King Masinissa of Numidia (Algeria) accepted a gift of colossal ivory tusks from his admiral. Aghast when he later learned that the admiral had stolen the famous giant tusks from the Temple of Juno on Malta, the pious king inscribed the ivories with his apologies and shipped them back to Malta. Only 130 years later, however, Verres, governor of Sicily and a notorious Roman looter, stole those same venerable tusks, with no qualms.

Augustus's Fossil Collections

In 31 B.C., the future emperor Augustus plundered the great tusks of the mythical Calydonian Boar from the Temple of Athena in Tegea, Greece, and installed them in Rome. I think that these trophies were most likely prehistoric elephant tusks dug up in Pleistocene beds near Tegea. The contemporary poet Ovid apparently viewed them in Rome: he compared the Calydonian Boar's tusks to those of an Indian elephant (thought to be larger than the African species in antiquity). Some 200 years later, the "keepers of the wonders" at the Sanctuary of Dionysus in the Emperor's Gardens in Rome informed Pausanias that one Calydonian tusk had crumbled away, but he admired the surviving one—it measured about 3 feet (almost a meter) long.

Some 400 years after Pausanias, the Byzantine Greek historian Procopius viewed another great pair of tusks labeled "Calydonian Boar" at Beneventum, Italy. He described them as "well worth seeing, measuring not less than three spans around and having the form of a crescent." A circumference of three hand-spans, about 27 inches (70 cm), and the distinctive curvature or "crescent" shape suggest that the Calydonian tusks at Beneventum were a pair of woolly mammoth tusks, common remains in Italy.[34]

As emperor, Augustus (63 B.C.–A.D. 14) established the world's first paleontological museum at his villa on the island of Capri. According to his biographer Suetonius, it housed "a collection of the huge limb bones of immense monsters of land and sea popularly known as giants' bones, along with the weapons of ancient heroes." This offhand statement is an important milestone in paleontology, because it shows that Suetonius, writing in the early second century A.D., was aware of the *animal* origin of the prodigious remains conventionally ascribed to humanoid giants (we'll return to this museum in chapter 4).[35]

Some fossils collected by emperors were spoils of war like the Calydonian tusks, and others were sent to Rome from the far outposts of the empire, much as mammoth bones from the American colonies were shipped to London during the British Empire. But enormous remains turned up in Italian soil too. The poet Virgil observed that farmers often plowed up vast bones in their fields, where paleontologists now find mammoth remains. Large Pleistocene fossils also occur on the island of Capri and in the environs of ancient Rome itself, so people digging wells or building foundations in those places could have come across the giant bones of several Ice Age species.

In the Roman era, interest in giant bones coincided with a growing fascination with freaks of nature: some bizarre specimens were displayed in a sideshow atmosphere. For example, besides his display of the Calydonian tusks in Rome and his museum of giant bones on Capri, Augustus exhibited the bodies of a giant and giantess in a vault at Sallust's Gardens in Rome. According to Pliny, the giants, named Pusio and Secundilla, were each 10 feet, 3 inches (over 3 m) tall. It's not clear whether these were mummies or skeletons, but they may have been composites of human bones and bones of extinct mammals. At least two writers of Augustus's reign, Manilius and Diodorus of Sicily, mentioned sensational displays that combined animal and human limbs, and later Aelian referred to "artificers of nature" who fabricated fakes by blending dissimilar animal remains.[36]

Tiberius's Paleontological Investigations

Augustus's successor, Tiberius (42 B.C.–A.D. 37) immersed himself in Greek history and mythology. He spent part of his boyhood in Sparta, where he would have heard the tale of Orestes' huge bones, and later he lived in Rhodes, where locals referred to the giant skeletons buried in the eastern part of the island as the "Demons of the East." The diversion of the Orontes River in Syria, where the huge skeleton was reported by Pausanias, was probably Tiberius's project. He retired to Capri, the site of Augustus's paleontological museum, and he took an active interest in zoological curiosities himself.

As emperor, Tiberius received reports of strange remains observed in Gaul (France), according to Pliny. Pliny's statements are maddeningly vague, but the events have paleontological possibilities, so I asked Eric Buffetaut (Centre National de la Recherche Scientifique, Paris) to help me decipher them. One report, from an unnamed island off northwestern France, concerned about three hundred monstrous remains of various sizes and shapes exposed by the sea after a storm. Pliny says that "the recognizable remains included elephants, rams with odd horns, and Nereids [mermaids]." It is not clear what the rams and Nereids really were, but "remains of elephants" strongly suggests an exposure of jumbled bones in which the distinctive parts of proboscideans stood out.

The islands off northern France are granite or metamorphic, barren of fossils, except for one. The Channel Island of Jersey has Pleistocene deposits containing the well-preserved tusks, skulls, and limb bones of woolly mammoths. The Romans would recognize the mammoth tusks, molars, and skulls as elephantine. Buffetaut and I believe that this report to Tiberius is the earliest documented sighting of the Jersey mammoth fossils.

Another report concerned an "equally large array of sea monster remains on the coast of Santones" (Saintes), north of the Gironde River (France). This may have been a mass stranding of contemporary cetaceans, but if fossil remains were involved, then the skeletons of Jurassic or Cretaceous reptiles that occur along the coast of Saintes might have inspired the report. Perhaps an exposure of *Halitherium* skeletons was seen in the area south of the Gironde estuary. These early sirenians were large Miocene ancestors of the dugong (a marine mammal related to manatees). "Sirenian bones are conspicuous fossils, especially the pachyostotic ribs and bizarre skulls," observes Buffetaut. The thick, mineralized ribs of Miocene dugongs were collected along the Gironde by Paleolithic hunters some 15,000 years ago to make their distinctive arrows (chapter 4). Living sirenians apparently inspired ancient lore about merpeople known as Nereids and Tritons. The strange remains of the large prehistoric sirenians "might well have been interpreted as remains of sea monsters, since they are found in sands that are also rich in fossil seashells."

The intriguing dispatches from Gaul preserved by Pliny show Tiberius's interest in cryptozoological mysteries. Suetonius tells us that Tiberius enjoyed quizzing learned scholars on obscure points of mythology and philosophy. It's tempting to imagine the emperor needling the philosophers about inexplicable natural wonders from his own imperial archives of anomalies.[37]

A unique record of purposeful paleontological research carried out by Tiberius appears in a little-known account, published by Phlegon of Tralles who lived during the reign of Hadrian (A.D. 117–38). Phlegon, a contemporary of the biographer Suetonius (both men served on Hadrian's staff), had access to the imperial archives, storerooms, and many other writings now lost. Phlegon's *Book of Marvels* devotes an entire section to discoveries of "Giant Bones."

During Tiberius's reign (A.D. 14–37), writes Phlegon, a series of "devastating earthquakes utterly destroyed many notable cities of Asia Minor," especially in Pontus (southeastern Black Sea coast, Turkey). According to Pliny, this was "the worst earthquake in human memory, destroying twelve Asiatic cities in one night." In cracks of the earth, vast skeletons appeared. Phlegon reports that the survivors "were hesitant to disturb the bones, but as a sample they sent a tooth from one of the skeletons to Tiberius." Ambassadors from Pontus carried the tooth, which was just over a foot long, to Rome. They asked the emperor if he wished to see the rest of "the hero." Eager to know the creature's full size and form, but anxious to avoid desecrating a hero's grave, Tiberius devised "a shrewd plan." He hired a geometer named Pulcher to make a model of the giant based on the tooth. According to Phlegon, Pulcher sculpted a head proportionate to the size and weight of the tooth, and from that the mathematician estimated how large the entire body would have been. Pulcher's replica, presumably a grotesque humanoid bust of clay or wax, pleased Tiberius, who sent the hero's tooth back to its home soil for reburial.[38]

This narrative is more than just an anecdote about Tiberius's clever response to a provincial embassy. The facts are historically and scientifically sound: The Black Sea region is prone to severe earthquakes that do expose gigantic mastodon, steppe mammoth, and *Elasmotherium* skeletons. *On Earthquakes*, a lost work by Theopompus of Sinope (in Pontus), described a quake on the Asovian shore of the Taman Peninsula (between the Black Sea and Sea of Azov) that "tore open a ridge and disgorged immense bones" whose "skeletal structure measured 24 cubits [34 feet; 10 m]." Interestingly, the size of the giant tooth (about 12 inches) that was sent to Tiberius corresponds to the length of the molar of a woolly or steppe mammoth, although a large mastodon molar might look more "human." (See fig. 3.12 for the relative sizes of human, mastodon, and mammoth teeth.)

Phlegon cited a historian named Apollonius (works now lost), but he may also have consulted official correspondence about earthquake aid. According to Suetonius, the normally stingy Tiberius was generous to provinces hit by quakes and rebuilt the destroyed cities at his own expense. In gratitude, the ambassadors

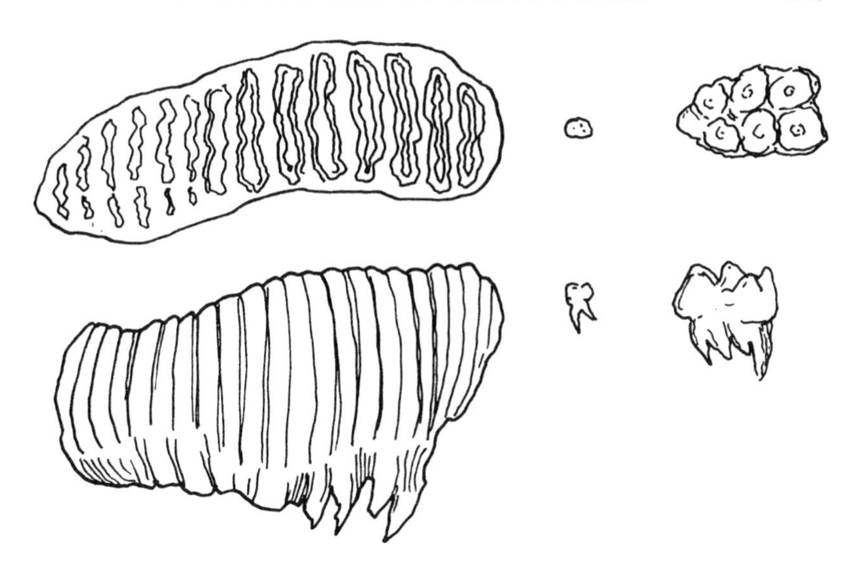

3.12. Relative scale of human and prehistoric elephant teeth. *Left* (*top* and *bottom*): mammoth molar, top and side view. *Center*: human molar, top and side view. *Right*: mastodon molar, top and side view. Drawing by author, approximate sizes based on scanned images provided by Arthur H. Harris, Laboratory for Environmental Biology, University of Texas at El Paso.

from Asia Minor dedicated a colossus (a bronze statue of a giant) in the Temple of Aphrodite in the Roman Forum. Did they also present the giant tooth at that time? Unusual relics from the provinces were routinely sent to Rome, and a hero's tooth would be an especially suitable tribute at the dedication of a giant statue. Emperors often commissioned replicas of items of interest (recall Septimius Severus's restoration of the beached whale skeleton), and colossal statues and busts were extremely popular in Rome at that time. Even the mathematician's extrapolation of the hero's stature from a single tooth is not far-fetched, since a lower first molar does give a good indication of a mammal's total body size and weight. Phlegon's account of the giant tooth is the earliest written record of a scientific reconstruction of a life-size model from prehistoric animal remains!

How did such a momentous event slip by historians of science in general and of paleontology in particular? Phlegon and other paradoxographers have "not been treated kindly by modern critics," in the words of classical folklorist William Hansen (who published the first English translation of Phlegon in 1996). Scholars discount the genre for failing to measure up to scientific philosophy or objective history, and that attitude has kept works like Phlegon's obscure. But Sally Humphrey's recent remarks on paradox as a real tool of discovery and Hansen's realistic view of paradoxography—as popular literature about current events that engaged the curiosity of the general public and emperors alike—allow us to sift out nuggets of ancient history and natural knowledge available nowhere else.[39]

Phlegon's Accounts of Paleontological Finds in Greece and North Africa

Viewed from a paleontological perspective, then, Phlegon's narratives of giant bones are no longer a miscellany of dubious marvels. They come into focus as some of the first descriptions of the remains of the largest extinct mammal species of Asia Minor, Greece, and Egypt—remains that would not be scientifically recognized until the nineteenth and twentieth centuries.

Messenia, Greece

In Messene (a city in the southwestern Peloponnese, south of Megalopolis), "not many years ago," relates Phlegon, a powerful rainstorm broke apart a big stone jar in the ground. Inside was a huge skeleton with three craniums and two jawbones. The inscription read "Idas," the name of a Homeric giant-hero of Messenia. The citizens of Messene "prepared another stone jar at public expense and reburied the hero with honors." Idas, the strongest hero of his day, had helped slay the Calydonian Boar, and he was killed by a divine thunderbolt, like the other fallen giants of Arcadia (chapter 5).

Large storage jars were used as coffins in early antiquity (archaeologists have excavated heroic burials in amphoras in Messenia). Phlegon's account looks like another case of superhuman-size prehistoric animal bones discovered during the eighth or seventh century B.C. and interred as relics of a local hero, and then rediscovered centuries later. But what about those extra bones? According to Homeric-Hesiodic lore, strongmen of myth were often said to have multiple heads or limbs. Their inclusion in the jar is unique material evidence of a literal belief in that archaic folk image of giants.[40]

Wadi Natrun, Egypt

If anyone still doubts that giant skeletons really do exist, says Phlegon, they should go to Nitria in Egypt. There, one can view an exhibit of "huge bones, unencumbered and plain to see. The bones are not concealed in the earth and jumbled together in disorder, but arranged in such a manner that the person viewing them can recognize the thigh bones and shin bones and so on with the other limbs." Phlegon seems to be describing articulated skeletons eroding out on the surface of the desert. The contrast to the fragmented bone assemblages normally seen in Greece and Asia Minor is unmistakable.

Since the era of giant beings, Phlegon explains, all life-forms have waned over successive generations. The dimensions of the bones at Nitria are "sure proof that nature is running down." The notion that animals and humans became progressively smaller over the ages was a commonplace in antiquity (see chapter 5).

Nitria, now called Wadi Natrun (Natron Lakes, a below-sea-level valley of sodium carbonate deposits, known as *nitrum* or *nitron* in antiquity), lies about 40 miles (64 km) south of Alexandria. A desolate landscape of drifting dunes broken by limestone cliffs, Wadi Natrun was a stop on the ancient caravan routes going south and west to the oracle of Ammon at Siwa. Phlegon's details suggest an eyewitness account. He served on Hadrian's staff—did he accompany the emperor to Alexandria in A.D. 130, shortly after Hadrian reburied the huge bones of Ajax at Troy? We know from other writers that Hadrian made excursions into the desert south of Alexandria. The oasis at Nitria would have been a logical destination, especially given Hadrian's interest in remarkable remains.

Phlegon's account is the first written description of the fossils of Wadi Natrun, where modern paleontologists have discovered Egypt's most signficant Pliocene exposures. The conditions in this desert basin are totally unlike the situation in the Aegean, Italy, and Anatolia where farmers plowed up big bones or earthquakes disgorged broken skeletons. Here, scouring winds uncover the complete skeletons on the surface, as in the windswept dinosaur beds of the Gobi Desert (fig. 3.13). Since Wadi Natrun was the source of the natron required for mummification in ancient Egypt, we can guess that prospectors of the second millennium B.C. were probably the first humans to see these impressive remains. Nitria was not an easy place to visit, yet by Phlegon's day travelers went out of their way to view the spectacle of massive, articulated skeletons of mastodons, such as *Gomphotherium angustidens* (fig. 3.14), or the huge giraffids *Sivatherium maurusium* and *Libytherium*. Nowadays, paleontologists search for Egyptian fossils by sweeping with brooms after severe sandstorms. It's easy to picture the ancient Egyptians using the same technique for their display at Nitria.[41]

Phlegon's account brings up an interesting question: Did large fossils influence early Egyptian beliefs about giant deities? Some myths about Egyptian giants and the gods Osiris and Set have suggestive features. According to Diodorus of Sicily (ca. 30 B.C.), when life was first originating on earth, giant beings flourished in Egypt, but the Egyptian name for them has been lost. These huge creatures were completely exterminated by Osiris and other gods. Diodorus also described a great mound visited by travelers in the western desert, at a place named Zabirna (near Siwa), where the god Dionysus had killed and buried an enormous monster called Campe. There is evidence of a "cult of fossils" for several periods

3.13. Typical fossil mammal exposure in the Egyptian desert. Here, a man uncovers a large *Arsinoitherium* skull in the Fayyum Basin, 1907. A similar scene was described the second century A.D. by Phlegon of Tralles, at the Pliocene fossil beds of Nitria (Wadi Natrun). Neg. 18298, photo Granger. Courtesy Department of Library Services, American Museum of Natural History.

in Egypt, and archaeologists have discovered two ancient shrines to Set where worshipers dedicated large fossil bones by the ton. Set was a composite "typhonic" monster-god of the desert (similar to Typhon of Greek myth)—and the desert, where strange and immense skeletons emerge from the sand, was the traditional habitat of Egyptian monsters. Paleozoologist Herbert Wendt has suggested that the unknown animal head of Set in Egyptian art might have been based on the skull of the *Libytherium* (a large giraffid similar to *Samotherium*).

In Egyptian mythology, Set trapped Osiris in a large coffin and

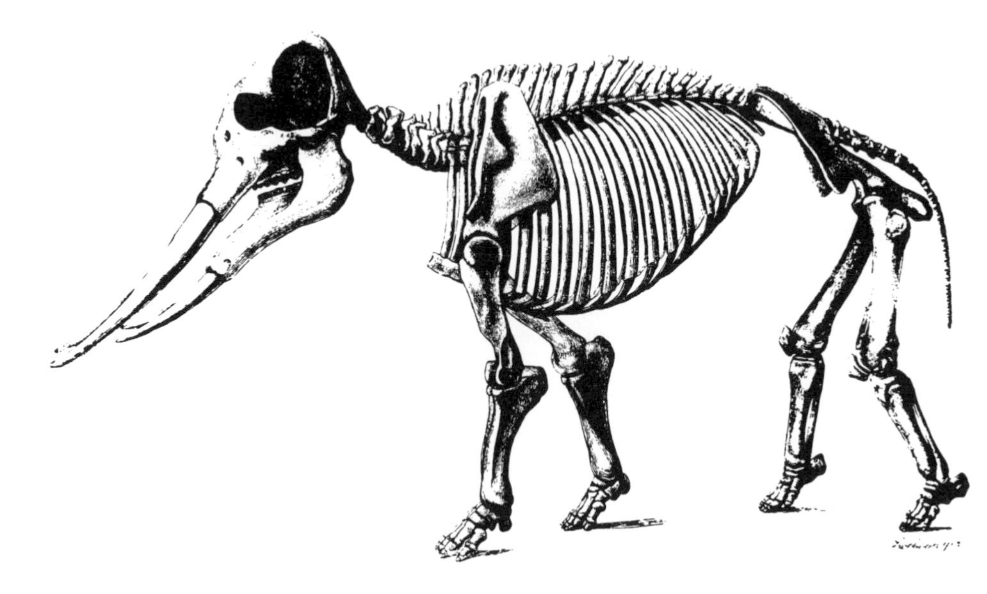

3.14. *Gomphotherium angustidens.* The fossil remains of this large mastodon are found in southern Europe and North Africa. From Gaudry, in O. Abel, *Rekonstruktion vorzeitlicher Wirbeltiere* (Jena, 1925), fig. 228.

then dismembered and scattered his body across the land. The goddess Isis gathered up the bones and placed them (in some versions, wax models) in shrines all around Egypt; we are also told that she buried the real reconstituted skeleton in a secret grave. Various places in Egypt had either a part or the true body of Osiris. Saïs and Philae claimed to possess the complete body, while Busiris claimed Osiris's spine, Abydos his head, Bigah his left femur, Thinis his right femur, and Nubia his shin. Were some of these Osiris relics large prehistoric bones? I think it's possible. The Egyptians regarded the Fayyum depression, a region rich in a variety of Tertiary fossils of large and bizarre mammals, as the origin of all life-forms, and the Fayyum was also said to be the place where Isis reassembled and secretly buried Osiris's limbs. In the third and second centuries B.C., Wadi Natrun (Nitria) was known as "the secret burial mound of Osiris." It's fitting that the largest bones to be seen at Wadi Natrun belong to *Anancus osiris*, a mastodon named in honor of Osiris.[42]

Osiris in Attica

Claims to possess the body of Osiris were not limited to Egypt. Phlegon reported that when the Athenians fortified a tiny unnamed island off Attica, men digging foundations for walls found a coffin 100 cubits long (140 feet!) with an equally tremendous body inside. The epitaph declared that the corpse was "Makroseiris," Greek for Great (or Long) Osiris. Isis-Osiris cults were established in Greece before the third century B.C., around the time that Wadi Natrun was called the burial mound of Osiris. Isis worship was centered in Piraeus, the harbor of Athens. I suspected that the Makroseiris tale circulated among Athenian Isis worshipers anxious to link the Osiris myth to their home territory of Attica. What might have helped them justify such a claim?

I consulted geological maps and located a small island (Fleves, ancient Phabra) less than a mile off Cape Zoster, Attica, and about twelve miles south of Piraeus. It guards the southern sea approach to Piraeus, so it is a likely candidate for the tiny island fortified by Athens (Phlegon does not say when). The entire islet consists of Neogene sediments identical to those of Pikermi, so diggers might well turn up big bones. To an Isis worshiper from Piraeus who was familiar with the Osiris shrine at Wadi Natrun, the resemblance between the great bones there and on the island of Phabra would have been a striking coincidence. The story of the stupendous coffin and the Makroseiris epitaph can best be understood as fanciful elaborations of actual fossil finds spread by Greek Isis worshipers who associated extraordinary bones with the body of divine Osiris.[43]

Carthaginian Giants

Men digging fortifications in North Africa also found immense bones. The Hellenistic geographer Eumachus (third to second century B.C.) reported that "when the Carthaginians were surrounding their territory with trenches," they came upon two enormous skeletons, 24 and 23 cubits long (about 34 feet; 10 m).

Ancient Carthage, on the Gulf of Tunis, dug trenches (60 feet wide) around the city limits in the fifth century B.C. More trenches and earthworks marked the fluctuating territory controlled by Carthage until its defeat by the Romans in the third and last Punic War (146 B.C.).

By the time Phlegon repeated the story, the dimensions of the skeletons were obviously exaggerated (unless they refer to mixed fossil assemblages). The exact boundaries of Carthage are unknown, but clustered within its territory near the ancient cities of Utica, Sufetula, and Theveste are rich and varied Neogene fossil beds containing mastodons (*Tetralophodon* and *Gomphotherium angustidens* [fig. 3.14], both about 10 feet, or 3 m, tall), deinotheres, and mammoths (*M. africanavus*). Men excavating trenches in those areas would be likely to strike bones of prodigious size. The finds described by Eumachus and Phlegon are the oldest references to the discovery of prehistoric proboscidean remains of Tunisia and Algeria.[44]

The Last Vestiges of Classical Giants

Saint Augustine (A.D. 354–430) lived in Carthage 550 years after the last Punic War, in the twilight of the classical era, on the cusp of the Byzantine age with its new Christian paradigms. In *The City of God*, Augustine discussed the reality of huge creatures of the remote past. He was defending biblical dogma, but to confirm the historical existence of giants he turned to the old Greek and Roman authorities and the long history of paleontological discoveries since Homer. He invoked Virgil's verses about farmers plowing up massive skeletons, and he cited Pliny's observation that the aging earth produces ever-punier life-forms—a "historical fact regretted by Homer." "Some people refuse to believe that bodies were so much larger than they are now," wrote Augustine, "but skeptics are generally persuaded by the evidence in the ground." The "frequent discovery of incredibly large bones revealed by the ravages of time, the violence of streams, or other events" is "tangible proof" that ancient life-forms were enormous—and that "some of those even towered far above the rest."

On the shore of Utica (Gulf of Tunis), declared Augustine, "I myself—not alone but with several others—found a human molar so immense that we estimated that if it were divided up into the dimensions of one of our teeth it would have made a hundred of them. I believe that molar belonged to some giant." Bones and teeth are "very long-lasting," he noted, allowing rational people of "much later ages" the opportunity to visualize the magnitude of beings from eons ago. In past times, such wondrous teeth had been the prized possessions of emperors and temple treasuries.[45]

We might end with Augustine's powerful experience on the beach at Utica, holding in his hands unassailable evidence of the ancient races of giants. His essays, composed as belief in the old Greek myths was fading, have been called "landmarks in the abandonment of Classical ideals." But as his respect for Greek and Roman learning and history shows, the old traditions of giants, heroes, and monsters were not yet extinguished. So let's conclude this chapter with Claudian, the last great pagan writer in the classical spirit. Written as barbarian armies were advancing on the Roman Empire, the two men's overlapping perspectives on the vestiges of primeval creatures strike a haunting chord at the close of an era.

Born in Alexandria, Egypt, in A.D. 370, Claudian was Augustine's younger contemporary. In a poem begun about the same time as Augustine's memoir, Claudian evoked the trophies of the Gigantomachy displayed in a sacred grove in Italy. "Here hang the gaping jawbones and monstrous hides affixed to trees, their horrible visages still threatening. And heaped on all sides bleach the bones of slaughtered monsters." His poem was never finished, for Claudian perished during the Gothic invasion of Rome. A few years later, Augustine died as the Vandals swept through North Africa—and his own bones were eventually enshrined in Italy as sacred relics of the Christian church.[46]

From North Africa to the English Channel, from the Mediterranean and Black Sea to the foothills of the Himalayas, the Greeks and Romans preserved eyewitness accounts of their discoveries of extraordinary remains and people's efforts to explain them. The writers tell us that large bones, teeth, and other fossils were transported over land and sea and venerated in sanctuaries and museums all around the ancient world. The next chapter investigates a variety of material evidence uncovered by archaeologists to complete our picture of ancient fossil hunting.

CHAPTER

4

Artistic and Archaeological Evidence for Fossil Discoveries

The Monster of Troy

IN A GLASS CASE in the Boston Museum of Fine Arts, a strange creature lurks on an ancient Greek vase. This vase painting, made in the famous pottery center of Corinth in the sixth century B.C., is known to art historians as the oldest illustration of the story of the Monster of Troy, a creature described in Homeric legends. But the animal on this particular vase troubles specialists in Greek art, because it doesn't fit the typical monster image.

The Monster of Troy was already an old tale when Homer retold it in the eighth century B.C. In that legend, a fearsome monster suddenly appeared on the Trojan coast after a flood. It preyed on the farmers in the neighborhood of Sigeum. The king's daughter, Hesione, was sent as a sacrifice to appease the beast, but Heracles arrived in time to kill it. The vase painting shows Hesione and Heracles confronting the monster; she hurls rocks from a pile at her feet while Heracles shoots arrows. Two of Hesione's rocks have struck home—there is one just under the eye and another lodged in the creature's maw—and one of Heracles' arrows is stuck in the jaw (fig. 4.1).

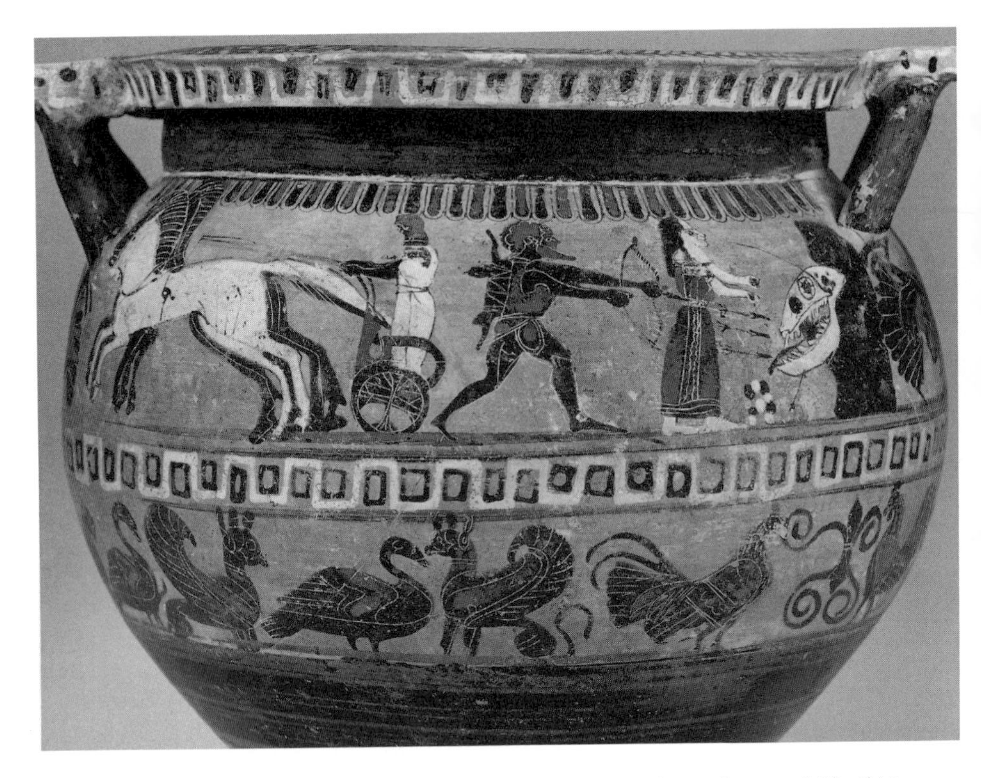

4.1. The Monster of Troy vase, late Corinthian column-krater, 560–540 B.C. Heracles and Hesione confront the legendary monster that appeared on the coast of Troy, near Sigeum. The artist has depicted the monster as a large fossil animal skull eroding out of an outcrop. Museum of Fine Arts, Boston. Helen and Alice Colburn Fund, 63.420.

The human and other animal figures on the vase (horses, geese, leopards, and griffins) are well rendered, in contrast to what seems to be a crudely drawn monster. In his definitive 1987 article on sea monsters, art historian John Boardman singled out the poor artistic quality of this creature's "shapeless, unworthy head" gaping "balefully from a rocky cave." Karl Schefold calls it a "hideous white Thing," painted by a naive artist deficient in creative imagination. Vase scholars have assumed that the Monster of Troy should look like a typical sea monster—even though the Homeric legend described it as a land creature. Most scholars take the dark background behind the odd, disembodied head to be a sea cave, although in a careful study of the scene, vase specialist John Oakley observed that the head "protrudes from a rocky cliff," and Schefold noted that it is "stuck to the cliff like a gargoyle."[1]

This creature is certainly not a typical sea monster. Monsters of the deep were depicted by Greek artists as scaly serpents, lizards, or seahorse-like dragons with undulating bodies, ears, large staring eyes, a dorsal crest, and an upturned snout resembling that of a crocodile, shark, dog, or pig. But if we pay attention to the details and geography of the Homeric legend, and if we view this atypical Monster of Troy in its historical context, then its peculiar appearance begins to makes sense.

The vase painter lived in Corinth, a hub of Greek trade routes where people kept abreast of events in the eastern Mediterranean. The vase was painted in 560–540 B.C., right around the time when the Spartans announced their discovery of the gigantic bones of the hero Orestes in Tegea. The artist created this image in the thick of the ancient bone rush, when Greeks were searching out large, unusual remains that emerged from the ground all around the Aegean. The vase shows the famous Monster of Troy, which appeared at Sigeum on the Trojan coast. That same coast was the scene of another Homeric legend about a pair of monsters that emerged from the cliffs of a tiny island (Tenedos) across from Sigeum. And it was at Sigeum that the colossal bones of the Homeric heroes Ajax and Achilles suddenly appeared in the Roman era (reported by Pausanias and Philostratus; see chapter 3). The sediments around Sigeum contain rich fossil deposits that continually erode out of the collapsing bluffs by the sea.

If we look at the vase again with these facts in mind, instead of a poorly drawn sea monster peeking out of a sea cave we suddenly perceive a monstrous animal *skull* poking out of a cliffside. The great size of the white head, its shape, and the anatomical features suggest that a large fossil skull eroding out of an outcrop was the model for this earliest illustration of the Monster of Troy. Intrigued by this surprising possibility, I showed photographs of the Corinthian vase to several vertebrate paleontologists and asked them for their professional impressions (fig. 4.2).

The paleontologists immediately noticed realistic details missed

4.2. The Monster of Troy, detail. Note the articulated jaw, bony plates around the eye socket, broken premaxilla, and the extended occiput of the monstrous skull. Photo courtesy of John Boardman.

by classical art scholars whose search image had led them to look for traditional artistic motifs instead of ancient natural knowledge. The paleontologists noted the jaw articulation, the hollow eye socket, the extended back of the skull, the forward-leaning teeth, and the natural detail of a broken-off premaxilla (upper jaw and nasal structures). They agreed that the general size and shape of the skull was consistent with that of a large Tertiary mammal emerging from an outcropping.

But two nonmammalian features—the ring of bony plates around the eye socket and the mouthful of jagged teeth—also caught the paleontologists' attention. The numerous teeth and lack of molars reminded some of them of a reptile or toothed whale skull. Eocene whale skulls are exposed around the Mediterranean, and the skulls of the giant Miocene giraffids, including *Samotherium* and *Helladotherium*, have large, forward-projecting teeth, which would be especially noticeable if the premaxilla were missing (fig. 4.3). Sclerotic eye rings are features of birds and dinosaur skulls, never mammals. Dinosaur remains are unlikely to

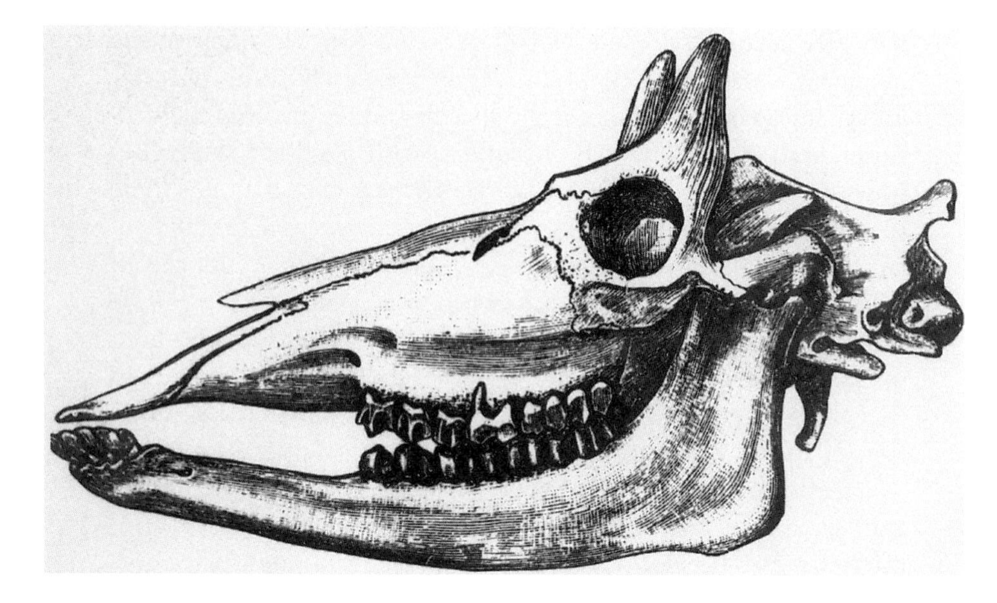

4.3. *Samotherium* fossil skull, giant Miocene giraffe of Samos, Greece. (Horns are absent in female specimens.) After Lydekker, ca. 1890.

have been observed around the Mediterranean, but the ancient Greeks were certainly familiar with the skulls of living birds. And they may have encountered fossil skulls of giant ostriches in Tertiary and Quaternary deposits across Anatolia and Greece. Ostriches are remarkable for having the biggest eye sockets of any terrestrial vertebrate.

"The artist probably saw a large skull appearing on an outcrop in profile, a typical exposure of Tertiary sedimentary deposits," suggests Sevket Sen. "Then he stylized the skull to give an impression of a monster." Dale Russell speculates that the artist combined features from several unrelated creatures, such as teeth and eye rings, to create a fearsome "chimerical" effect. Indeed, we know that the ancients did visualize monsters as composites of beasts, birds, and reptiles. But what is so striking about this particular monster is that (except for the tongue) the artist drew realistic features of various animal *skulls*, instead of living creatures.

But, as Eric Buffetaut comments, we need not identify the specific components of the monstrous skull on the vase. Its extraordinary size relative to Heracles and Hesione conveys a convincing image of a large fossil skull emerging from a cliff. The scientific judgment was clear: whatever the specific species involved, the Corinthian artist must have been familiar with fossilized skulls. But how would the art historians react to the new information? Would this be just one more case in which the sciences and the humanities failed to communicate across disciplinary lines? Not at all! When I relayed these new paleontological perspectives to John Boardman in Oxford, he readily agreed that earlier identifications of the monster as a poorly drawn sea serpent were in error, and that the strange head appears to be the earliest artistic representation of a fossil discovery in antiquity.[2]

Corinthian vases are distinguished by naturalistic depictions of animals and by their complex treatments of myths. This Corinthian vase painter clearly expected his audience to associate a grotesque skull emerging from a bluff with the traditional Homeric tale of the Monster of Troy. Did he intend to offer graphic proof that monsters definitely did once exist by having an immense, chimerical skull stand in for a living monster? Or was the artist making a sophisticated comment on the origins of archaic myths?

Two other vase paintings, also from the sixth century B.C., help put the unusual Monster of Troy vase in perspective. An amphora painted about the same time (ca. 575–550 B.C.) by a different Corinthian artist shows a very similar scene, of Perseus rescuing Andromeda from the monster at Joppa. Perseus throws stones (from a pile at his feet) at the monster's head. Hovering above a dark ground, the white head is about the same size as the Monster of Troy and has a lolling tongue and staring eye. But this monster is fleshed out as a large, scary beast, as described in the mythical narratives (fig. 4.4). The Monster of Troy vase can be seen as a practical exposé of the more typical Perseus vase. The Monster of Troy artist seems to be employing realism to reveal the bone beneath the flesh. He suggests that his fellow artists' depictions of Homeric monsters were originally inspired by discoveries of bizarre animal skulls emerging from the earth.

A black-figure cup painted some years after the Monster of Troy vase (ca. 500 B.C.), now in the National Museum of Copenhagen,

4.4. Perseus rescues Andromeda from the monster of Joppa. Black-figure amphora, Corinthian, early sixth century B.C. Berlin F1652. Drawing by author, after Boardman 1987, plate 24, fig. 11.

shows Heracles dragging a huge monster out of a cave—or a rocky cliffside—with ropes. In this vignette, the monster is depicted as a primitive human giant with a protruding tongue (fig. 4.5).

Monsters of Greek myth were perceived in the popular imagination and portrayed by artists either as huge *beasts* or as giant *humans.* The artist of the Copenhagen vase has opted for the latter. The Perseus vase and the Copenhagen vase therefore illustrate the two traditional branches of mythical interpretation of monsters. But the unparalleled depiction of the Monster of Troy as a large fossil animal skull on the Boston vase points to a natural basis for the two branches of monster and giant images in art and literature. Here is powerful evidence that fossil remains of prehistoric animals influenced ancient ideas about primeval monsters![3]

This amazing artifact, painted some 2,500 years ago, is the earliest drawing of a large animal fossil to survive from classical antiquity. The artist represented that fossil as a specific monster of myth,

4.5. Giant dragged from cave by Heracles. Black-figure cup, ca. 500 B.C. National Museum, Copenhagen, Department of Classical and Near Eastern Antiquities, no. 834.

a monster that appeared on a coast where the remains of large, extinct animals are continually exposed. And further, the depiction is evidence that some ancients were conscious of the myth-making processes that informed the imagery of monsters in the other two vases. The anonymous Monster of Troy artist's paleontological "commentary" anticipates a later literary deconstruction of the way real elephant teeth inspired an archaic myth about dragon teeth (by the writer Palaephatus; see chapter 5). The ability of some ancient Greeks and Romans to consciously link the monsters and giants of their own myths to unusual skeletal remains observed in their own Mediterranean lands stands in sharp contrast to their imported beliefs in Scythian griffins and Indian dragons, fossil monsters of exotic cultures' folklore and landscapes.

The paleontological significance of the vignette on the Monster of Troy vase went unrecognized, despite numerous published descriptions by classicists, archaeologists, and art historians since the

Boston Museum acquired the vase in 1963. It seems incredible, in view of the literary record of avid fossil collecting in antiquity and the rich Eurasian and circum-Mediterranean prehistoric bone deposits known to modern paleontology, that scholars have not sought out ancient art objects like this one.

Nor, as we shall see, have they paid much attention to archaeological evidence for ancient "paleontological" activity. Because the interdisciplinary study of Mediterranean archaeology and paleontology is still in its infancy, this chapter can present only random "snapshots" of the ways physical evidence found by archaeologists can illuminate the ancient engagement with fossils. The following evidence has been gleaned from accidental, little-remarked modern discoveries of fossils in ancient temples, tombs, and other sites around the Mediterranean. Much important evidence has been lost or discarded, through accident or lack of sustained scholarly attention. I hope the following collage of archaeological finds will inspire a more systematic study of fossils excavated in ancient sites and will alert archaeological excavators to the cultural and historical relevance of fossil remains collected in antiquity.

Archaeological Evidence of Fossil Collecting

Fossils have always been objects of human curiosity and acquisition. In Mongolia in the 1920s, for example, fossil hunter Roy Chapman Andrews discovered jewelry made of dinosaur eggs in Stone Age sites near the rich Cretaceous bone beds of the Flaming Cliffs. In the Congo in the 1980s, paleontologist William Sanders found a large *Stegodon* tooth from the Pliocene in a human habitation site of the Pleistocene epoch. The out-of-place tooth of a mastodon that went extinct millions of years before humans appeared "could only have gotten there if some African hunter had brought it home as a curio 21,000 years ago," observes Sanders.

But the topic is considered marginal. Fossils discovered in ancient sites "are rarely saved or studied," laments David Reese, one of the few zooarchaeologists to specialize in identifying prehistoric remains from Mediterranean sites. In 1965, the British paleontologist Kenneth Oakley called his own wide-ranging interest in fossils collected in antiquity an indulgence in "a category of useless knowledge."

Yet Oakley's self-professed empty erudition produced pioneering studies in paleontological history, showing that from Paleolithic times and earlier, people selected compact and intriguing fossils, especially teeth, shells, and marine creatures, to use as ornaments, amulets, and curios. For example, Jurassic ammonites (coiled mollusks) were pierced for suspension by Cro-Magnons in southern France, and an attractively ridged fossil from a Triassic lungfish was brought by a hunter to a cave in Switzerland some 15,000 years ago. These prized items were often transported great distances by primitive cave dwellers. One of the oldest examples of long-distance fossil trade is a Pliocene gastropod found in the Lascaux Cave of central France, which could have come only from Ireland or the Isle of Wight. Trilobites from Germany and other distant sites have also been found in Paleolithic French caves.

Paleolithic hunters in the Gironde Valley in France encountered abundant fossil skeletons of *Halitherium*, a large Miocene ancestor of dugongs. They collected the thick, flinty ribs to make their distinctive arrowheads and blades. Meanwhile in the Pyrenees, Paleolithic hunters searched for *fossilized* canine teeth of formidable cave bears for their amulets (much safer than obtaining teeth from live cave bears, which stood over 7 feet tall!). Translucent blue, green, red, and brown triangular teeth of sharks of the Jurassic through the Cretaceous, still glossy with enamel, were favorite fossil curios in a wide range of archaeological sites, collected by humans as early as 30,000 B.C. Fossils were included as funeral offerings, too. For example, among the grave goods of Saka-Scythian nomads of Kazakhstan (ca. 400–200 B.C.), mourners placed Jurassic *Gryphaea* shells and other marine fossils that they had gathered from outcrops near the Caspian Sea, as well as exotic camel bones, which could only be found 600–1,200 miles to the south.[4]

We can only guess how Paleolithic cave dwellers or steppe nomads explained fossils, both big and small, because they left no writings. But we have seen that the ancient Greeks and Romans recorded their own and other cultures' paleontological observations, along with their interpretations. We know that Greeks and

Romans recognized stony shells and fishbones found far inland as evidence for vast past seas and they interpreted enormous bones in the earth as the remains of long-extinct giants and monsters. The same ancient sources tell us that such bones were collected and displayed in public and sacred places. Has any archaeological evidence survived to corroborate the literary record?

Bothersome Bones

What happened to all the remarkable remains in sanctuaries and heroes' shrines around the Mediterranean? Will we ever recover giant bones and teeth like those hefted and measured by Herodotus, Pausanias, Pliny, Plutarch, Philostratus, Phlegon, and Augustine? Fossilized bones millions of years old that weathered out of the ground in antiquity would have been extremely fragile, especially when further exposed to the elements and human handling. Then, too, some relics, like Pelops's big shoulder blade and the Calydonian Boar's tusks, had already crumbled away by Roman times, according to Pausanias, who tried to search them out. As for relics that escaped destruction or looting in the classical era, they may have been smashed or misplaced during the barbarian invasions of the late Roman Empire and subsequent upheavals in Mediterranean history. And, finally, over the millennia up through the present day, ancient graves and occupation sites have been continuously disturbed and rifled by robbers. Bronze sarcophagi, fine tombs, and temple treasuries were primary targets. Anyone grubbing for treasure simply tossed aside mouldering bones—whether recent or ancient, fossilized or not—to grab up gold, gems, metal to melt down, fine painted pottery, and other valuables with obvious resale value.

Unfortunately, until recently, archaeologists in the Mediterranean ordinarily did the same. In their own zeal for datable artifacts and museum-quality art, they misidentified confusing and bothersome bones and threw them away as ancient rubbish. A picture of the chaotic treatment of countless bones unearthed by excavators in Greece comes through in Carla Antonaccio's study of the archaeology of hero cults in early Greece. In a countryside "saturated" with heroes' graves, excavators exhumed a profusion of animal bones, teeth, horns, and tusks from burial sites of the Bronze Age and later. In the area of Messenia (southwestern Peloponnese), for example, archaeologists unearthed antlers of extraordinary size, tusks, a large animal skeleton in a tomb ascribed to a Homeric hero, unidentified large animal bones given ceremonial burials, and graves containing nothing but animal remains. Yet none of these remains were retained for study; none received the detailed publication that would allow them to be identified by zoologists or paleontologists. Until the advent of zooarchaeology, untold numbers of the remains of living and extinct animals recovered from ancient burials were discarded in archaeological "dumps" and lost forever to science.[5]

Zooarchaeology

Zooarchaeology, the formal study of faunal remains discovered in archaeological excavations, emerged as a discipline in the 1970s. Since then, archaeologists have been more careful to record animal remains. But zooarchaeology has not contributed as much as one might hope to the study of ancient fossil collecting. Zooarchaeologists tend to focus on issues of ancient demography and economics, rather than culture and meaning-production. They evaluate human skeletons and identify living domestic and wild animal skeletons to determine what ancient people ate and what kind of animals they sacrificed. The occasional prehistoric fossil or tooth that turns up in an ancient site is an anomaly. Many anomalies are simply ignored because they don't fit the search image for edible or sacrificial animals. Even when archaeologists do take note of fossils, such finds are often misidentified and misplaced. A fossilized bone or tooth might be recorded in the back pages of field notes, and if published in the excavation report at all, it appears as miscellany in a footnote or appendix.

According to David Reese, it's not uncommon for fossils excavated by classical archaeologists to be mistaken for modern species. Paleontologists rarely have a chance to examine vertebrate fossils unearthed by their archaeological colleagues from ancient sites. One fortuitous exception was Barnum Brown's chance discovery of an extinct elephant molar in the rubble of the ancient medical school of Kos in 1926, discussed at the end of this chapter. Brown imagined that the great doctor Hippocrates had contemplated the specimen. That, of course, can never be proven, but as Eric Buffetaut comments, "There is no doubt that the tooth had been considered suffiently valuable to be carried from one of the vertebrate [fossil] localities on Kos to the Asklepeion by the ancient Greeks."[6]

Teeth and marine fossils, especially when they show signs of human handiwork, are the most common prehistoric remains to be studied by archaeologists. However, a few large fossil remains of extinct megafaunas have been recognized in ancient sites. The fine Spartan tomb containing Orestes' giant skeleton and the fancy Athenian coffin holding the enormous bones of Theseus will have been destroyed ages ago, of course. Yet even if those spectacular paleontological relics are long lost, and even though zooarchaeologists mostly record only living animal species, it's worth sifting through the back pages of zooarchaeological reports, lists of unusual items dedicated in ancient sanctuaries, and unpublished material collected by scholars from various disciplines. Enough scattered proof emerges to show that fossils, including the bones and teeth of large extinct mammals, were actually collected and enshrined just as the Greek and Roman authors said they were.

False Fossil Leads and Dead Ends

A Giant Shoulder Blade in Ancient Athens

I know it's too much to hope that archaeologists will ever locate the sunken ship carrying Pelops's massive shoulder blade home from Troy. But two intriguing finds, one involving a huge scapula in an ancient site and the other a sunken ship carrying fossils, make the story of Pelops's bone plausible. In 1998, I learned that a shoulder blade longer than my arm had been excavated from a ninth-century B.C. context in the Agora of Athens. My first thought was that the early Athenians had discovered a fossil mastodon scapula in the bone beds around nearby Pikermi. But the Agora archaeologists believe that the big bone came from a contemporary whale, whose mammalian scapula resembles that of a giant human. A square hole was cut into one end; this may have allowed the big bone to be suspended for display. Whether modern whale or fossil, this intriguing find means that sometime in the ninth century B.C., when interest in the relics of larger-than-life mythological heroes such as Pelops was first stirring, someone expended considerable toil to lug a colossal shoulder blade to Athens.[7]

Shipwrecked Fossils in the Aegean

In 1973, while working on Cyprus, David Reese heard that a classical shipwreck had been discovered off Cape Kormakiti (on the northwest coast of the island). The hold of the sunken vessel was said to contain fossil animal bones and teeth. Archaeologist Emily Vermeule of Harvard, who was excavating a Bronze Age site that summer, also heard about the wreck. She was given "a fossil tooth of a large ruminant," thought to be some species of camel, from the sunken ship. The man who gave her the fossil tooth is now dead, and the tooth itself is lost. Reese, Vermeule, and I have tried in vain to obtain more information about this highly significant find, which was never reported to the Cypriot antiquities authorities. Relations were strained between Turkey and Greece that summer, and war broke out in 1974, resulting in the Turkish occupation of northern Cyprus and its waters. Cape Kormakiti is still occupied by the Turkish army. No further word has ever emerged about the rumored ancient cargo of fossils, possibly heroes' relics, lost at sea in antiquity, and then lost again in a modern war.[8]

Archaeological Proof of Fossils Collected in Antiquity

A ship passing north of Cyprus may well have been carrying fossils to or from Asia Minor. Cyprus lay along the ivory shipping route from Syria to Greece. Heavy ivory tusks were considered "profitable ballast" in antiquity. Given the interest in the acquisition of fossil relics, heavy mineralized bones might have been another kind of profitable ballast. A thoroughly studied shipwreck discovered in 1982 off Ulu Burun (near Kas, Turkey) is famous for its cargo of ivory tusks and other Bronze Age luxury items. Recently, I learned the little-known fact that the ship, which sank in the late fourteenth century B.C., also carried a cache of fossil seashells. Excavation director Cemal Pulak believes the beach-polished shells were collected by a sailor in Syria or Palestine who planned to make them into pendants during the voyage, as items for trade.[9]

Evidence from numerous ancient sites shows that marine, plant, and mammal fossils were imported from afar and gathered close to home as valuable ornaments, relics, and votive offerings. In 1888, for example, archaeologist Sir Arthur Evans acquired a beautiful pendant from a sixth-century B.C. Etruscan tomb (in Ascoli Piceno, Italy): the large translucent gray and red Miocene shark tooth, a common fossil in the western Mediterranean, was mounted in gold filigree. In the same era, another Etruscan individual placed a pleasingly patterned black stone in a loved one's grave. A modern botanist has identified it as a chunk of a fossilized prehistoric tree (*Cycadeoidea etrusca*) that once graced Etruria. Fossilized flora from various places were certainly recognized as petrified plants in antiquity. They were admired and transported as curiosities. For example, Pliny described stones from Munda, Spain, which when split open contained the exact likenesses of palm branches. In the fourth century B.C., Theophrastus remarked on petrified reeds from faraway India.[10]

Maltese Fossils

In the sixth century B.C., the philosopher Xenophanes saw fossil marine creatures embedded in rocks on Malta and concluded that the island was once covered by the sea (chapter 5). Cicero remarked that Malta's ancient Temple of Juno was renowned for its treasure of ivory tusks of prodigious size, possibly fossil mammoth tusks brought there from Italy or North Africa (see chapter 3). Several archaeological finds on the island confirm that the ancient Maltese had a strong interest in local fossils. As early as the fourth millennium B.C., they dedicated heaps of Miocene fossil shark teeth at a sacred site. Maltese potters of 2500–1500 B.C. used the serrated teeth of the gigantic Miocene shark *Carcharodon megalodon* to decorate ceramic bowls with distinctive grooves. Helicoid gastropods (screw-shaped snails) of the Miocene were also gathered by Neolithic Maltese and placed in temples.

The archaeologists who found the latter were surprised to find a number of oversize gastropod replicas carved from limestone and modeled in baked clay. The models recall the imitation shark vertebrae made of gold and marble found in several Minoan sites of the same era on Crete. These are the earliest datable replicas of marine fossils ever made. Do they represent efforts to figure out how the mysterious natural fossils had been formed? Were the gastropods and shark vertebrae so valuable that forgeries were worthwhile? We know that in twelfth-century China, for example, well-preserved fossil fish impressions were in such demand that locals produced counterfeits. All we can say for sure is that from earliest antiquity Malta was a place where fossils were highly prized.[11]

The Giant Bones of Capri

In 1905, workmen digging the foundations for the posh Quisisana Hotel on the island of Capri came upon an assemblage of massive fossil bones and teeth. Mixed in with the remains were stone arrowheads and blades. Over the next few years, Italian paleontologists determined that the flint weapons were made by Ice Age hunters, and they identified the bones and teeth as those of extinct Pleistocene mammals, including mammoths (*P.* or *E. primigenius*, *M. chosaricus*), rhinoceroses, and giant cave bears (*Ursus spelaeus*). The fossil specimens and weapons were placed in a museum in Capri Town (figs. 4.6, 4.7, 4.8).

Incredibly, this significant find has been passed over by historians of the early Roman Empire. It was on Capri that the emperor

4.6. Molars of a mammoth (*Mammuthus chosaricus*), found by workmen digging the foundations for the Quisisana Hotel in 1905, Capri, Italy. Stone weapons were mixed with the Pleistocene mammoth, cave bear, and rhinoceros bones. A similar find may have provided the "giant bones and heroes' weapons" displayed by the emperor Augustus at his museum on Capri. Photo courtesy of Filippo Barattolo, © Centro Caprense Ignazio Cerio.

Augustus established his museum of giant bones and the weapons of ancient heroes (chapter 3). Scholars question whether or not the emperor really included antique weapons with the big bones he collected, and if so, why. I think the solution to this old puzzle lies in the discovery by the Quisisana workmen in 1905. Their find almost certainly recapitulates a similar discovery on Capri in Augustus's time, perhaps during the building of his villa. The coincidence of strange, enormous bones together with skillfully crafted but antiquated stone weapons would strike ancient observers as evidence of mythical heroes slaying giants and monsters. It would explain why Augustus, an emperor obsessed with reclaiming the

4.7. Rhinoceros (*Stephanorhinus hemitoechus*) molars and fragment of jaw, found during excavations at the Quisisana Hotel, 1905, Capri, Italy. Photo courtesy of Filippo Barattolo, © Centro Caprense Ignazio Cerio.

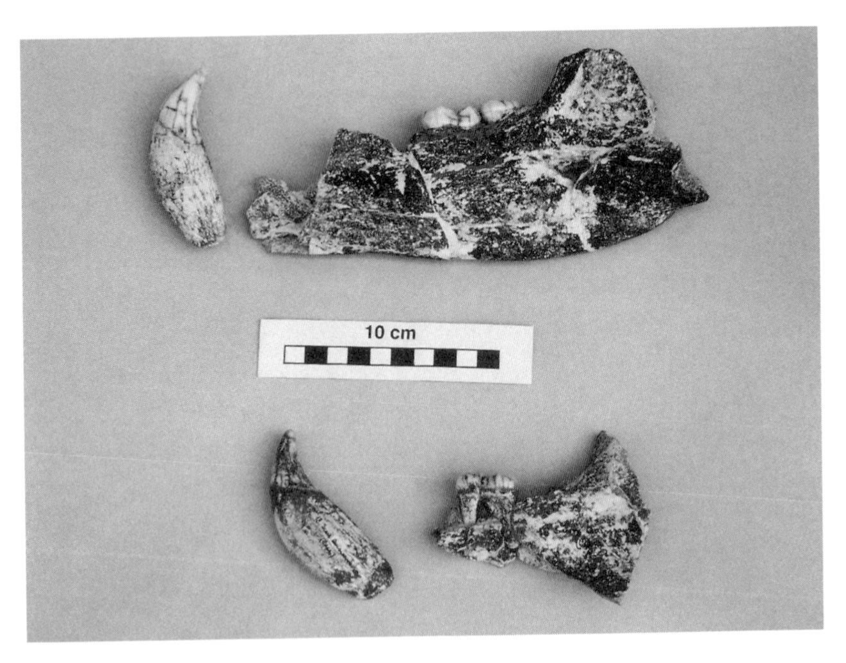

4.8. Cave bear (*Ursus spelaeus*) jaw fragments and canine teeth, from the excavations at the Quisisana Hotel, 1905, Capri, Italy. Photo courtesy of Filippo Barattolo, © Centro Caprense Ignazio Cerio.

glorious Italian past, decided to display the "weapons of heroes" along with the remains of monsters at Capri. Moreover, we can now visualize the prehistoric specimens that were displayed in the world's first paleontological museum: tremendous woolly mammoths, rhinoceroses, and cave bears (fig. 4.9).[12]

Fossil Wonders in Ancient Egypt

In about 3100 B.C. a resident of Maadi, near Cairo, drilled a hole in a green fossil shark tooth of the Middle Eocene so it could be worn as a pendant. Similarly perforated shark teeth from the Tertiary have been unearthed in a Neolithic site (ca. 4500 B.C.) in the Fayyum Basin. Plutarch, who lectured in Egypt, observed that people found petrified mollusks "in Egyptian mines and on the mountains, proof that all the land was once sea." Plutarch's report is confirmed by a treasury of fossil votives excavated from the ruins of the mining temple at Timna (Sinai). Ancient miners had gathered numerous echinoids, gastropods, seashells, and serpulid worm tubes from gravel scree in oases near the Egyptian copper mines and stone quarries in the late Bronze Age and dedicated them as offerings at their local temple.

In 1903, a unique record of one such fossil find came to light in Heliopolis. There, Italian archaeologists excavated a round building that housed a remarkable array of inscriptions and small artifacts from different epochs. In that library-museum they found a fossil sea urchin engraved with the history of its own discovery. Hieroglyphs inscribed around the face of the Eocene echinoid indicate that it was found by a New Kingdom miner or scribe-priest named Tjanefer south of a certain stone quarry, probably in the Sinai mining area. In 1998, Egyptologist John Ray reflected that this long-forgotten stone sea urchin preserved in an Egyptian treasury of valuables represents the age-old sense of wonder that inspires scientific curiosity. This marvel-turned-artifact is a pinpoint of light beckoning us through the "labyrinth" of ancient thought.[13]

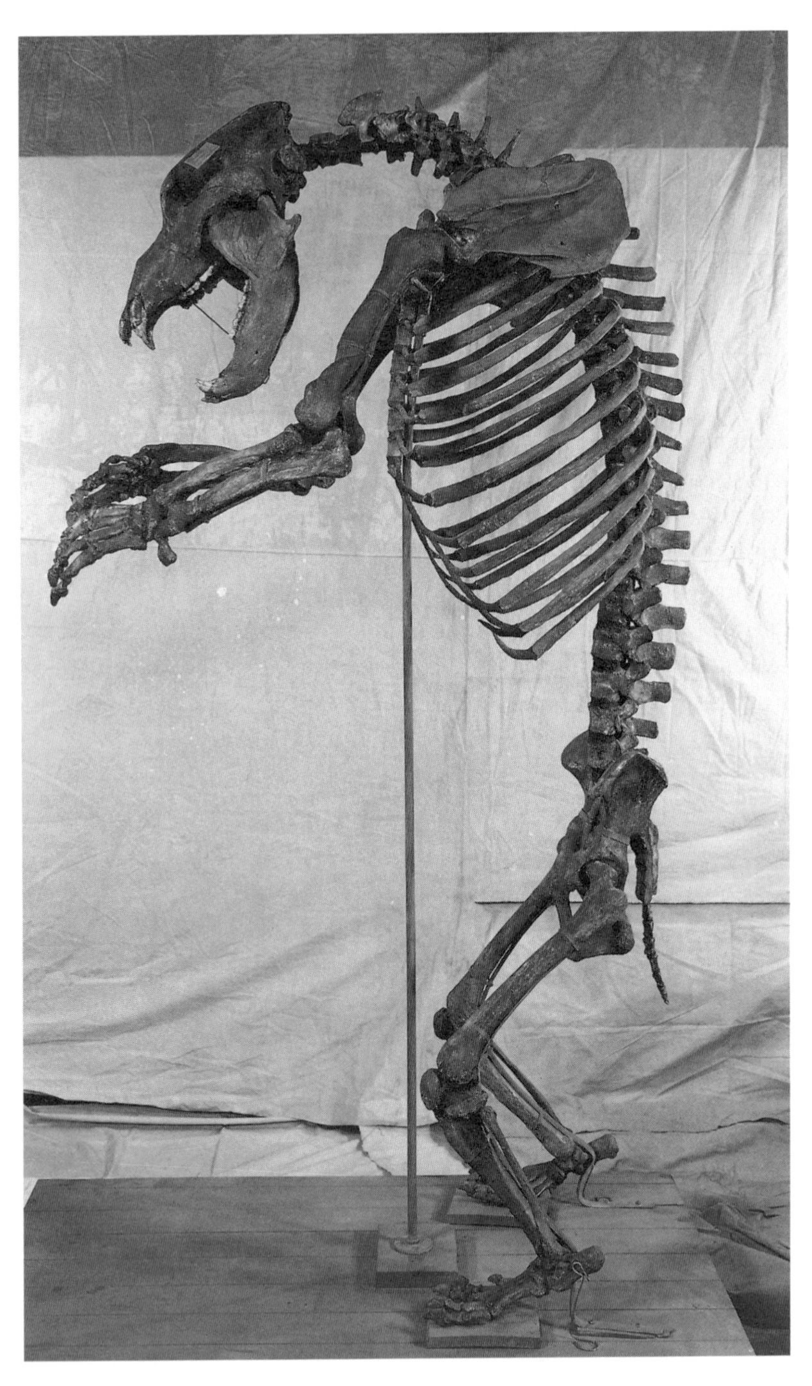

4.9. Cave bear skeleton (*Ursus spelaeus*). Cave bear remains were probably included in the museum built by the emperor Augustus for his collection of "the huge bones of monsters popularly known as giants' bones." Neg. 320463, photo A.R. and R.E.C. Courtesy Department of Library Services, American Museum of Natural History.

The Black Bones of Set

Between about 1300 and 1200 B.C., Egyptian worshipers of the typhonic god Set (Osiris's enemy; see chapter 3) collected nearly *three tons* of black, river-polished fossil bones. They brought the fossil bones to two sites where Set was venerated a few miles south of Asyut on the Nile.

In 1922–23, the archaeologist Guy Brunton was the first to discover the immense quantities of fossils heaped in Set shrines at Qau and Matmar, stunning evidence that large prehistoric bones were revered as sacred relics and ceremonially reburied. The next year (1923–24), Brunton's colleague Sir Flinders Petrie discovered another hoard of fossils heaped in tombs at Qau. He was particularly struck by several fossil bones carefully wrapped in linen and placed in rock-cut tombs. Most of the fossils were large hippopotamus bones, but remains of extinct crocodile, hartebeest, boar, horse, and gigantic buffalo were also collected, along with some fossilized skulls and limbs of human beings. The fossils were very heavy, dark colored, and highly polished by river sand. Brunton suggested that Egyptians revered the black bones of stone as the remains of Set, the god of darkness. Kenneth Oakley surmised that the ancient priests of Set recognized some of the mineralized bones as human and deliberately combined them with the animal fossils to represent the human and bestial aspects of the god.

Brunton's team spent six weeks sorting through the mass of fossil bones at Qau. In 1925, Petrie expressed his hope that geologists would discover the original deposits of the fossils. In 1926 the geologist K. S. Sandford conducted explorations within a five-hundred-mile radius of Qau but failed to locate the source of "this strange collection of animals," apparently of the Pliocene-Pleistocene. He wrote, "We await the verdict of the paleontologist with the greatest interest." In 1927, Brunton stated that the tons of fossil remains would be "the subject of a special memoir"; in 1930 he again promised a forthcoming volume devoted solely to the bones. Unfortunately, that was the last word about the extraordinary black fossils of Set. It seemed that they were lost to the scientific community.

In 1998, I contacted Andrew Currant, curator of Quaternary mammals, Department of Paleontology, British Museum, Natural History, to inquire if there was any record of the Qau fossils collected by Brunton and Petrie. To my surprise, Currant tracked down a "substantial but largely undocumented collection" of fossils from Qau stored at the museum's warehouse in Wandsworth. The material is still packed in the original crates that Brunton shipped from Egypt in the 1920s! These unique fossil relics from Qau, languishing in unopened crates in South London, certainly deserve further study by zooarchaeologists, paleontologists, and Egyptologists.

But the saga of the fossils wrapped in linen found by Petrie is even stranger. David Reese finally traced their whereabouts in January 1999. Some years ago, he learned, the Petrie Museum in London sent a large collection of ancient Egyptian textiles to the Victoria and Albert Museum. Some of the earliest textiles of uncertain provenience were culled out; they escaped destruction when they were acquired by the Bolton Museum in Lancashire. Angela P. Thomas, senior keeper at the Bolton Museum, noticed that some very ancient linen items of peculiar bulk had been inadvertently included with the textiles (fig. 4.10). Original labels scrawled on the backs of envelopes in Petrie's handwriting confirmed that these bundles were the long-lost linen-wrapped fossil bones discovered at Qau in 1923–24![14]

A Fossil Relic at Troy

In the course of his famous excavation of the burned ruins of Bronze Age Troy in 1870–80, Heinrich Schliemann came upon "a very curious petrified bone" near an area of human burials in earthenware urns. The fossil had been carried to the city in the thirteenth century B.C., the time of the legendary Trojan War, and during Troy's earliest contacts with the rest of Anatolia, the Aegean, and northern Syria. Schliemann sent the bone to William Davies of the Fossil Department of the British Museum in London. Davies identified it as the mineralized vertebra of an extinct

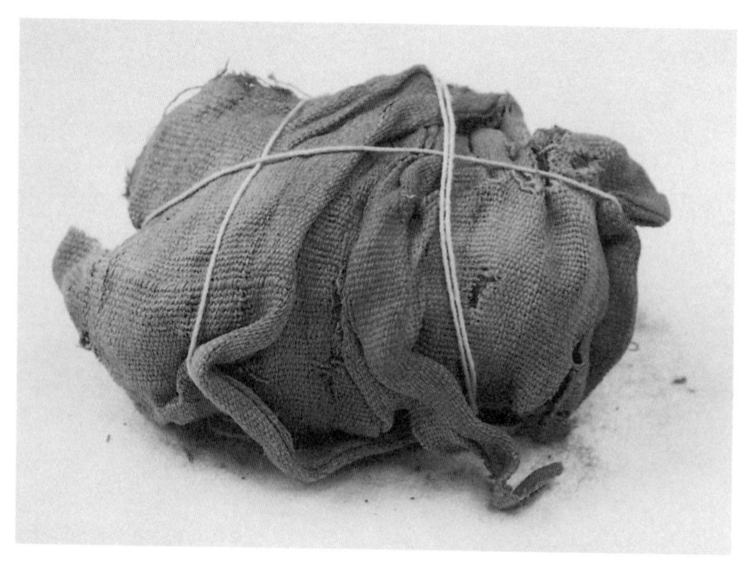

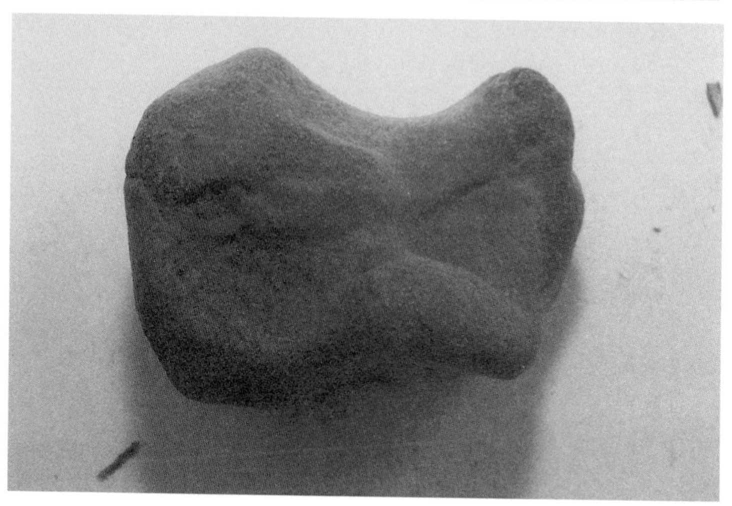

4.10. Linen-wrapped fossil bones from Set shrine at Qau, Egypt. Discovered in a rock-cut tomb by Sir Flinders Petrie, 1923–24. This linen bundle is about 4 inches long. Photos courtesy of the Bolton Museum and Art Gallery.

Miocene cetacean. Noting that the ancients often transported such "objects of attraction" long distances, Davies concluded that the bone relic was obtained "from a Miocene tertiary deposit either in the Troad" or perhaps from further afield.

Schliemann's discovery that ancient Trojans collected an unusual petrified bone was eclipsed by his spectacular discoveries of gold jewelry and other precious Bronze Age artifacts. But for us, the material evidence of fossil gathering in Bronze Age Troy adds a new dimension to Pausanias's story of the enemy seer captured by the Greeks during the Trojan War (chapter 3). The Trojan seer predicted that the Greeks' fossil relic of their hero Pelops would prove to have magical power. Schliemann's find suggests that the concept of numinous fossil relics was indeed familiar to the Trojans of that era.

More than a century after Schliemann shipped the fossil vertebra to the British Museum's Fossil Department for identification, David Reese suggested to me that the Trojan relic might still be stored there. Sure enough, curator Andrew Currant confirmed that the bone was in London, available for further study. Examination of the fossil might reveal the origin and species, allowing us to know whether it came from the vicinity of Troy, where huge Miocene fossils excited interest in antiquity, or from farther away, perhaps recovered from a bone bed near the Black Sea or Ionian coast, an Aegean island, or even from the Orontes Valley in Syria during Troy's earliest contacts with those places.[15]

Natural Wonders at Samos

Some 10 million years ago, Samos was the geographical-zoological crossroads of Africa, Europe, and Asia. In the seventh and sixth centuries B.C., the island of Samos was still a crossroads, a cultural nexus of Greece, Asia Minor, and Egypt. The great bones of Miocene mastodons and samotheres that emerge from the soil of Samos were shown to ancient travelers as the vestiges of Neades or Dionysus's war elephants (chapter 2). Travelers also visited Samos's magnificent Temple of Hera, the largest temple of its day.

The sanctuary's treasury buildings and precincts have been excavated by the German Archaeological Institute since 1910. The Heraion excavations are yielding a fantastic assortment of exotic faunal remains, first published in 1981 and 1983.

Samos emerges as an extraordinary ancient museum of natural wonders both native to Samos and from afar. The temple complex was a place where local villagers and sophisticated travelers alike could contemplate zoological and geological marvels, from the petrified bones of the monstrous Neades and the skeletons of enormous reptiles to carvings of baboons and lifelike models of Scythian griffins. It's possible that there was a menagerie of strange living animals, too. Among the exotic faunal remains unearthed at the Heraion were hippopotamus teeth of record size and ostrich eggshells available only from the Levant or North Africa, gigantic *Tridacna maxima* clamshells imported from the Red Sea, skulls and horns of a hartebeest (*Alcelaphus buselaphus*) found only in North Africa, and the impressive skull of a 16-foot (5-m) crocodile from the Nile.[16]

The rare natural wonders transported from Africa and the Middle East to Samos were donated by wealthy worshipers, who also commissioned artists to create costly bronze cauldrons decorated with fierce griffin heads (the same ones that first inspired my pursuit of fossil monsters; chapter 1). But German archaeologist Helmut Kyrieleis points out that the modest but imaginative offerings of the "common man" at the Heraion are too often overlooked by ancient historians. Ordinary folk brought homemade miniature cauldrons with griffin heads, simple wood carvings, and numerous "small oddities, or simple *naturalia*" such as rock crystals, pinecones, bits of coral, and stalagtites.

Anything paradoxical, rare, or intriguing was a fitting gift to the gods *and* a valuable contribution to a temple's "cabinet of curiosities." Although Samos is particularly rich in such dedications, all around the Mediterranean fishermen, farmers, quarrymen, shepherds, and villagers dedicated beautiful or odd fossilized shells and larger petrified remains to sanctuaries. Ancient authors say that children searched for small stone cylinders (fossil crinoid stems) along the Meander River and near Mount Sipylos (Turkey), to place in local temples of Mother Earth. On the banks of the Eurotas River near Sparta, people gathered "stones shaped like little helmets" (so-called helmet echinoids?). These little military marvels were known as Thrasydeilos (Rash and Timorous) because they were said to leap out of the river at the sound of a trumpet but sink in fear at the word "Athenians"! The helmet stones were dedicated in great numbers in the Bronze Temple of Athena at Sparta (the helmeted goddess was often associated with war; the joke seems to refer to the Peloponnesian War). Hellenistic verses relate that fishermen dedicated the colossal bones they netted at sea to rustic sanctuaries. Precious and curious things—even giant heroes' bones—could be found by any man, woman, or child who carefully observed the natural world.[17]

Teeth of phenomenal size were favorite curios in sanctuaries. Some of the worked hippo and elephant tusks found in archaeological sites were obviously used for ivory carving, but the unmodified large tusks and molars from living or extinct species kept in temples were probably "curiosities, trophies, or heirlooms." Archaeologist P. J. Riis believes that the very large hippopotamus molar found in the sixth-century B.C. temple built by Greeks at Tell Sukas on the Orontes River (south of Antioch) was a hero's relic. Riis explains, "Greeks who happened to find a tooth of huge dimensions, but otherwise resembling a human molar, would, no doubt, treat it with great respect, assuming that it was a disjected member of a heroic burial." Indeed, Riis's find recalls the huge hero's tooth found after the devastating earthquake in Pontus and sent to Tiberius in Rome (the emperor returned it for proper burial; chapter 3). It also brings to mind the giant skeleton discovered in the Orontes riverbed (described by Pausanias; chapter 3). Large elephant and hippo teeth have been recovered from sacred sites at ancient Mycenae (Peloponnese), Knossos (Crete), Cyprus, Syria, Israel, and Turkey, among other places. As mentioned above, a number of extra-large hippo teeth were among the exotic votives at the Heraion at Samos, too: one found near an altar measures almost 20 by 3 inches (50 cm by 7.5 cm).[18]

At Samos, German archaeologists are still uncovering unexpected faunal remains from the Temple of Hera. In 1988, in a heap of votives, pottery, and thousands of animal bones dated to the seventh century B.C., a conspicuously oversize fossil femur caught Helmut Kyrieleis's eye. Kyrieleis paints a by-now-familiar scene: "Imagine a Samian farmer who finds a bone of an Amazon in his field and reverently offers it to the great sanctuary" of Hera, where it awed visitors to Samos over centuries. We can hear the echoes of temple guides expounding on the farmer's fabulous find and directing curious travelers, perhaps even the historians Euphorion or Plutarch, to the bone beds near Mytilini where they could view for themselves a profusion of large bleached skeletons eroding out of the earth, skeletons we now know belonged to great mastodons and samotheres.

Thanks to Kyrieleis's sensitivity to the literary record and his knowledge of the rich Neogene bone beds in the center of the island, he was able to recognize the special historical significance of an enormous fossil in a heap of miscellaneous bones. The placement of the relic in the temple in the seventh century B.C. confirms the written tradition that huge bones were displayed in Samos, and other ancient accounts of giant bones placed in sanctuaries.

In 1988, the Heraion fossil relic was provisionally identified in Athens as the femur of a "pleiocene [*sic*] hippopotamus." But Nikos Solounias, who specializes in the paleontology of Samos, points out that hippopotamus remains are not found in Samian fossil deposits. Solounias agrees with Kyrieleis's hunch that the bone was discovered by a Samian farmer, and its light color suggests that it is a fossil from Samos (fig. 4.11). If it is a local fossil, the femur probably came from a Miocene mastodon or rhinoceros, rather than a Pliocene hippopotamus. If the fossil femur is from a hippopotamus, it would have been brought to Samos from elsewhere, as was the huge hippo tooth also found in the Heraion. As I write, arrangements are being made for further analysis to determine the bone's species, provenience, and date.[19]

A Hero's Bone in Messenia

The ancient Messenians, who lived on the southwest promontory of the Peloponnese, showed a marked interest in collecting large bones. According to Pausanias, relics identified as those of the

4.11. Large fossil femur found in the ruins of the Temple of Hera, Samos, Greece (matchbox at left for scale). The bone relic was placed in the temple in the seventh century B.C. or earlier. Photo courtesy of Helmut Kyrieleis.

seventh-century B.C. hero Aristomenes were shipped from Rhodes to Messenia, probably during wars between Sparta and Messenia. Perhaps around the same time, the huge bones of another ancient Messenian hero, Idas, were given a fine burial in a stone urn, only to spill out of the ground during a severe storm centuries later. Phlegon of Tralles said that the citizens of Messene reburied those bones with a new inscription in the second century A.D. (chapter 3). Archaeologists have found evidence for heroic burials in Messenia, including burials in large jars, dating from the Bronze Age to Hellenistic times. But, as mentioned above, before the 1960s, countless faunal remains, some unexpectedly large, were briefly noted and thrown away by excavators in Messenia.

By coincidence, the first archaeological excavation to scientifically analyze faunal remains in a Mediterranean site took place in Messenia. In 1978, George (Rip) Rapp, Jr., and S. E. Aschenbrenner published the first results of the Minnesota Messenia

Expedition's long-term excavations of an ancient town near modern Nichoria, a few miles southwest of the main ancient settlement of Messene. The site at Nichoria (the ancient name of the place is unknown) proved to have been occupied from the middle Bronze Age to Byzantine times. The archaeological community was impressed by the Minnesota team's groundbreaking methods, despite the relatively modest finds.

I was riveted by the last sentence in the final section of the excavation report, labeled "Invertebrates and Miscellaneous Animal Remains." The excavators found a large fossilized femur from an extinct animal on the acropolis (citadel sanctuary)—"clearly an item carried to the acropolis as a curio." The first archaeologists to pay systematic attention to the nonhuman bones from an ancient Greek site had found a hero's relic! Although the archaeologists recognized the oversize fossil bone as a relic deliberately placed in a sacred/public place, they did not connect it with the literary accounts of heroes' bones that were venerated in ancient Messenia. Since the archaeologists also found "heroes' graves" at Nichoria, this big bone may have been a relic gathered as part of the cult of heroes. One of the heroic burials identified at Nichoria was a *pithos* burial, in which bones were placed in a large jar, recalling Phlegon's account of the reburial of the huge bones of the hero Idas in a stone jar at nearby Messene (chapter 3).

The Minnesota team did attempt to identify the large femur. In 1978, a dinosaur specialist unfamiliar with Mediterranean paleofauna, Robert Sloan, assigned the broken-off thigh bone to a "Pliocene or Miocene fossil elephant." If the place of this relic in the history of ancient paleontology is to be understood, an accurate identification is important. If Miocene-Pliocene, the evidence of the bone would suggest that the ancient occupants of Nichoria had imported the relic. Nichoria is underlain by Pliocene deposits, but they consist of marine silts, so a Miocene-Pliocene mammal bone would probably come from a Pikermi-type deposit in, for example, Attica, Euboea, an Aegean island, or Anatolia. The nearby bone beds of the Peloponnese contain mostly Pleistocene remains of elephants, mammoths, and rhinoceroses (chapter 2). Ancient authors tell us that large and unusual faunal remains were sometimes transported great distances, like the bones of Aristomenes from Rhodes. Indeed, several other exotic rarities (such as elephant ivory and fossil shark teeth) turned up at the Nichoria dig. So, was this giant bone evidence of long-distance trade in relics? It would all depend on an accurate paleontological identification.

Reevaluating the Messenia Expedition's invertebrate finds in 1992, David Reese learned that the zooarchaeologists' focus on food sources had led them to misidentify other fossils collected by the ancient Nichorians. For example, the Minnesota zooarchaeologists mistook a trove of Pliocene oyster and scallop shells for edible living species. Reese determined that these jawbreakers were fossil curios, not the debris from ancient seafood dinners. The stone shells were probably gathered locally, from the surrounding Pliocene rocks.

Another "surprising and notable" faunal find discovered by the Minnesota team came from the entrance to a tomb of the thirteenth century B.C.: a very large horse molar. Sloan identified the molar as that of the equine breed known to historians as the "Great Horse." Nearly twice the size of Greek horses used at the time of the tomb's construction, the Great Horse (1.6 m, over 5 feet, at the withers) would not be introduced to Greece until ten centuries later. Did someone carry this unusual horse tooth from an exotic land (perhaps Libya or the Russian steppes) to Nichoria in the Bronze Age? Or was the molar a fossil curio found closer to home? The remains of the extinct large-bodied (1.8 m high) Pleistocene horse are common in the Megalopolis basin. Teeth are highly diagnostic: reexamination of the tooth by a Mediterranean paleofaunal expert would solve this mystery. Unfortunately, the tooth has been misplaced.

The Nichoria bone (the distal end of a right femur, 5–6 inches, or 15 cm, wide) is exciting, because it and the Samos bone (also the distal end of a large femur) are the only large fossil bones collected as relics by ancient people to be identified so far by field archaeologists working in Greek sites. A photograph of the Nichoria fossil was published in 1978. More than two decades after the bone's discovery I endeavored to locate its current whereabouts. My first thought was to obtain more photographs that might help a vertebrate paleontologist check Sloan's provisional identification. In keeping with the theme of lost evidence, however, I learned from the codirector Rip Rapp (Professor of Geoarchaeology, University of Minnesota at Duluth) that out of the thousands of uninventoried photographs and negatives stored in the Messenia Expedition archives in library basements in Minneapolis and Duluth, the original photo of the big fossil bone was missing.

If the photo was gone, what about the bone itself? The faunal remains from Nichoria were placed in labeled bone bags and stored in the Kalamata Museum in Greece—all, that is, *except* for the big fossil bone. Hopelessly lost, maybe even discarded, guessed Rapp. But the initial verdict of "hopelessly lost" was not the end of the trail. In the summer of 1998, the long-forgotten femur of a huge animal that lived and went extinct millions of years ago, discovered by ancient Messenians, and carried to their sunny acropolis for safekeeping, was unexpectedly rediscovered by Rapp himself, gathering dust in a storeroom in Duluth. He agreed to send the fossil relic to me for study. A few weeks later, the awe-inspiring stone bone of a Greek giant dominated my desk in Princeton.

On a clear, windy morning in September 1998, I boarded the train to New York City. In my backpack was the five-pound bone, wrapped, I hoped, as carefully as Pelops's shoulder blade on its way to Troy. Nikos Solounias, the specialist in Greek fossils, had agreed to meet me at the American Museum of Natural History to see if we could learn more about the Messenian hero's bone. We handled it gingerly, since the fossil was friable, and every touch brought away ancient powder. The rusty-black color of the heavy fossil immediately indicated to Solounias that it had come from the lignite deposits of the Megalopolis basin (about 35 miles or 55 km north of Nichoria) or some similarly lignitiferous bone bed. The sharp break told us that the ancient Messenians had probably carried the entire femur, twice the size of a man's thigh bone, to their acropolis. It was time to identify the most likely prehistoric species.

Now, Solounias began to pull out drawer after drawer in the

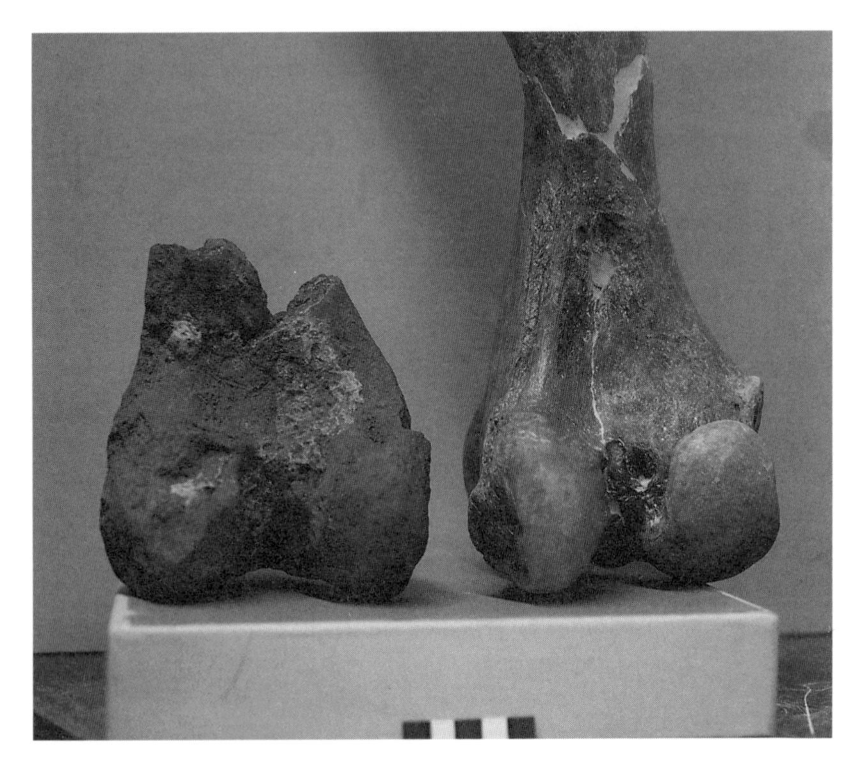

4.12. *Left*: large fossil femur (probably from a Pleistocene woolly rhinoceros) collected as a relic in ancient Messenia, Greece, excavated from the acropolis at Nichoria by the Minnesota Messenia Expedition. *Right*: fossil femur of a Pleistocene rhinoceros, for comparison. Photo courtesy of Nikos Solounias.

museum's vast collection of bones to compare specimens to our relic. Over the hours we eliminated large extinct deer, antelope, and huge oxen species. The bone was only slightly larger than the *Samotherium*'s femur, but the fossils of Samos are chalky white and the morphology of this blackened femur was different. Our fossil was from a very large adult mammal, but it was not quite big enough for a mastodon or mammoth. The size suggested a *Chalicotherium* or rhinoceros. We climbed to the rhino floor and opened drawers laden with limb bones. Eureka! Our ancient Greek giant was almost certainly one of the three rhino species (possibly *Coelodonta antiquitatis*, woolly rhino) that lived and per-

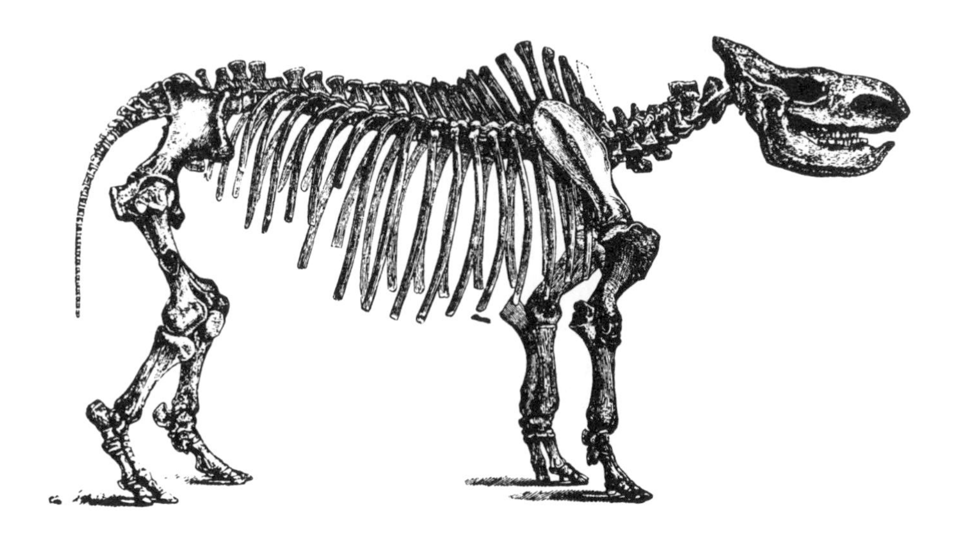

4.13. Woolly rhinoceros, *Coelodonta antiquitatis.* During the Ice Age, this rhinoceros roamed as far south as the Peloponnese in Greece. The fossil relic stored on the ancient acropolis at Nichoria has been identified as a rhino femur from the Megalopolis bone beds. Karl Zittel, *Handbuch der Palaeontologie* (Munich, 1891–93), fig. 240.

ished in Megalopolis during one of the Ice Ages (figs. 4.12, 4.13). Just the right conditions had transformed the huge beast's bone into stone, a natural wonder for ancient Greeks to ponder.

The natives of Nichoria had not obtained their relic from overseas after all but had acquired a local Peloponnesian fossil, perhaps during the ancient bone rush or during the later search for heroic bones after Messenia was freed from Spartan control. They carried their hero's bone overland from the celebrated Arcadian burial grounds of lightning-struck giants and heroes, the same big bone beds of Megalopolis and Olympia that had yielded the huge shoulder blade of Pelops, the giant skeleton of Orestes, and so many other Peloponnesian heroes' bones (chapter 3).

Only a tiny portion (14 percent) of the total faunal collection recovered at Nichoria has yet been studied. Further study may reveal that other fossil wonders, perhaps even more large bones, were collected in antiquity. But the exotic and fossil remains identified thus far show that even a small village like the Nichoria settlement maintained a store of natural wonders and heroic relics. If the settlement was typical, the determination of the origin of its bone relic and other fossils may point to special features that defined an ancient city's self-identity and vision of its past, its trading habits, and status: namely, the possession of local heroes' bones and natural marvels enshrined on the acropolis.[20]

Remarkable or anomalous evidence from the past that seems to resist the expected patterns can be made to reveal hidden meanings when humanities scholars and scientists combine their specialized knowledge and creative thinking. The happy resolution of the mysterious Monster of Troy vase and the saga of the Nichoria bone relic show how much can be learned when students of different disciplines pool skills and resources—and I like to think the outcomes point the way to a productive future for archaeopaleontological studies.

Fossil Legacies

Long before archaeologists began to pay attention to faunal remains in ancient Greek sites, and at the close of the modern bone rush in Greece, the great fossil hunter Barnum Brown ventured into zooarchaeology and ancient history. After his successful excavations in Samos, Brown turned his attention to fossil exposures on the island of Kos, the birthplace of Hippocrates. One day in 1926, Brown decided to visit the ruins of the Asklepieion, Hippocrates' famous medical school. Rummaging in the jumble of broken marble statues and pottery fragments, Brown picked up part of a molar of the prehistoric elephant *P. antiquus.* Someone had obviously carried the fossil tooth to the sanctuary in antiquity.

Thrilled by the historical possibilities, Brown imagined that the molar had been "handled and discussed by Hippocrates himself." Brown went too far in announcing that he'd found "the earliest known fossil collected by man," but the tooth is certainly tangible evidence that fossils were valuable items stored at temples. (Not surprisingly, given the fate of so many artifacts and antiquities of paleontological significance, the famous molar of Kos has long since disappeared.)

Brown fancied that the tooth "seemed to come from the invisible hand of the master Hippocrates," a legacy passed on to Brown "as a problem unsolved by [classical] philosophy."[21] Here Brown inadvertently put his finger on a major puzzle of ancient paleontology: the natural philosophers' failure to explain large prehistoric remains. The learned thinkers who intuitively grasped the meaning of small shell fossils as evidence for past seas did not address the enigma of extraordinarily large bones, despite sensational discoveries and public displays like those described by numerous ancient writers and confirmed by archaeological excavations. Why? The next chapter explores the conceptual resources provided by popular myths as well as the mysterious silence of the philosophers.

CHAPTER

5

Mythology, Natural Philosophy, and Fossils

Tension between Popular Traditions and Natural Philosophy

When the ancient Greeks and Romans encountered perplexing fossils around the Mediterranean, what concepts did they draw on to explain them? As we've seen, myths and folklore about giants and monsters offered a strong interpretive model for huge, weird bones buried in the earth. But because modern classical scholars tend to read myth as fictional literature, not as natural history, the significant contributions of popular traditions to ancient paleontological thought have not been appreciated.

Moreover, scholars both ancient and modern consigned descriptions of giants and monsters to the realm of fantasy and superstition. Since discussions of remarkable remains are missing in the "objective" writings of the best-known classical historians, like Thucydides, and natural philosophers such as Aristotle, most modern historians and scientists have simply assumed that large prehistoric remains went unnoticed in antiquity. But, as we have seen, the empirical experiences narrated by travelers, mythographers, ethnographers, geographers, natural historians, compilers of natural wonders, and other hard-to-classify ancient writers tell a very different story—a story confirmed by archaeological evidence. The world of ancient paleontology is preserved in these works of less exalted aim but surprising erudition.

The writers who described fossil discoveries in antiquity consistently referred to mythical paradigms to explain extraordinary remains. Greek myth is a complex skein of tales about the origin of the natural world and the history of its earliest inhabitants. The oldest oral narratives were first written down in the eighth century B.C. in works attributed to the poets Hesiod (*Theogony*) and Homer (*Iliad* and *Odyssey*). Later Greco-Roman mythographers embroidered on these myths and on other, now lost, epic poems, constantly adding new stories to the stock of old lore about gods, giants, monsters, and heroes. The natural philosophers inherited these early cultural traditions about giants and monsters, but they rejected mythological explanations for earth's history. So, while ordinary people and writers tried to identify extraordinary bones using myths and legends, philosophers apparently declined to participate in such vulgar pursuits, preferring to focus on the normal and normative rather than the exceptional and anomalous.

The fourth-century B.C. philosopher Plato expressed the prevailing intellectual attitudes toward popular beliefs in scattered remarks about giants, Centaurs, and other monsters. Plato's sole allusion to remarkable bones occurs in the *Republic*, where he refers to folklore about Gyges, a seventh-century B.C. tyrant of Lydia (Asia Minor). "The story goes," says Plato, that a violent storm and an earthquake broke open the ground, revealing a hollow bronze horse containing a gigantic skeleton and a magic ring. Since Plato often manipulated folk genres for his own purposes, we can't know whether he was citing actual lore, but the anecdote does capture the texture of the later reports of large fossil discoveries by Pausanias and others.[1]

Geomyths of Primeval Giants and Monsters

The word "geomythology" was coined in 1968 by Dorothy Vitaliano, a geologist interested in folklore, to describe etiological legends that explain—in poetic metaphor and mythological imagery—volcanoes, earthquakes, and landforms. Her term would also apply to lore about conspicuous prehistoric fossils. As Vitaliano points out, geomyths can perpetuate popular misconceptions, but they can also contain insights about natural phenomena. Historian of geology Mott Greene suggests the phrase "natural knowledge" for the descriptions of natural phenomena conveyed in Greco-Roman mythological literature. He observes that classicists unversed in science often miss natural knowledge embedded in myth and popular literature.

Myth was considered historical in antiquity, even among writers who disparaged its fantastic elements and unreliability. But not everyone took myths and legends literally; many saw them in a symbolic or metaphorical light, sensing that oft-told and embellished stories had coalesced around kernels of truth. Pausanias admits that he once regarded old giant myths as "rather silly," but after visiting Arcadia with its huge bones, he came to appreciate the tales as realities couched in poetic language. As the geographer Strabo observed, even if a historian is not fond of myths, they are worth examining "since the ancients expressed physical notions [and] facts enigmatically . . . and added the mythical element to their accounts. It is not easy to solve with accuracy all the enigmas, but if one studies the whole array of myths, some agreeing and others contradictory, one might be able to conjecture the truth."[2]

In the myths told by Hesiod and Homer, long before the appearance of present-day humans the primeval Earth produced a series of monsters and a race of giant humanoid Titans. The Titans, led by Kronos, slaughtered their father Uranos, whose blood produced the next generation of Gigantes (Giants). "Earth brought forth the giants, . . . who were matchless in the bulk of their bodies and invincible in their might, with terrible aspect. . . . Some say they were born at Phlegra, but according to others in Pallene." In the chaotic, primal era, when the "origin of life was still recent," generations of giants and strange creatures proliferated, such as the Cyclopes, the Centaurs, and Typhon, a horrendous monster whose bellows sounded like the dissonant shrieks of myriad beasts. Some of these generic creatures—among them, giants, Centaurs, and Cyclopes—apparently "bred true," producing more of their kind, but other intermonster matings spawned monsters different from their parents. The Greek for monster, *teras*, means "strikingly abnormal offspring," which carries the implication that their distinctive characteristics were not hereditary. In one version of Typhon's complicated family tree, for example, his offspring included Geryon, the grotesque giant who raised wondrous cattle, and Orthos, a giant dog, who mated with the flame-belching monster Chimera to produce the great Nemean Lion and the Sphinx.[3]

The Gigantomachy

The Titans Rhea and Kronos gave birth to Zeus and his siblings, the future Olympian gods. With the help of some giants, Zeus overcame his father, Kronos, and became king of gods. The myths climax in the cosmic wars for supremacy that pitted the older Titans or Gigantes (later poets conflated the two), and Typhon and the other monsters, against Zeus and the new, more human, gods. This era of earthshaking battles was known as the Gigantomachy, celebrated in ancient poetry and art. Even in antiquity, however, the war between the gods and the giants was seen by some writers, such as Claudian (b. ca. A.D. 370), as a metaphor for geomorphic change over eons. During the Gigantomachy, he wrote, "islands emerged from the deep and mountains lay hidden in the sea; rivers were left dry or changed course," and mountains collapsed as the earth sank. His words offer an uncanny description of the geological events described in chapter 2.

Zeus, hurling his bolts of thunder and lightning, destroyed legions of giants and Typhon. Heracles killed numerous giants and monsters, including Geryon, while the other gods, goddesses, and heroes and heroines slew still more giants and monsters all around the Mediterranean. The defeated giants and monsters were buried in the earth where they fell. As we've seen, the geographical distribution of the fallen giants and monsters conforms to localized fossil beds discovered in modern times. For example, Zeus slew armies of giants in Arcadia, Crete, and Rhodes, and the gods heaped the island of Kos on top of giants: all four places are now known to have fossil elephant remains. An eroding cliff between Corinth and Megara was said to bear the petrified bones of the giant Skiron. According to myth, this ogre threw victims over the cliff to a monstrous turtle waiting below, until the hero Theseus threw him off the same precipice. One can't help wondering whether gigantic tortoise and mastodon remains like those found at Pikermi were observed in archaic times in the crumbling Neogene cliffs west of Megara.[4]

At Phanagoreia (near Sennoi on the Taman Peninsula, Black Sea coast), a temple commemorated Aphrodite's destruction of giants there. The goddess drove the giants into a cave, where Heracles slaughtered them. This legend may have been influenced by rich remains of ancestral mammoths and other Ice Age mammals there. According to paleontologist Alexey Tesakov, the peninsula's high cliffs and sand-clay pits bear incredibly rich Neogene remains continually exposed by the sea. The most famous exposure is Siniaya Balka (Blue Ravine) on the Azov coast, where the skeletons of *Elasmotherium caucasicum* (a colossal rhino) and *Mammuthus meridionalis tamanensis* erode out by the hundreds. This is the place where Theopompus of Sinope described the exposure of many immense bones by an earthquake in antiquity (chapter 3).[5]

The era of the Gigantomachy was a favorite subject for monumental public sculpture and vase paintings. In early Greek art, giants were imagined as quadruped monsters, or as warriors, huge ogres, or primitive strongmen armed with tree trunks and boulders; some later artists added serpent legs to symbolize their earth-born nature. It's important to keep in mind that giants were not necessarily visualized as human. In the words of Manilius (first century B.C.), the giants were broods of "deformed creatures of unnatural face and shape" that appeared and were destroyed in the era "when mountains were still being formed." Note that Manilius, like Claudian, associated the giants with vast geophysical upheavals of the prehuman past.[6]

Some giant monsters, such as Geryon and Typhon, and even giant heroes like Idas, were said to have multiple heads or an unusual number of limbs. This is a widespread folk motif denoting extraordinary strength. In the ancient Greek mythical metaphor, extra limbs and mixed animal-human features also indicated the composite nature of strange creatures whose bodies lay buried under the earth. An ancient description of Typhon says that the monster's thighs were of human shape. In view of Typhon's consistent associations with fossiliferous locales, and the widespread tendency to see the femurs of large extinct mammal species as human, this detail hints that prehistoric skeletons contributed to the image of Typhon.[7]

Zeus's cosmic lightning decisively ended the era of giants and monsters, and a new epoch began. The carcasses of the defeated creatures ended up in the ground all around the Mediterranean world, where they were later revealed by natural forces or human digging. The possession of such links to the mythical past became in many places a matter of local pride. For example, Anatolian Lydia, Spanish Cadiz, Thebes, and Olympia each claimed to have the body of Geryon, and his huge oxen were said to have been dispersed around mainland Greece (Arcadia, Attica, Epirus), in Italy, in Asia Minor, and on the shores of the Black Sea. Typhon's lightning-blasted corpse was variously thought to be buried in Syria, Cilicia (southern Turkey), Phrygia (north-central Turkey), and Sicily.[8]

The ancients identified several sites notable for their concentrations of enormous bones as major battlefields of the Gigantomachy. In Arcadia, Pausanias visited places where the battles supposedly raged, around Megalopolis, Trapezus (Bathos), and Tegea where the giants had made their last stand. As we saw in chapter 3, Pallene (Kassandra Peninsula, Chalkidiki) was regarded as the giants' headquarters during the Gigantomachy. Several ancient authors referred to the continual exposures of gigantic bones around Pallene, also known as Phlegra ("Burning Fields"). Numerous fallen giants were also located in Phlegra in Italy (the Phlegraean Fields, Bay of Naples) and in southern Spain around Tartessus. These eerie battlegrounds were distinguished by another marvelous sign. The soil was still burning from the awesome barrage of Zeus's lightning bolts!

Was the image of burning earth just a vivid fantasy? I looked for a pattern in the geology of killing fields of giants and places of burning earth. It turns out that the fossil beds associated with fallen giants and burning soil—at Megalopolis, Trapezus, Tegea, Pallene, Phlegra, Tartessus, Cilicia, Syria—also contain combustible lignite deposits or volcanic phenomena. Lignite is a low-grade coal or peat formed by decaying plants of earlier eras, such as Miocene forests of sequoia trees. Struck by lightning or spontaneously ignited, lignite can smoulder indefinitely, sometimes fed by volatile natural gases. At Phlegra (at the base of Mount Vesuvius), the earth is volcanic, with smoking fumaroles venting flammable gases. The discovery of gigantic, blackened, burnt-looking petrified bones embedded in soil that also mysteriously burns might well evoke scenarios of huge creatures incinerated by bolts of lightning in the deep past.[9]

Along the Alpheios River, where dense concentrations of Pleistocene mammoth remains were excavated by Professor Skoufos in 1902, Pausanias conversed with locals about their Gigantomachy battlegrounds. He learned that the Arcadians "sacrificed to violent thunder and lightning storms" at Trapezus, where the ground still smoulders. "They say that the legendary battle of gods and giants took place here, and not at Thracian Pallene [the giants' legendary headquarters in Chalkidiki]." According to Pausanias's contemporary Appian, lightning was also worshiped in Syria, where Typhon was killed by Zeus's thunderbolts, and where burning earth and large fossil remains of mammoths coexist.

The notion that fossil-monsters were killed by lightning is not unique to classical antiquity. "Lightning bones" were traditionally gathered in the fossil beds of the Siwalik Hills, and the North American Sioux Indians identified the huge fossils revealed by storms in the South Dakota badlands as lightning-struck "Thunder Beasts." The lightning motif reflects the natural fact that violent thunderstorms expose big fossils to view, but it also reflects the attempt to imagine a force powerful enough to destroy monsters of such size and strength. Today, many believe that a giant asteroid impact was the great cataclysm that destroyed the dinosaurs.[10]

Evidence of Giant Populations of the Past

Not all fabulous creatures of myth were inspired by fossils, of course; some were imaginary. And it's impossible to know which came first, the idea that the youthful earth was populated by giants the likes of which were no longer seen on the aging earth, or early peoples' speculations about prodigious bones weathering out of the ground where they had obviously lain for ages. But it seems safe to say that abundant prehistoric skeletons of uncommon stature and form influenced the earliest Greek myths that gigantic beings once existed and suffered mass destruction. In turn, those traditions were used to explain each new discovery of big bones throughout Greco-Roman antiquity.

People of antiquity were well acquainted with animal and human anatomy. Hunting and butchering animals, animal sacrifices, and ritual cremation and inhumation of the bones of their dead made skeletons very familiar objects. Evidence from Herodotus shows the ancients' deep interest in examining any bones they came across for unusual features. For example, in about 440 B.C. Herodotus visited the battlefield in Egypt where the Persians had crushed the Egyptians in 522 B.C. The bleached skeletons of the armies were still scattered over the surface of the ground where they fell. Herodotus spent considerable time comparing and recording the differences between the skulls of the two races. In Boeotia, for years after the battle of Plataea (479 B.C.), people combed the battlefield for valuables and unusual bones. Herodotus reports that among the finds were a seamless skull, a bizarre jawbone, and a 7.5-foot (over-2-m) skeleton. Such accounts reveal the ancient fascination with examining, comparing, and measuring human and/or remarkable bones.[11]

It was a commonplace in the classical era that all living things were progressively diminishing over generations. The earth's energy was waning, and it no longer produced life with the same vigor. This idea may have arisen from observations of fossil skeletons whose dimensions exceeded those of living humans and

5.1. Man and mammoth bone. An average ancient Greek male stood about 5 feet tall. Drawing by author.

known animals (fig. 5.1). In this view, men of the distant past would have towered over the puny men of the present day. Humans were thought to be the smaller, weaker descendants of this superior race of "heroes," whose oversize bones were revered as relics. (The Greek word *hero* was applied to human ancestors of the distant past as well as to mythical superheroes.) As Pausanias put it, the tallest human beings were the "so-called heroes and whatever race of mortals may have existed in the heroic age before the humans of this age." Quintus Smyrnaeus (third century A.D.) expressed the classical image of heroes' stature in his description of the great warrior Achilles' burial at Troy: "His bones were like an ancient Giant's."[12]

The ideal man's height in classical antiquity was about 4 cubits (5.5 feet; 1.7 m); the average man probably stood just over 5 feet. The traditional height of a mythical hero was about 10 cubits (15 feet; 4.5 m), about three times the size of an average man. I think that the traditional hero's stature was derived by comparing the human-looking femurs of the largest extinct mammals and an average man's thigh bone, which would be about 15 inches long (about 38 cm; see figs. 2.7, 2.8, 2.9). A mineralized mastodon or mammoth thigh bone is two or three times that length (and is much heftier), leading people to visualize a giant about three times as big as themselves. Figures 2.11, 3.4, and 3.5 illustrate how the ancients may have perceived a mammoth skeleton as a two-legged giant. Since no living giants of that magnitude were ever seen or reported, the great bones were taken as vestiges of the era when the earth was young. Viewing such bones, Pliny observed, "It is obvious that the whole human race is becoming shorter day by day!"[13]

It followed that ancient heroes had hunted correspondingly bigger animals. This conception was fortified by relatively recent memories: for example, the wild cattle of the Bronze Age (*Bos primigenius*; chapter 2) were in fact much more imposing than the smaller domesticated cows of the classical era. The intuitive idea of biological "decline" has some basis in reality. Giantism usually does fail to maintain viability: the most titanic prehistoric faunas did die out or evolved smaller forms. And in some Mediterranean islands, the largest species, mammoths and hippos, tended to become smaller over time, while certain tiny species, like mice, grew bigger, as though striving for some Aristotelian mean.[14]

Skepticism about the past existence of giants usually centered on their relative height, but persuasive evidence supported the "shrinking" hypothesis. Who else but supersize men could have piled massive boulders into the Bronze Age fortifications—already unthinkably ancient by the classical era—known as Cyclopean (Cyclops-built) walls? Thucydides, the fifth-century historian who ordinarily discounted folklore, is actually our earliest historical source for the belief that two prehistoric races of giants once inhabited Sicily, the Cyclopes and the Laestrygonians. In the fourth century B.C., Plato's anecdote about giant bones in Lydia makes him the second-earliest writer (after Herodotus) to refer to the discovery of a giant skeleton from the remote past. Later writers pointed to heroic "throwbacks"—the occasional births of very tall men and women in the present—as strong proofs that a race of much larger humans had existed in the past.[15]

The most spectacular evidence, of course, was the procession of colossal femurs, shoulder blades, vertebrae, and teeth that emerged from the earth, conventionally known as "heroes'" or "giants' bones" but sometimes attributed to monsters. Each new fossil exposure excited interest, as people measured themselves and known animals against the impossibly big bones and tried to match them to mythohistorical beings and events. The family trees of heroes and the geographical distribution of giants stirred controversy as villagers and religious oracles debated the identities and names of the beings who had left newfound bones. As Pausanias remarked, "When ancient legends about local heroes and their genealogy had no mythological poem to follow, the stories became quite inventive!" (It's amusing to compare the words of an early critic of Charles Darwin's family trees for extinct animals: those pedigrees, he said, are "worth no more than the genealogies of Homeric heroes!")[16]

The Greeks and Romans read the vast dimensions and worn condition of large, petrified bones as signs of their great age. Unlike the Gobi *Protoceratops* remains, which resembled the sun-bleached skeletons of recently deceased creatures (chapter 1), most fossilized bones around the Mediterranean are dark, mottled, disarticulated, and fragmented. I think that this aged appearance, along with their depth in rock layers exposed by earthquakes and collapsing cliffs, led to the idea of their extreme antiquity. The Greeks realized that creatures of the deep past were different from and bigger than present species, and that they had all been destroyed long ago. And this brings us to the idea of extinction. Do the Greco-Roman geomyths contain the germ of the concept of past species' extinction?

Creation, Transmutation, Extinction, and Petrifaction in Popular Traditions

The geomyths envisioned the birth, reproduction, transmutation, and disappearance of giants and monsters in a long-past era. Another branch of folklore described how once-living creatures were transformed to stone. Taken together, these traditions formed an internally coherent model that accounted for the observed facts of mineralized bones of remarkable dimensions. The following section looks at the implications of those popular beliefs; thereafter we'll consider the alternative theories proposed by the natural philosophers.

Origin and Change over Time

How successive, different faunas emerged over geological time is one of the most difficult puzzles of paleontology. From the Middle Ages until Darwin, science was dominated by the biblical doctrine of creationism, which holds that the world's species were created all at once by God and have never changed or gone extinct. It still finds adherents. Modern "creation scientists" hold that only an "intelligent design" can plausibly explain the complexity of earth's origin and biology, and that evolution itself is a new "creation myth" devised by the "scientific elite."

It is a common misconception that creationism and fixity of species held sway in classical antiquity too, and that extinction was unthought-of. For example, Jack Horner writes, "As far as anyone can tell, the concept of extinction was completely unknown [in classical antiquity]. From the dawn of humanity . . . the standard belief was that all the plants and animals on Earth had been created at the same time and that all of them had continued to exist in the same way from that day onward." Horner says that this idea went unchallenged until the "1600s, when a few European naturalists speculated that since living animals could not be found to match certain skeletons, the animals must have perished sometime in the past." This is the generally accepted view of paleontological historians.[17]

But it turns out that creationism and fixity of species were not dogmatic principles in classical antiquity. Natural changes in the environment and in life-forms over immense periods of time were important concepts in pre-Christian myth and philosophy. In the Greco-Roman mythic tradition, and in ancient Hebrew traditions as well, all species were not created in one fell swoop, and novel life-forms appeared and then disappeared over time.

The myths preserved by Hesiod and Homer visualized a succession of new types of strange mortal creatures bred over several generations, long before the appearance of present-day humankind. These creatures arose through matings between different sorts of monsters. Some of the offspring of these unions were unique monstrosities, but strange hybrids whose offspring *did* resemble the parents, such as Centaurs, were accorded a kind of species status. Titans and giants were also species that reproduced themselves over generations. We have the names of numerous giant pairs and their descendants, including immature giants destroyed in the era of the Gigantomachy. For example, the god Apollo slaughtered the giant children Otus and Ephialtes and buried them on Naxos. The presence of "young giants" on Naxos might represent an interesting natural observation, since the prehistoric elephant remains of Naxos are in fact smaller versions of the very large ancient elephant *P. antiquus* found on the mainlands.[18]

Popular Knowledge and Beliefs about Extinction

Extinction is a crucial concept in modern paleontology. It is generally assumed that the idea of extinction of whole groups of animals did not develop until the seventeenth century. But some 2,500 years ago, notions of extinction, both catastrophic and gradual, were developed by the Greeks and applied to remarkable fossil bones. The paleontological legend of the Neades of Samos clearly states that the huge remains belonged to real animals that had all died out before the first humans arrived on the island (chapter 2). Several authors placed the era of the long-vanished giants in the prehuman age when mountains were still being formed and the "origin of life was recent."

The geomyths described past epochs marked by the destruction of entire populations of viable creatures. Unlike the Scythian nomads' scenario of lurking griffins in the Gobi, the Gigantomachy narrative led the ancient Greeks and Romans to perceive the *extinct* status of the beings whose strange skeletons emerged from the earth. They knew that the bones were all that remained of colossal creatures that appeared, reproduced, and then were destroyed long ago in specific locales where their skeletons petrified in the ground. Once the gods' *Blitzkrieg* ("lightning war") ended the epoch of giants and monsters, those creatures were gone forever. This narrative closely parallels the modern concept of catastrophic extinction.[19]

The idea of the extinction of whole groups of creatures before the appearance of current humans occurs in several classical myths. For example, Heracles destroyed the Centaurs in Arcadia and/or Thessaly, completely killed off all the large predators of Crete and North Africa, and rid the Mediterranean of sea monsters. Orion, a giant son of the Earth and the greatest hunter of all time, slew untold numbers of wild beasts. On Chios, he exterminated the aboriginal fauna of the island, then fell exhausted on the beach. When Orion boasted that he would kill all the animals on earth, the gods destroyed him, burying his giant body on Delos or Crete. The settings of this extinction myth are revealing. Chios, an island between Samos and Lesbos, was part of the Anatolian mainland in the Pleistocene. Paleontologist Sevket Sen, who excavated mastodons and other extinct species on Chios in 1991 and 1993, tells me that along the western sea cliffs big bones continually weather out on the beach, where Orion slept. Delos and Crete, rivals for the giant's grave, also have prehistoric elephant fossils. In Pliny's day (first century A.D.), an earthquake on Crete revealed a huge skeleton, 69 feet (21 m) long, which "some people thought must be that of Orion." (Others argued that it was Otus, the young giant killed by Apollo.)[20]

Other myths told of the destruction of oversize versions of known animals. Some examples are the great Nemean Lion that ravaged Mycenae; the monstrous canines Cerberus and Orthos; the giant Teumessian Fox of Boeotia; the ferocious Cretan Bull that terrorized Crete, Arcadia, and Attica; the gigantic Erymanthian and Calydonian boars and the Sow of Crommyon; the awesome cattle of Geryon; and the huge carnivorous birds and hulking wolves of Stymphalia killed by Heracles. Trophies of some of these super-beasts were displayed in antiquity. For example, the Temple of Apollo at Cumae, Italy, boasted the great tusks of the Erymanthian Boar, and the tusks of the Calydonian Boar were the pride of the Temple of Athena in Tegea. (The tusks may have belonged to Pleistocene proboscideans whose remains are abundant around Cumae and Tegea; chapter 2.) Pausanias's comment that some people believed that the Nemean Lion and other huge beasts of myth "came out of the ground" is interesting. I think that some stories of supersize animals and birds might have been influenced (and then continually confirmed) by familiar-seeming but unexpectedly large fossil remains of extinct sabre-toothed tigers, giant hyenas and pigs, cave bears and lions, ostriches, gigantic tortoises, and immense cattle.[21]

Some myths may reflect historical disappearances of species, such as the wave of extinctions (of mammoths, cave bears, giant cattle, etc.) in the late Pleistocene and Holocene epochs. J. Donald Hughes, a historian of ancient ecology, discusses several examples of the ancient awareness "that wildlife might be totally extirpated" from certain areas. Several ancient writers remarked that "animals are no longer found where they were once abundant." Pliny observed that "a number of birds had not been seen for generations" where they once thrived. Within living memory, people knew that lions had become extinct in Greece, Asia Minor, and Libya. Bears disappeared from Attica and the Peloponnese, and ostriches vanished from Arabia. People knew that elephants no longer inhabited North Africa and Syria, and that the tiger had gone extinct in Armenia and northern Iran. They knew that certain horse breeds had disappeared in Greece and Anatolia, and that giant bovines no longer roamed Crete, Greece, and Italy.[22]

The Flood Myth

The classical myth of Deucalion's Flood describes extinction by catastrophe. Unlike the biblical Flood myth in which a breeding pair of every species created by God was saved from extinction in

Noah's Ark, the Greco-Roman tradition described a deluge that wiped out all animals and ushered in an epoch of new fauna. In the classical myth, earth's life-forms passed through several ages. In one of the later eras, Zeus decided to drown all life (some versions speak of five successive deluges). The narrative seems to collapse into one event the long sequence of sea transgressions and emerging land masses in the Miocene discussed in chapter 2. Earthquakes, collapsing mountains, and violent storms accompanied the flood. The watery deep covered all but the highest peaks. In a scene reminiscent of paleontologists' reconstructions of the Miocene disaster at Pikermi (chapter 2), the Roman poet Ovid imagined flocks of birds falling exhausted into the sea; schools of fish dying entangled in submerged forests; and lions and wolves and sheep, boars and stags all desperately seeking refuge together on high ground before being swept away.

In the Greek myth, two Titans, Deucalion and Pyrrha, escaped in a boat, which ran aground on a high mountain. As the waters receded, they found a muddy earth devoid of living things. The goddess Themis told them how to repopulate the land. "Throw the bones of your Mother behind you" was her mystical command. Deucalion and Pyrrha decided that the "bones of Mother Earth" must be stones. They cast some boulders over their shoulders and watched in awe as the mud clinging to the stones became flesh and the stones themselves became bone. Some of the new creatures reproduced shapes previously known, says Ovid, while other animals were novel and strange.[23]

Petrifaction Folklore

Identifying mud with flesh and stone with bone is a widespread folklore motif. Since bone formation (ossification) *does* occur through mineral deposits (calcification) in living skeletons, the folk "comparison of bones and rocks is not entirely metaphorical," notes paleontologist Jack Horner. The transformation of bones, the hardest part of the body, into stone, the hardest part of earth, appears in many Greco-Roman tales. One story tells how the bones of the giant Atlas petrified in the ground; another describes the skeleton of the nymph Echo gradually desiccating into stone; a man of the heroic era is transformed into a rocky islet off Euboea in still another. The early inhabitants of the island of Seriphos were lithified en masse, and in other tales the bodies of a giant wolf, fox, hound, and humans were turned to stone in the mythical past.[24]

Deucalion and Pyrrha's creation of new fauna after the flood is a fantastic *reversal* of the mysterious process of petrifaction. "Bones of Mother Earth" could also mean the petrified bones of Earth's offspring, the giants and monsters of yore left buried in the ground. Observing prehistoric remains that had somehow hardened into stone may have suggested the idea of reversing the process to bring lost species back to life. So, in the myth, instead of the normal sequence of flesh rotting away from bones that then turn into stone, the flesh reappears on the stones, magically transforming them back into living creatures.

Petrifaction Science

The metamorphosis of bones into solid stone was a process of interest to the natural philosophers of Aristotle's school. They correctly suspected that infiltration of water and dissolved minerals played some role. By the fourth century B.C. it was common knowledge that mineral-laden hot springs coated things, including bones, with a hard film and then petrified them. The petrifying force was seen as a kind of "vapor" that somehow consumed moisture as it caused shriveling and hardening into stone. In a remarkable passage about animal remains found in mines in Spain, Pliny recognized the relationship between crystal precipitation and petrifaction of bones (chapter 2). Aristotelians referred to "drying exhalations of the earth" that produced solidified *fossils* (meaning dug-up curiosities, such as crystals and buried ivory). In *Parts of Animals*, Aristotle alluded to petrifaction folklore: "Once the soul departs, what is left is no longer an animal. Nothing remains except the configuration, like the animals in folktales that are turned to stone."[25]

Aristotle's successor as head of the Lyceum, Theophrastus (ca. 372–287 B.C.), was born in Eresos, a village that lies in the midst of the extensive petrified forest of western Lesbos. Some 18 million years ago, hundreds of Miocene sequoias, palm trees, and other vegetation were engulfed in volcanic ash and mineralized by rainwater and hot springs. The stone trees, many still standing, with roots and pinecones intact, have been declared a Greek national monument. In his treatise *On Stones* (on rare and unusual materials dug from the earth), Theophrastus discussed certain minerals' capacity to change objects into stone, and described petrified reeds. According to Pliny, he also described mottled "fossil ivory," and wrote about "stones resembling bones that are produced from the earth." Pliny was referring to Theophrastus's two-volume treatise *On Petrifactions.* Alas—this unique work about ancient paleontology has disappeared into the abyss of time. Pliny's lone statement, some fragments of another book by Theophrastus about fish embedded in rocks near the Black Sea, and the title of Theophrastus's lost work are the only evidence we have that an ancient natural philosopher seriously addressed the problem of prehistoric bones.

Since Theophrastus invented botany and mineralogy and wrote a lost treatise on petrified fish, it seems likely that one volume of the lost *Petrifactions* described petrified plants like those of his native Lesbos, and the other volume discussed petrified animal skeletons. Some modern scholars maintain that Theophrastus thought bone-shaped stones were produced by plastic forces in the earth, an idea that would hold sway in the Middle Ages. But I think his grasp of the petrifaction process in *On Stones* suggests that *On Petrifactions* might have approached an understanding of how bones of unfamiliar and oversize animals happened to mineralize in the earth.[26]

Natural Philosophy

In the sixth century B.C., philosophers began to counter the authority of the myths to explain natural phenomena. As an alternative to the "impossible" actions of the gods, they sought natural causes that conformed to physical laws as they were then understood. Natural philosophers advanced insightful theories about the origin of life, transmutation, extinction, and petrifaction, and they cited simple marine fossils as evidence of former seas. Yet no surviving writings by the pre-Socratics, or by Aristotle, Lucretius, or other natural philosophers, mention the bones of surprising dimensions that caused such a sensation among their countrymen. Their silence seems to reflect an eternal tension between academic science and mass culture (a tension exemplified by Plato's ironic anecdote about big bones, discussed at the beginning of the chapter). In what we might call the "archives of anomalies," however, a few fragments presented at the end of this chapter hint that some Aristotelian followers and other philosophically aware writers took an interest in mysterious remains.[27]

Floods and Marine Fossils

It is generally accepted that marine fossils found far inland influenced the myth of Deucalion's Flood. As the historian Solinus wrote, the receding waters of the flood "left behind shells and fishes and many other things, making the inland hills resemble the seashore." Convinced by the same evidence, the earliest natural philosophers agreed that there had been extensive former oceans, but they rejected the role played by gods.

The first writer to recognize the organic nature of fossil shells was the Greek philosopher Xenophanes (only fragments of his works survive). He left his native Ionia in about 545 B.C. to wander the Mediterranean, living for a time in Sicily. Perhaps on the basis of his observations of severe coastal changes in western Anatolia, Xenophanes proposed cyclic encroachments of sea and land, and even maintained that life was alternately destroyed and then generated anew in grand cycles. All land was once covered by sea, he said; therefore hard rock was once soft mud. In a rare example of empirical proof in philosophy, Xenophanes enumerated examples of bivalve shells (Mesozoic-Tertiary mollusks) on mountains, impressions of fish and seaweed in quarries on Sicily, myriad sea creatures embedded in rock on Malta, and oodles of small fish (probably Oligocene-Miocene *Prolebias* species) in slabs of rock on the island of Paros.[28]

Xenophanes' geological evidence for receding seas was universally accepted. For example, in the fifth century B.C., Xanthos of Lydia arrived at the same conclusions after observing cockle and scallop shells in inland Asia Minor, and Herodotus discussed marine fossils in the deserts of Egypt. Eratosthenes (285–194 B.C.) of Cyrene (north coast of Libya) concluded from the vast numbers of shells miles from the coast that North Africa had once been under water. Some historians of ancient science have called these examples of deductive reasoning based on observations of various types of marine fossils "remarkable and impressive," "unusually scientific," even "astounding."[29]

But historians of paleontology view the accomplishment differently. They place common marine fossils at the very "easiest end of the spectrum" of fossil evidence, in the words of Martin Rudwick. He points out that it took no profound effort to recognize the organic nature of simple fossil shells, because they are complete and relatively unchanged from their living state. Indeed, the Tertiary fish and shells so closely resembled living fish and shells familiar to the Greeks and Romans that no theory of extinction was needed to explain them. The drying mud of receding waters accounted for their stony state. Rudwick's comments help us appreciate how mythological explanations for the much more difficult phenomena of disarticulated prehistoric mammal remains bridged a gap between empirical evidence and theory, a gap that the philosophers seemingly avoided.[30]

The Enigma of Giant Bones

For philosophers, the Flood myth was successfully countered through the visualization of a natural, rather than a divine, deluge to explain out-of-place, petrified—but not necessarily extinct—marine creatures. Yet the "easy" theories accounting for shells

5.2. Philosopher and a giant bone. If an ancient Greek philosopher ever contemplated the fossil femur of a Miocene *Samotherium* (giant giraffe of Samos), this sketch shows the relative size of the thigh bone. Drawing by author.

could not explain how so many bones of colossal land creatures no longer seen on earth came to be in the ground. The stature of humans of the distant past and the viability of monsters were topics of philosophical debate. But even if the former existence of giants and monsters was not in question, their origin and mass extinction were. The problem for the philosophers was the lack of a naturalistic theory to successfully counter the myths of the Theogony and the Gigantomachy. We have to wonder why the ancient philosophers (with the possible exception of Theophrastus) ignored the most difficult paleontological evidence of giant bones (fig. 5.2).

"Explaining silence is tough. How confident can we be that natural philosophers actually knew about big bones as a phenomenon of nature rather than legend?" asks classical scholar Brad Inwood, who specializes in ancient natural philosophy at the University of Toronto. "I would expect them to be less interested in these facts than historians, travel writers, geographers, and natural history writers," says Inwood. The philosophers were "motivated by an argumentative relationship with tradition, rather than the explanation of evidence." Geoffrey Lloyd, historian of ancient science, points out that the community of philosophers did not judge each other's theories on the amount of empirical evidence they could present. Historian John Burnet reminds us that our ancient sources are extremely fragmentary, and what we do have focuses on philosophers' results, not on their research data.[31]

Were the natural philosophers aware of giant bones? Signs point to yes. It's hard to believe that investigators of nature were oblivious to the tangible evidence at the root of the myths of giants and the Gigantomachy. We know that the philosophers lived and traveled in regions where discoveries of "giant" bones were sensational events, and fossil relics were publicly displayed throughout the Greco-Roman world. Did the academic community of philosophers avoid the topic as a way to transcend the unseemly hoopla over marvels? There is a hint of this in Plato's dialogue *Phaedrus.* When someone asks Socrates whether legends are really true, Plato has him reply that philosophers do normally reject or rationalize away popular myths. But what philosopher has the ingenuity or time to account for the overwhelming multitude of mythical monsters? "That's why I don't concern myself with them," says Socrates, "I have no time for such things. I accept what is generally believed, and pursue more serious matters."

Another possibility is that the incomplete nature of the vertebrate evidence discouraged scientific investigation. Perhaps natural philosophers "recoiled from the difficulties [presented by] the imperfections" of large fossil bones, as suggested by paleontologist Henry Fairfield Osborn in 1921. According to Georges Cuvier, early naturalists passed over the impressive "fossil bones of quadrupeds" because they were "frightened" by the chaotic evidence of "isolated bones, confusingly intermingled and scattered, and almost always broken and reduced to fragments." They would not even "hazard giving a name to them," and so vertebrate paleontology remained the "least cultivated part of fossil history."[32]

Maybe, as Burnet suggested, the philosophers' silence is just an accident of capriciously preserved ancient texts. In the end, we can only speculate. Although none of their surviving works explicitly related their theories of the earth's past to unusually large skeletons in the ground, the natural philosophers were pursuing a concept essential to paleontology, that "nature has a history." Their theories about life's origin, change, and extinction may have indirectly influenced the way giant bones were interpreted by other ancient writers who *did* report such discoveries. Some of those authors tried to reconcile the silence of the established philosophers with the widespread popular awareness of remarkable bones. And in an interesting parallel to competing geological and paleontological theories of the modern era, the ancient geomyths painted a picture of mostly catastrophic change, while the natural philosophers leaned toward a more gradualist view.[33]

Creation, ™Evolution,ʃ and Extinction Theories

Anaximander of Miletus (ca. 611–547 B.C.) was one of the first to understand that the Mediterranean was a vestige of vast seas (known as Tethys to modern paleogeologists). According to scraps of his enigmatic writings, the sun's heat interacted with the primal ooze to generate sea creatures. Some of these progressed to a sort of "chrysalis" stage from which the first primitive humans emerged onto dry land and sustained themselves by reacting to difficult new circumstances. Should Anaximander be called the first evolutionist? Scholars disagree, but his ambiguous scenario does suggest that humans originated from or within a biologically different species that no longer exists, and that humans survived by adapting to a new environment. Marine fossils probably fostered Anaximander's conviction that all life originated in water. He resided in Miletus, which has abundant Miocene animal fossils observed in antiquity, but we will never know if his thinking was influenced by extraordinary bones. We do know, however, that when Pausanias discussed the giant bones found at the Orontes River (chapter 2), he alluded to Anaximander's theory to explain them.[34]

The eccentric philosopher Empedocles of Acragas (Sicily) deplored literal belief in myths. His mysticism and spectacular death in ca. 432 B.C. made him legendary in his own day (it was said that he leaped into Mount Etna's crater). We know that he traveled several years in the Peloponnese, where the giant bones of the Gigantomachy battlefields were famed far and wide. He attended the Olympic Games in 440 B.C., at Olympia on the Alpheios River northwest of Megalopolis. The giant shoulder blade of the hero Pelops had been discovered nearby and was displayed in a shrine at Olympia. The enormous bones of the hero Orestes had been unearthed at Tegea with great fanfare in about 550 B.C. That famous incident was reported in detail by Herodotus, who may have met Empedocles in Italy.

So Empedocles had ample opportunities to see and hear about the bones of giants on his travels in Italy and Greece. He knew of Xenophanes' explanation of the marine fossils of Malta and Sicily. Was he also familiar with the conspicuous elephant fossils of Sicily? A persistent *modern* legend still perpetuated by historians of paleontology claims that Empedocles not only observed the elephant skulls but related them to the myth of the one-eyed Cyclops in his writings (for the origin of this false legend, see the introduction). But in the fragments that survive of his poems, there is no word of unusual bones, elephants, giants, or Cyclopes.[35]

One fragment of Empedocles *does* discuss extinct monsters, however. The inclusion of mythical monsters in his cosmology has long puzzled scholars. In a notoriously obscure passage, Empedocles applied his mystical Love/Strife theory to the creation of monsters. In the epoch before humans, "myriad kinds of mortal creatures were brought forth, endowed with all sorts of shapes, wondrous to behold." Nature produced disembodied heads and limbs that assembled into various weird hybrids, such as "human-headed oxen and ox-headed humans." This imagery recalls the human-beast hybrids used by archaic artists to represent all sorts of different mythical monsters. Empedocles argued that these bizarre products of chance arose naturally, but were ultimately unviable, unable to defend, feed, or reproduce themselves, and so died out forever, by natural means, while fit animals survived.[36]

"It is hard to escape the conclusion that Empedocles was here seeking to provide a scientific explanation of composite creatures like minotaurs, centaurs, and so on, which featured so heavily in mythology," says Sue Blundell, a historian of ancient philosophy at University College, London. W.K.C. Guthrie concurs: Empedocles "was always glad to show that his . . . system accounted for phenomena known or believed in by his countrymen." Did those misunderstood phenomena include the remarkable bones that his countrymen attributed to mythical beings? It is tempting to think so, but we have no proof in his writings.[37]

Lucretius, an Epicurean philosopher writing in Rome in the first century B.C., refined Empedocles' theory of extinct monsters. In *On the Nature of Things*, Lucretius provided the clearest expression of extinction and "survival of the fittest" in ancient literature. Without referring to anomalous bones but obviously seeking to counter the myths' picture of fertile monsters and catastrophic destruction by the gods in the Gigantomachy, Lucretius wrote that nature produced "many monsters of manifold forms" and "bigger animals" in ages past, but these gradually died out when they could not find food or reproduce. Larger humans and strange creatures once existed, but hybrids of physically incompatible species, such as human-horse Centaurs, were biologically impossible.

"Everything is transformed by nature and forced into new paths. One thing dwindles . . . another waxes strong," wrote Lucretius. "In those days, many species must have died out altogether and failed to multiply. Every species that you now see drawing the breath of life has been preserved from the beginning of the world by cunning, prowess, or speed." Those "without natural assets fell prey to others, entangled in the fatal toils of their own being, until nature brought their entire species to extinction." From these cycles of destruction, humans learned that "nothing endures forever," that everything "will have its day of doom." It's hard to resist comparing Lucretius to Jack Horner speaking of evolution and extinction two thousand years later: "Vegetation disappears.

Habitat changes. . . . the number of niches . . . dwindles." "Environmental stress . . . was so intense" that "entire populations were decimated," and "their fortunes and ultimate fate echo throughout all creation."[38]

Aristotle, Fixity of Species, and Anomalies

In the fourth century B.C., Aristotle denied, on physiological grounds, Empedocles' notion that mixed species ever existed, just as Lucretius would later do. But, unlike Lucretius, Aristotle also rejected Empedocles' insightful "random nature" hypothesis of the past origin and extinction of monsters, in favor of a "designing nature" theory. Aristotle's "theistic" view does seem to preclude mutability of species. But the static idea of divinely created, unchanging species became dogma only after Christian theologians took it up in the Middle Ages.[39]

Fixity of Species

It was at Assos, on the fossil-rich Anatolian coast, that Aristotle began his investigations of natural history, and he often visited his friend Theophrastus on nearby Lesbos. In Athens, Aristotle and his students at the Lyceum developed a zoology based on scientific observations, but they also gathered folk wisdom from around the ancient world. Pliny claimed that Aristotle's student Alexander the Great ordered the shepherds, hunters, farmers, and fishermen of Greece and Asia Minor to inform the philosopher about any unusual animals they came across in the empire.

In classifying an ascending scale of 540 species of fishes, birds, and mammals, Aristotle's method was to look for regular combinations of characteristics. His search image focused on describing predictable, living forms in nature; unique "monstrosities" were noted but excluded. Geoffrey Lloyd points out that Aristotle's approach problematically allowed "normative statements" to be "deduced from descriptive ones." The assumption that nature's design was to produce forms that were essentially unchanging tended to dampen philosophical speculation about creatures that did not fit categories of known animals. Oversize bones with no living counterparts were not classifiable in this system.[40]

Some historians of paleontology assert that "Aristotle delivered the deathblow to evolutionary ideas in natural philosophy" with his "dogma of immutability of species." But Geoffrey Lloyd cautions that "one should not exaggerate the extent to which the Aristotelian idea of fixity of species prevailed" in antiquity. "With those for whom it did, however, there was not going to be much speculation about now-extinct animals." Lloyd compares the situation to ancient astronomy. Because the Greek philosophers assumed that the heavens were stable, they were not adept "at observing novae, sun spots, and other phenomena that might be thought to contradict that assumption." Transient heavenly events were recorded by the ancient Chinese scholars, on the other hand, because their search image led them to expect irregularities in the sky.[41]

"Aristotle's species conservatism" was not "a central theme of his biology," agrees Brad Inwood. Aristotle's "views did not strictly entail that species cannot die out (which is all that would be required to explain big bones), only that living species do not evolve." Yet in no extant writing by Aristotle do oversize skeletons appear, even in contexts where we might expect them. For example, the fragment about the proverbially noisy Neades of primeval Samos includes no mention of their famous bones (chapter 2). The Aristotelian system, like Linnaeus's, was based solely on animals currently alive: it had no provisions for species that no longer existed. In developing his view of nature's intention to reproduce standard species, Aristotle did acknowledge occasional, irreproducible "mistakes in nature." He also recognized the existence of exotic, uncategorizable *living* creatures. In *History of Animals*, for example, he devotes a passage to contemporary fishermen's experiences with bizarre sea monsters, creatures whose "rarity makes them unclassifiable." By calling attention to anomalies only to label them "monstrous" and irrelevant, Aristotle may well have intended to transcend wrongheaded popular interest in the bizarre, including gigantic old bones.[42]

The Anxiety of Anomalies

According to Thomas Kuhn's influential *Structure of Scientific Revolutions* (1962), the history of science reveals a series of peaceful interludes punctuated by conceptual revolutions that replace old worldviews. When scientists are engaged in gathering data on normal phenomena, the resistance to noticing discrepancies is strong. Kuhn observes, "Normal science does not aim at novelties of fact or theory and, when successful, finds none." If anomalies begin to be perceived in great enough numbers, however, they cast doubt on expected patterns, and anxiety arises over the failure to make them conform to the accepted paradigm. Then, when sufficient anomalies accumulate to induce repeated crises within the scientific community, the scientists struggle to adjust, and revolutionary scientific advances can occur.

Using Kuhn's criteria for scientific revolutions, we might say that in antiquity, Aristotle and other philosophers were intent on recording with precision, for the first time, a vast body of normal phenomena that presently existed in the world. Discoveries of remarkable bones identified by the populace as vestiges of mythical giants or heroes fell short of the kind of crisis of anomalies that would be required to divert the attention of the ancient scientific community from its normative task. That point was not reached until eighteenth-century scientists perceived an undeniable crisis of anomalies, as riddling skeletons emerged around the world in such sheer numbers and variety as to overwhelm the biblical Genesis and Flood paradigms.[43]

Of course, ordinary people in classical antiquity certainly *did* notice sensational irregularities such as gigantic skeletons. While the circle of academic philosophers declined to discuss such observed phenomena, the rest of Greco-Roman society hashed out the meaning of big bones discovered all around the Mediterranean and even farther afield, drawing on the coherent and flexible paradigms in myths, legends, and folklore. We know this because a group of educated intellectuals, like Pausanias and Philostratus, knowledgeable about myth, history, and philosophy but also attuned to current events and curious, even paradoxical, phenomena, preserved the evidence in their writings. Their narratives made up the body of ancient paleontological knowledge discussed in chapter 3.

Philosophical Archives of Anomalies

According to a biographer of the third century A.D., Aristotle supposedly wrote two works titled *On Composite Animals* and *On Mythical Animals.* No such works have survived. If the titles are spurious, as historians believe, then perhaps they reflect wishful thinking on the part of some later scholar who saw gaps in the Aristolian oeuvre.[44] But some tantalizing evidence suggests that natural anomalies, including giant bones, did in fact capture the attention of certain of Aristotle's successors and other writers interested in reconciling brute facts with popular belief and philosophical "truth." Theophrastus's vanished works on petrifactions and "stones that resemble bones" fall into these dossiers of anomalous evidence, as does his other lost treatise, on Empedocles. The difficult-to-classify writings of Pseudo-Aristotle, Palaephatus, and Philostratus also attempted to come to terms with problematic evidence and popular beliefs.

Pseudo-Aristotle

The Aristotelian corpus contains some works by unknown authors. *On Marvelous Things Heard* (attributed to Pseudo-Aristotle) is a collection of nature lore that may have been compiled by Aristotle's followers or attached to his works because someone thought the topics fit the kinds of things explored at the Lyceum. One of the entries expresses the compilers' rationale: "This report may strike us as legend, but . . . one must not pass over it without record, when making a catalog of events in specific places." The list of marvels includes several items of paleontological interest, such as the burning battlegrounds in southern Italy and in Megalopolis; resin hardened into stone (amber); a place near Cumae

(Italy) where things were petrified; large footprints in rock in the heel of Italy; petrified human bones in Lydian mines; cylinder-shaped stones (crinoid echinoderm fossils) dedicated in a shrine near Mount Sipylos (west-central Turkey); and little stones shaped like beans (nummulites or *Camerina* fossils) gathered for medicine in Egypt. *On Marvelous Things Heard* is an agenda of natural mysteries to be investigated. Was it compiled by someone who hoped to reconcile the philosophical exclusion of anomalies and the keen public interest in the anomalous?[45]

Palaephatus

Another work, usually dismissed by modern scholars as clumsy and contrived, falls into a gray area at the margins of philosophy and popular knowledge. *On Unbelievable Tales* was written by a friend and follower of Aristotle known as Palaephatus (a pen name that roughly translates as "ancient tales"). Palaephatus rationalized traditions about heroes and monsters by tracing how actual events blurred over time into myth. Instead of rejecting myths outright, he tried to strip away impossibilities to reveal an underlying historical core.

Centaurs, those half-human, half-horse creatures supposedly wiped out by Heracles in the mythical past, were a focus of tension between the logic of popular belief in marvels and philosophers' belief in immutable principles of nature. Myth accorded Centaurs the status of a viable species. Empedocles suggested that Centaur-like creatures once existed but died out as monsters unfit for survival; later, Aristotle and Lucretius vigorously denied that such hybrids could ever exist.

Plunging into the Centaur debate, Palaephatus repeats Aristotle's notion that incompatible natures cannot coexist. Then he articulates a principle of unchanging species: "If there ever were such animals, then they would exist today." But not only does his statement contradict ancient knowledge that some real animals had gone extinct, his wording leaves the door open for relict Centaur "sightings" and atavistic births that would prove their existence in the past, much as abnormally tall men and women proved the former existence of giants. And indeed, live Centaurs *were* reported in the Roman era. These sightings of relict creatures that had supposedly gone extinct eons ago can be compared to modern-day sightings of Bigfoot or Yeti, the Loch Ness monster, and other unknown species or supposedly extinct animals investigated by cryptozoologists. Belief in relict or crypto-creatures thrives in the fertile ground between science and fantasy, nourished by imagination and by ambiguous evidence (chapter 6).

Next, Palaephatus tackled the myth of the dragon's teeth sown in the soil by the hero Cadmus, which sprouted into armed men. He suggested that instead of a dragon, Cadmus killed a king in Boeotia named Draco ("Dragon") who "owned the various valuable possessions that kings usually own, in particular elephants' teeth." Elephants had only recently become known to the Greeks; Palaephatus realized that before his day, their teeth would have been identified as those of giants or monsters. Cadmus stored the precious molars in a temple, says Palaephatus, but Draco's allies seized the teeth and fled to Attica and the Peloponnese, where they raised armies.

Palaephatus's rationalization may be contrived, but he neatly sums up the high value placed on exotic relics such as elephant molars, which could turn up in Greek soil and were in fact stored in archaic temples. Palaephatus is the only ancient author to explain a monster myth as a misunderstanding of real animal remains. His paleontological exposé brings to mind the Corinthian vase painter's depiction of the Monster of Troy as a large fossil skull (chapter 4).[46]

Philostratus and Apollonius of Tyana

Apollonius, a philosophically inclined sage of the first century A.D., is a controversial figure for historians. His biography, by the sophisticated writer Philostratus (ca. 217–238 A.D.), recounts the sage's journey to India to gather wisdom and his later travels around the Mediterranean. For us, Apollonius and his biographer

Philostratus are crucial figures in ancient paleontology, because they demonstrate a practical approach to contentious natural history questions in antiquity, such as the appearance of griffins and dragons and the former existence of giants.

We first met Apollonius in chapter 1, when he offered a reasonable explanation for griffin "wings." In chapter 2 he described fossil ivory; in chapter 3, he gave an account of dragons in the fossiliferous hills of northern India. Here, his opinion on the accumulating evidence of giant bones seems radically rational in view of the natural philosophers' failure to confront the problem of oversize remains. After listening to local "vulgar myths" in Sicily about Typhon and the giants destroyed by the gods, Apollonius offers a "more plausible conclusion, one more in keeping with philosophy." "I agree that giants once existed," states Apollonius, "because gigantic bodies are revealed all over earth when mounds are broken open." "But," he continues, "it is mad to believe that they were destroyed in a conflict with the gods."

Pay attention to the evidence, he counsels, but look for a natural explanation. Apollonius's statement sounds like a challenge to philosophers, and it shows that at least one thinker was capable of conceiving that natural causes could explain the demise of giant creatures whose skeletons lay in the earth. Philosophers had already found a natural cause to explain stranded seashells, proposing cycles of flooding and receding seas to replace the myth of a divinely sent deluge. The enigma deepens: why did no natural philosopher proceed from the simple, rational standpoint expressed by Apollonius to explain the widely observed evidence of unexpectedly big bones?[47]

Ancient Paleontology

"The correct interpretation of fossils is a difficult human endeavor. To gaze upon a fossil is not necessarily to comprehend its significance as a biological entity," remarks paleontologist Peter Dodson. Neither are the extreme antiquity and extinction represented by fossils easily grasped by observers. But ancient Greco-Romans apprehended all three difficult paleontological concepts when they encountered the petrified bones of Miocene-Pleistocene species around the Mediterranean. Colin Ronan, a historian of scientific ideas, poses a logical question. Why, armed with this popular knowledge and experience of fossils, and building on the frameworks provided by Anaximander, Xenophanes, Empedocles, and even Plato, did no ancient philosopher ask the right questions about Aristotle's ascending scale of zoology, to develop a "truly dynamic" idea of evolution some two millennia before Darwin? Ronan blames the "static belief in fixity of species and the mistaken notion that they were all created at once" for stifling ancient paleontological thought. But as this chapter shows, fixity of species and onetime creation were not monolithic notions in either Greco-Roman myth or philosophy. Even Plato, for example, suggested that environmental forces caused life-forms to change over time. Humankind "existed for an incalculably long time from its origin," wrote Plato, "and various changes in climate have probably stimulated a vast number of natural changes in living beings." Pausanias affirmed the concept when he remarked that "animals take different forms in different climates and places."[48]

Other formidable obstacles blocked the development of formal scientific theories to account for oversize animal fossils in the classical era. Philosophical theories and popular beliefs about extinction and transmutation passed through centuries of intellectual dialectics, and the natural philosophers apparently sidestepped the challenging evidence of big bones; they never made the leap from their theories of vast changes in earth's climate and landforms to explain the geologically embedded skeletons of unusual size. No one sensed the true immensity of the earth's history or the long chronology of biodiversity. And the Mediterranean fossil record itself is extremely fragmented owing to the geological factors surveyed in chapter 2.

It was only after people began to comprehend the real magnitude of geological time that species mutability could be understood. "Evolution is nothing more, and nothing less, than change through time," says Jack Horner, yet without an appreciation of billions of years of life on earth, the true significance of the fossil record must remain elusive. Not until the nineteenth century had investigators gathered enough fossil material from all the continents to understand our relationship to the populations of bizarre creatures that had flourished and mutated or died out over eons.

The "contorted and broken stratigraphy of most Mediterranean lands did not provide well-ordered sequences of fossil remains" in layers of earth that might have "suggested a sequence of animal types" to ancient observers, explains classicist E. D. Phillips in "The Greek Vision of Prehistory." The fossil database was narrow and arbitrary: bones and teeth were randomly revealed by erosion, earthquake, or human grubbing. No fossil exposure in antiquity could begin to approach what Horner calls the "sufficient resolving power" provided by the systematic excavation of a thick and rich "geological column" comprising orderly strata of diverse evolutionary material laid down over distinct ages.[49]

Knowing these impediments to modern paleontology makes the Greco-Roman insights about fossils all the more extraordinary. The history of ancient paleontology is not a story of revolutionary leaps but a story of the remarkable stability of plausible ideas about big fossil bones, ideas which flowed from a mythical paradigm that flourished in the face of philosophical indifference and even hostility over centuries. This raises some intriguing questions: What advances might have been realized in antiquity had the philosophical community participated in interpreting the accumulating evidence of giant bones, as Apollonius of Tyana urged? Popular imagination and insights might have benefited from philosophical logic, and natural philosophy might have been prodded to deal with anomalous facts.

In the Middle Ages and later, some people believed that large fossil skeletons were the bones of fallen angels or remains of giants swept around the earth by Noah's Flood, while others argued that fossils were sports of nature created by the earth itself, that they fell from the stars or could be pried out of toads. How the fossil insights of the classical era were lost is another story. But it seems that there are real cultural gains to be made if academic science and folk knowledge can find ways to trust each other. An estrangement between scientists and the popular imagination seems to encourage illogical beliefs and deliberate hoaxes. The increasing frequency of relict monster hoaxes and sightings in the later Roman Empire may reflect rising tensions between philosophical skeptics and popular assumptions regarding the existence and extinction of mythical creatures, the subject of the next chapter.

The natural philosophers turn out to be a dead end for the recovery of ancient vertebrate paleontology. Despite suggestive fragments of Empedocles on extinct monsters and Theophrastus on petrifactions, and the provocative questions of Apollonius, what survives of philosophical writings strongly suggests that, for whatever reason, the philosophers opted out of the "unknowable" problem of giant bones. But inquiry proceeded without them, resulting in natural knowledge based in experience and expressed in geomyths. The myths were not a formal theory in the modern sense, of course, but as paleontologist Niles Eldredge observes, neither is the modern theory of evolution a "fact." Like the mythical paradigm, our own modern paradigm is "an idea—a picture" that allows us to explain observed facts.[50]

Using the mythical paradigms to explain the observed evidence of oversize vertebrate fossil remains in the ground, the ancient Greeks and Romans

1. recognized the *organic* nature of unexpectedly large fossil bones, and tried to visualize the origin, appearance, and behavior of the creatures when alive;

2. perceived that large fossil bones were very old, assigning them to gigantic species that had lived in the remote past when mountains were being formed, before the current human era;

3. identified fossil bones as remains of *extinct* creatures, no longer seen alive, and speculated on the causes of their disappearance, by lightning, earthquake, flood, or other catastrophe;

4. described taphonomic conditions and created a speculative taxonomy, genealogy, and geography of large, extinct creatures;

5. collected, measured, compared, displayed, and—as the Tiberius episode of the huge tooth (chapter 3) shows—even attempted to reconstruct giant remains;

6. accounted for observed *anomalies*—gigantic remains incongruent with human or known animal anatomy were identified as creatures of past eras.

This set of notable accomplishments is not a formal *science* of vertebrate fossils, but the ancient intuitions certainly constitute the preconditions necessary for scientific paleontologicial inquiry. Moreover, these perceptions were based on popular understandings of the culture's most ancient myths, not on advances in natural philosophy. For a millennium before the medieval era, a complex geomythological model allowed Greeks and Romans to incorporate paleontological discoveries into a perceptive and consistent vision of prehistory.

Wishful Thinking

Like modern cryptozoologists searching for relict dinosaurs in unexplored lands, some Greeks and Romans imagined that a few supposedly extinct creatures of the mythical era might have eluded destruction and still survived. Living or preserved half-human hybrids, such as Centaurs, were especially sought after. Sightings and specimens of these impossible creatures pushed the ancient insights about the earth's most remote past, when strange and wonderful animals did in fact walk the earth, into the realm of fantasy. But as the next chapter shows, the desire to somehow bring vanished creatures back to life is an essentially human dream, as ancient as Greek myths about fabulous beings and as modern as Hollywood films about the lost worlds of dinosaurs.

CHAPTER

6

Centaur Bones: Paleontological Fictions

At ancient Sikyon, they keep an enormous sea monster skull, with a statue of the God of Dreams standing behind it.
—*Pausanias*

Dreams and Beasts are two keys by which we are to find out the secrets of our nature.
—*Ralph Waldo Emerson*

The Pickled Triton of Tanagra

PAUSANIAS, the intrepid traveler steeped in Greek mythohistory, went to Tanagra (Boeotia) in about A.D. 150 to have a look at the town's famous marvel, a pickled Triton. The connoisseur of relics and enormous bones had already seen a smaller Triton preserved among the wonders of Rome. "But," wrote Pausanias, "the Triton of Tanagra would really make you gasp!"

Tritons, wrote Pausanias, "are certainly a sight, with their sleek, froggy green hair and bodies bristling with very fine scales like sharkskin." The curators at the Temple of Dionysus told Pausanias that this half-man, half–sea creature had been killed long ago by

6.1. Tritons of Tanagra. A pair of Tritons on a painted terracotta figure, late sixth century B.C., found at Tanagra, Boeotia, where Pausanias saw a preserved Triton in the Temple of Dionysus in the second century A.D. Drawing by author, after Shepard 1940, plate 2, fig. 12.

Dionysus, the god of wine, and washed ashore. But Pausanias favored a more naturalistic tale that dispensed with divine agency and explained why the head of this otherwise well-preserved specimen was missing: some said the Triton had been lured onto the shore by men offering wine and then decapitated. Pausanias recalled that the Triton he saw in Rome had greenish-gray eyes, gills behind its ears, a humanlike nose, and a wide mouth filled with fangs. The fingernails of both specimens were encrusted like seashells, and the lower body resembled that of a dolphin (fig. 6.1).

This same pickled Triton of Tanagra had been examined by Demostratus, the author of a treatise on sea monsters, nearly two hundred years before Pausanias. Demostratus served on a Greek provincial council that investigated the Triton for one of the early emperors in Rome (perhaps Augustus, Tiberius, or Claudius; they were especially interested in natural wonders). Demostratus's treatise on sea monsters is now lost, but the natural historian Aelian preserved his record of the official investigation at Tanagra. According to Demostratus, the specimen resembled typical Tritons in art. But the body was extremely ancient and fragile even then. The head was "so marred by time that it was unrecognizable" (by Pausanias's day it had completely disintegrated). Hard scales fell off the body when Demostratus touched it. Another member of the council "removed a small piece of the scaly skin and burned it. A noisome stench assailed the nostrils of the bystanders. We could not determine if this creature was born on land or sea," Demostratus concluded.

In myth, Tritons were fish-tailed humanoids, offspring of the sea god Poseidon and Nereids, or mermaids. As contemporaries of Typhon, Geryon, giants, Centaurs, and other monsters of the era ended by the Gigantomachy, Tritons were destroyed by the gods and by Heracles when he cleared the seas and rivers of monsters. According to the mythic paradigm, then, Tritons and Nereids should have vanished by the time the current race of humans appeared. And according to natural philosophers, such hybrid human-animal creatures never existed. But the unexplored depths of the sea were a place of timeless mystery, a realm where unexpected wonders might surface at any time. What if the body of a Triton slain by Dionysus were miraculously preserved—or what if relict Tritons still survived in uncharted waters? Something was found or fabricated to create exactly that illusion in Rome and Tanagra.

The notion was not irrational. The world's oceans are still a vast, dark space hiding strange, unknown species that periodically come to light. Indeed, even today knowledge of at least one monstrous creature of the deep is based on rumor, sightings, and deteriorating, incomplete remains that wash ashore. The mysterious *Architeuthis*—giant squid—has never been observed alive by scientists. This stupendous monster, said to grow to at least 60 feet (18 m) long, with lidless eyes the size of dinner plates, has long inspired terror and awe, as storytellers (most famously, Jules Verne), artists, and zoologists try to visualize its appearance and behavior.[1]

The Triton body viewed by Pausanias was most likely a counterfeit monster, produced to fill a gap between curiosity about never-seen mythical creatures and natural knowledge in antiquity. But if we take a closer look at this and other ancient hoaxes and compare them to some modern hoaxes about hypothetical or extinct creatures, it becomes clear that paleontological fabrications can have deeper and more positive meanings beyond mere deception.

Over a millennium of classical antiquity, widespread evidence accumulated to confirm that the world had once been populated by immense and strange creatures, just as the ancient myths described. The mythical paradigm that explained the observed facts of giant bones remained stable through the Roman era, even as natural philosophers continued to evade the evidence of anomalous remains. Meanwhile, several writers, such as Pausanias and Philostratus, tentatively suggested that naturalistic explanations for the oversize bones were at least conceivable. The same group of writers who recorded the discoveries of remarkable remains and instances of paleontological restorations also preserved intriguing evidence of paleontological hoaxes, in the form of spurious remains of giants, Centaurs, Tritons, and other half-human creatures of myth. These artificial wonders, exhibited alongside giant bones and heroes' relics, are usually viewed today as the creations of opportunists playing on gullibility or collective delusions. But I think the man-made marvels tell us something important about the role of imagination in creative paleontology.

Hovering between fantasy and certainty, hoaxes of all eras reflect the tensions that arise between popular belief and established science. Reports of live Centaurs and Tritons and displays of their "remains" not only challenged the ancient philosophers' rejection of mythical hybrids but tested the limits of popular credulity. These hoaxes grew out of the estrangement of philosophical inquiry from popular knowledge, but they also reveal the age-old human longing to bring to life creatures of lost worlds.

Just as today's tabloid press regularly reports sightings of living dinosaurs or sea monsters, sightings of Tritons and Nereids and other supposedly extinct creatures of myth stirred excitement in antiquity. For example, Pliny says that a delegation from Olisipo (Lisbon) arrived in Rome to inform the emperor Tiberius that a Triton had been spotted in a cave by the sea. They also stated that a dying Nereid, covered with hair or fine scales even on the parts that looked human, appeared on the same shore. The governor of

Gaul reported to Emperor Augustus a mass stranding of Nereids on the Atlantic coast, and Pliny himself heard from reliable sources that a Triton was sinking ships at night in the Gulf of Cadiz.[2]

Triton remains were exhibited in Tanagra and Rome. What were they? Several theories have been proposed. Sailors' observations of dugongs (aquatic sirenian mammals related to manatees) may have inspired folklore about mer-people, but the details related by Pausanias (gills, scales, green hair, fingernails, and fangs) do not fit a dugong. Peter Levi, a translator of Pausanias, thinks that the Tritons were "genuine monsters preserved by pickling," possibly deformed animals or abnormal humans embalmed and exhibited as mer-people. Could they have been naturally preserved, unfamiliar extinct creatures? Prehistoric animals can be naturally embalmed in briny petrochemical solutions, as occurred at Starunia, Galicia (Poland), where perfectly preserved corpses of Pleistocene woolly rhinos and mammoths were trapped in a steep of salt, crude oil, and the mineral wax ozocerite. But even if extinct beasts were found in a similar steep in antiquity, it's hard to visualize an Ice Age land mammal that could be mistaken for a Triton.[3]

The most plausible explanation is that the Tritons were deliberate trompes l'oeil, illusions fabricated by taxidermists skilled in ancient embalming techniques. Several other authors of the Roman era allude to realistic fabrications of fabulous creatures of the primeval past. The motives for making such marvels are complex. We already know of emperors' projects to reconstruct extraordinary animals from actual skeletal parts—recall the beached whale skeleton restored in the Roman circus and the bust of a huge hero created for Tiberius on the basis of a single enormous fossil tooth. Other sideshow oddities, such as Augustus's pair of giants and the Joppa sea monster, may have been composites of extraordinary bones, possibly fossils, or mummified parts of unrelated species sewn together. Lifelike effigies could also be molded from wax, clay, and wood, with scales, hair, feathers, hides, and fingernails added for realism. It's not difficult to imagine the manufacture of a Triton figure from parts of a large dried fish and a human mummy in Pausanias's time. That possibility gains support when we compare the wide variety of fake wonders produced by medieval and modern artificers.

Artificial Wonders Ancient and Modern

The cottage industry of counterfeiting creatures, from unicorns and mermaids to basilisks and dragons, has a long and colorful history. Consider, for example, the autopsy of a mummified mermaid that created a lucrative splash in London in the early nineteenth century. Skeptics determined that the body was ingeniously constructed from the lower half of a large dried salmon attached to the torso and head of a mummified monkey, with the forehead, nose, and ears from a human cadaver superimposed. The eyes were artifical, tufts of hair were glued in the nostrils, and the nails were replaced with shapely fingernails carved from horn. The breasts were stuffed with batting, and a deep fold hid the seam stitching the fishtail to the monkey's waist. The whole thing was stained antique brown with ochre and reinforced with wood. From the Middle Ages on, similar mer-people and dragons, traditionally known as "Jenny Hanivers," were fashioned out of rays, skates, lizards, monkeys, and other natural and artificial materials (fig. 6.2). Certainly the technology for making similar counterfeits existed in antiquity.[4]

The most notorious modern paleontological hoax, Piltdown Man, was perpetrated by persons unknown in 1912 and not exposed until 1953. To fool British paleontologists eager to find a "missing link" in England, the conspirators created a cunning composite of human and ape skulls, stained it dark brown to resemble a fossil ape-man, and planted it in a bone bed. Pseudoscientific exhibits of human-animal wonders much like the ancient Triton of Tanagra have long been staples of sideshow attractions. In the late twentieth century, a film purporting to document the autopsy of an extraterrestrial humanoid at Roswell, New Mexico, gained wildly popular cult status. As these examples show, hoaxes fill in conspicuous blanks in official knowledge with current fanta-

6.2. Fake Tritons. *Top*: drawing based on a skeleton and mummies exhibited in the sixteenth century. *Bottom*: drawing of a "sea monster seen near Rome in 1523." Konrad Gesner, 1558–1604. Dover Pictorial Archives.

sies. That's one reason why hoaxes and scientific fictions seem credible at first: the logic of what is missing determines their shape.[5]

Evidence in Pausanias and other authors suggests that the practice of making artificial remains, such as the Triton of Tanagra, emerged in the Roman era. Diodorus of Sicily (ca. 30 B.C.) described his fellow Romans enthralled by naturalistic models or *tableaux vivants* of mythical monsters such as Geryon and Centaurs in the theater. Manilius (ca. 10 B.C.) mentioned composite "animals with human limbs" displayed in the time of Augustus. Pliny viewed the body of an embalmed Centaur exhibited in the imperial palace and an Egyptian Phoenix (a marvelous bird reborn every five hundred years) that drew crowds in Rome in A.D. 47. Lucian (A.D. 180) wrote a fascinating exposé of a cleverly faked human-headed serpent exhibited at great profit for years around the Roman Empire by a charlatan named Alexander of Abonoteichus. In that case a tame snake of uncommon size was fitted with a lifelike false human head fashioned out of stiffened and painted linen. Horsehair strings made the mouth open and close and controlled a darting forked tongue. A couple of decades later, Aelian (ca. A.D. 200) referred to "artificers who falsify nature," artisans who assembled wax figures of mythical creatures to make us believe that nature could have "blended dissimilar bodies into one."

How were these man-made "natural" wonders viewed in antiquity? Reactions ranged from blind faith to irony. Some people were fooled, while others scoffed; some admired the artistic talent that challenged accepted knowledge and belief, and still others, like Diodorus, perceived deeper meanings in the artifice. The Phoenix exhibited in Rome, for example, was apparently recognized as a fraud by everyone who saw it. According to Lucian, gullible crowds were taken in by the human-headed snake puppet, but "gradually many sensible men began to detect the trickery of the show." Pausanias seemed to rate the sensational Tritons as thought-provoking entertainment; the sight moved him to muse on the many other weird animals he had seen and heard about. Aelian's coy response to the Triton at Tanagra was to defer to the oracle of Apollo at Didyma: "If the god says Tritons exist, then who can doubt it?"[6]

Satyr Sightings and Relict Centaurs

Ancient sightings of relict populations of prehistoric creatures from the age of giants, and ancient hoaxes that sought to create the material evidence of their existence, typically involved human-animal composites, such as Tritons, Satyrs, and Centaurs. (Modern hoaxes often tend toward half-human composites too: mermaids, Piltdown Man, Bigfoot, and extraterrestrial aliens.) If Tritons were the prototypical "half-human" aquatic relicts, Satyrs and Centaurs were their terrestrial counterparts. Satyrs (and Silenoi) were primitive half-man, half-goat or -equine beings. Hairy and prognathous, with upturned noses, full lips, pointed ears, and tails, some Satyrs also had short horns and hooves. According to an ancient legend, Satyrs could father Centaurs. Like other denizens of misty prehistory, these creatures should have disappeared by the time current humans appeared in Greece, but like Tritons, they stubbornly persisted in report and hoax.[7]

Remarkable fossil remains resembling Satyrs were discovered inside slabs of rock on Paros and Chios in Roman times (chapter 3). In the fifth century B.C., the hide of Marsyas, the famous Satyr slain by the god Apollo, was a tourist attraction near the source of the Meander River (southern Turkey), according to Herodotus and Xenophon. But it was rumored that elusive Satyrs still survived in remote landscapes in Egypt, Libya, India, northern Greece, and certain desert islands. Pausanias remarked that a live wild Satyr from Libya was exhibited in Rome, and Plutarch described the capture of a Satyr in what is now Albania. In 83 B.C., the Roman commander Sulla was about to sail from Dyrrhachium to Italy when his soldiers surprised a Satyr asleep in a sacred meadow, a place where fire flowed from the ground. The creature looked just like Satyrs in art and drama, and, when captured and presented to the Roman commander, he uttered a harsh whinnying bleat.

Satyr sightings lasted into the early Christian era. Saint Jerome, a contemporary of Saint Augustine, stated that the emperor Constantine (d. A.D. 337) traveled to Antioch to view the remains of a Satyr that had been specially preserved in salt.[8]

Jerome's report suggests that, like the Tritons of Rome and Tanagra, phony Satyrs were manufactured and displayed as tourist attractions. The techniques for creating the illusion of a real Satyr had, in fact, been perfected much earlier, in the fifth century B.C., for Greek performances of a genre of plays featuring Satyrs. Actors in Satyr plays wore lifelike masks, and perhaps other props, to impersonate live Satyrs. Recently, German archaeologists and museum restorers produced several full-scale reproductions of the theatrical Satyr masks depicted in classical Greek vase paintings and sculpture. The museum fabricators used only materials available in antiquity, such as hair and skins. The remarkably realistic results bring to mind modern scientific illustrators' artistic restorations of primitive human ancestors (figs. 6.3, 6.4). I think that similar handiwork could account for "live Satyr" exhibits in Rome and for preserved corpses of Satyrs, like the one shown in Antioch. The salted Satyr viewed by Constantine may have been a human mummy fitted with a realistic mask, a tail, and hooves.[9]

Thessaly, in northern Greece, was the traditional haunt of Centaurs. Half-human, half-animal chimeras of various sorts (Sphinxes, Minotaurs, Harpies, etc.) originated in very early Greek art to represent an array of primeval monsters. But among these, Centaurs seem to exert a powerful and enduring hold on the imagination. The stylized Centaur, a composite of a man and a horse, became a familiar and appealing figure in literary versions of legends and in classical paintings and sculpture.

"Strangely enough," reflects art historian Peter von Blanckenhagen, the classical Centaur's appearance "does not look grotesque and impossible. The combination of the body of a horse with torso and head of a man results in something visually unified and acceptable, as if nature may indeed create such monsters." He attributes this natural effect to the formal similarity between the strong equine neck and the muscular human torso. What emerges is a "convincing image of a powerful being transcending the power of horse and man separately." Thinking back to Guthrie's theory of

6.3. Satyr masks reconstructed from ancient Greek vase paintings of realistic theatrical Satyr masks. Photo courtesy of Gérard Seiterle.

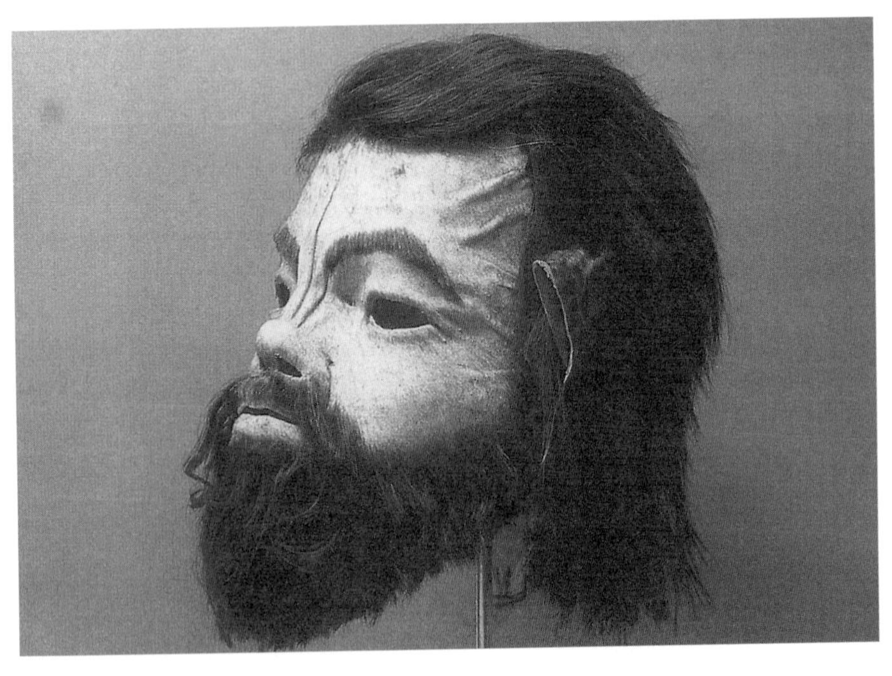

6.4. Satyr mask reconstruction, based on a fourth-century B.C. Greek vase painting. Photo courtesy of Gérard Seiterle.

the human urge to anthropomorphize (chapter 2), we may surmise that the Centaur was a particularly compelling way for the ancients to project human attributes back into the prehistoric past.[10]

But was the glorious Centaurs' existence limited to the mythical past? With expanding exploration in the Roman era, information about exotic lands and incredible animals, such as griffins and gorillas, was eagerly consumed in Rome. Not only was skeletal evidence piling up that giants and monsters had once populated the entire earth, but new zoological discoveries held out the possibility that some creatures of the mythic era—including the charismatic Centaur—might have escaped prehistoric destruction to roam unexplored landscapes.

As we saw in the previous chapter, hybrid beasts combining contradictory categories (including the bird-mammal griffins of chapter 1) were singled out by the circle of orthodox natural philosophers as impossible. But while Aristotle, his follower Palaephatus, and Lucretius heaped scorn on the viability of mixed species, especially Centaurs, writers like Aelian, Phlegon of Tralles, and others kept an open mind about seemingly incredible creatures, allowing the interplay of imagination and skepticism to fill in the blanks of the unknown. Aelian, for example, wondered whether time and nature might really have produced populations of such strange creatures, just as the myths claimed. If Centaurs were actually once prevalent in certain places and not just a figment of folklore, Aelian reasoned (echoing Empedocles) that they must have been at least a temporary fauna of the deep past.[11]

Palaephatus's authoritative assertion in the fourth century B.C.—that if Centaurs ever did exist, then they would still be seen alive—was given literal expression in a rash of Centaur sightings in the Roman period. During the reign of Claudius (A.D. 41–54), officials in Arabia declared that a small herd of Centaurs still inhabited Saune, a remote mountain wilderness infested by poisonous plants. Despite the danger, one of these "living fossils" was captured and transported to Egypt as a gift for the emperor. The Egyptians fed the wild Centaur a traditional diet of raw meat, but it could not tolerate the change in altitude and perished. The Egyptians had the corpse embalmed and shipped to Rome, where the emperor

Claudius exhibited the marvel in his palace. Pliny went with friends to see the spectacle: the Centaur was completely submerged in honey (a common preservative for transporting cadavers long distances). Immersion in honey would not only enhance authenticity but would prevent close inspection of the blurry illusion. Taking as an example Lucian's firsthand account of the phony human-headed snake exhibit, we can guess that the Centaur was shown in a dimly lit room, and that viewers were hurried along to the exit before they could examine the hoax.

Nearly a century later, through the reigns of nine emperors after Claudius, the embalmed Centaur of Saune could still be viewed, by special appointment, in the emperor Hadrian's imperial storehouse. Phlegon, the compiler of giant bone discoveries who served on Hadrian's staff (A.D. 117–138), examined the marvel himself. The Centaur was a bit smaller than what one might expect from classical Greek art, he observed, but it had a fierce face and hairy arms and fingers. The human rib cage and torso merged naturally with equine body and limbs, and its hooves were quite firm. The mane had originally been tawny, but the entire body had turned a very dark brown—owing, thought Phlegon, to the embalming process.[12]

In the Roman era, there were claims that ordinary mares could give birth to half-human colts, rare "throwbacks" that reverted to the forms of the long-vanished Centaurs. In about A.D. 50, for example, Claudius received a dispatch from provincial Greek authorities that a baby Centaur had been born in Thessaly. About fifty years later, Plutarch, perhaps inspired by that incident and by the Centaur of Saune, engaged in a provocative thought experiment about Centaurs and philosophers.

In his historical fiction "The Feast of the Seven Sages," Plutarch imagined how the world's wisest men would have reacted to the sight of a live Centaur in the sixth century B.C. On their way to dinner, the philosophers are interrupted by their host, Periander of Corinth, and asked to interpret the astounding birth of a Centaur to a mare in his stable. The sages peer at the newborn monster swaddled in leather, held in the arms of the stablehand. The half-human foal cries just like a human infant. One philosopher blurts out, "God save us," and turns his face away. Another worries about the symbolic meaning of the creature. But dinner is waiting, so another sage dismisses the newborn Centaur with a crude joke about the stablehand's unseemly romance with the mare. Notably, the three reactions—anxiety, pedantry, and mockery—that Plutarch attributes to the philosophers neatly recapitulate the attitudes of the natural philosophers toward popular geomyths and toward anomalous phenonomena (chapter 5).[13]

Centaur Excavations in Thessaly, Greece

In the 1980s, an ancient Centaur skeleton said to have been excavated in northern Greece traveled to museums in Wisconsin, Ohio, and Massachusetts. In 1994, the amazing remains were permanently enshrined in a faux marble showcase in the Library of the University of Tennessee, Knoxville. Peering into the glass-topped display, visitors gaze in awe at the complete skeleton of a Centaur embedded in a sandstone slab. A bronze arrowhead is lodged in the human rib cage. The label states that the specimen is one of three Centaur burials dated to about 1300 B.C., discovered in 1980 by Greek archaeologists at a site a few miles northeast of Volos, Thessaly (figs. 6.5, 6.6). The curator explains how Centaurs once roamed the forests of Thessaly until the appearance of humans, when they fled into more remote areas and gradually died out. "While Centaurs are considered by some scholars to be mythological, the discovery . . . at Volos forces us to reconsider this assumption."

The naturalistic fusion of the human and equine skeletons elicits an eerie sensation, a kind of "shimmering" between belief and doubt.[14] I think that the display allows us to spontaneously experience something like what many ancients must have experienced at the sight of the colossal, petrified bones of mammoths and samotheres, or the extraordinary bodies of Tritons and Centaurs, persuasively labeled as the actual remains of giant heroes and monsters of past worlds. The Centaur exhibit is just the kind of marvel that made Pausanias gasp and exasperated the natural philosophers.

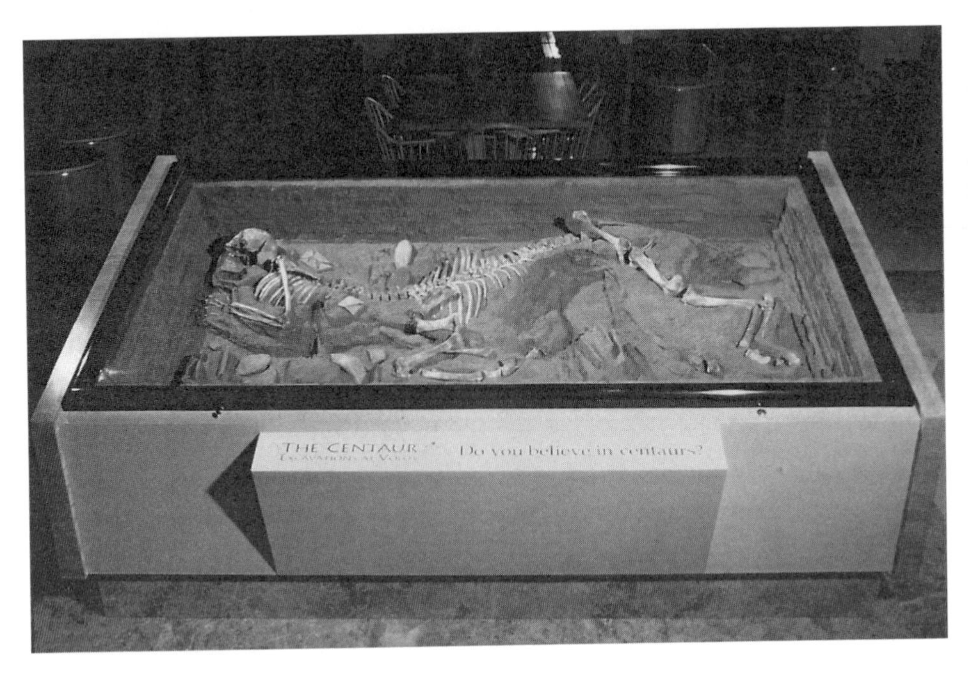

6.5. "Centaur Excavations at Volos." Artist: William Willers. Photo courtesy of Beauvais Lyons, Hokes Archives, University of Tennessee, Knoxville.

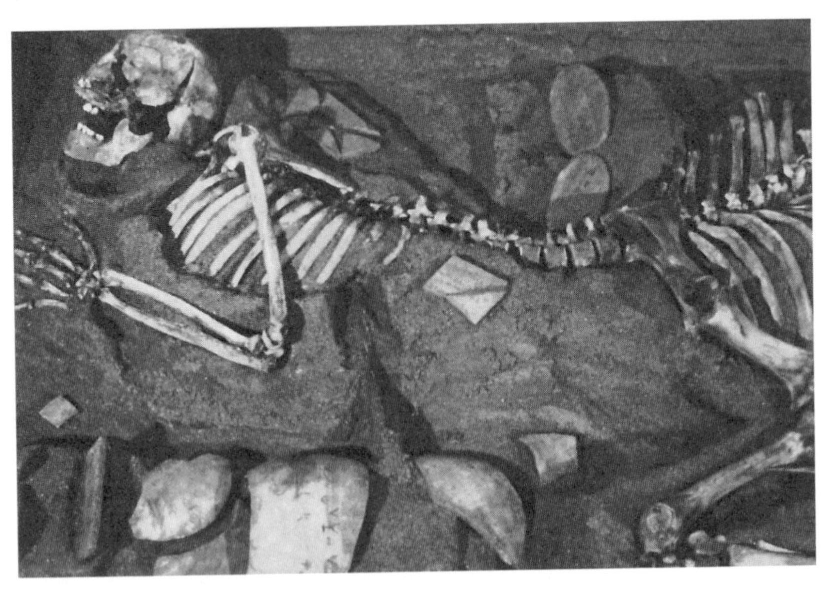

6.6. Centaur skeleton. Artist: William Willers. Photo courtesy of Beauvais Lyons, Hokes Archives, University of Tennessee, Knoxville.

The Centaur Excavation is an elaborate hoax, created by William Willers, a zoology professor and an artist. Combining the upper half of a deteriorating human skeleton from an anatomy classroom with the skeleton of a Wisconsin farm pony, Willers stained the bones dark brown with tea to make them look extremely ancient. He enhanced the effect of authenticity with pottery sherds and zooarchaeological labels, diagrams, and maps. Willers's intention was to present the mythical creature "at a level of realism that would allow a viewer to 'believe,' at least momentarily, in Centaurs." The physical evidence of a skeleton "elevates a mythical creature to an unprecedented level of reality in the mind of the viewer." "Whenever a person sees a skeleton," Willers notes, "a fleshing-out process occurs." "Fleeting images of the creature in the living condition pass before our eyes." Our imagination superimposes muscles, skin, and a face, and visualizes the creature in motion for an instant.

As an artist, Willers hoped that the ensuing imaginative effort would allow us to "gain some insight into the psychology of the ancients." As a scientist, Willers wanted to provoke people to question what their senses tell them and to treat authoritative texts with skepticism, adding a new twist to the old tension between popular beliefs and official knowledge. The multilayered impact of his Centaur assemblage stimulates dialogue among scientists, artists, and historians of popular culture.[15]

Indeed, the Centaur display embodies the point where age-old mythological imagination and scientific curiosity meet. Willers's combination of scientific and artistic sensibilities not only challenges his audience but complicates the very idea of the "hoax." He does not intend to deceive, but he does seek to startle. That jolt of uncertainty has a deeper function: it teaches us something profound about our sense of the possible.

Paleontological Fictions

Willers was not aware of the ancient Roman examples of fake Centaur remains when he created his hoax, which makes the coincidence all the more striking. The ancient and modern faux Centaurs demonstrate how complex hoaxes can transcend mere trickery or entertainment: they mark the intersection of imagination and the unknown. A better term for such hoaxing projects is "paleontological fictions," a phrase that captures the positive, creative, even nostalgic aspects of such improvised evidence. Indeed, every reconstruction of an extinct life-form—whether a scientific artist's drawing, a full-scale museum replica, or a computer-animated dinosaur in a Hollywood movie—is a form of paleontological fiction.[16]

If we view the sensational models of Tritons, giants, sea monsters, Geryones, and Centaurs as efforts to produce evocative replicas of creatures that either once existed and were destroyed in the mythological era or at least *might* have existed, then on some level these "hoaxes" emerge as ancient versions of modern artistic and scientific paleontological fictions: serious attempts to restore lifelike forms to the bare skeletons of extinct animals. Then as now, the gaps in what is known stimulate our curiosity and must be sketched in by audacious imagination.

Diodorus of Sicily expressed some of the complex motives for creating and appreciating paleontological fictions in his day. He acknowledged that even though no one believed that Centaurs or the monster Geryon ever really existed, nevertheless "we look with favor upon such products created for the theater" and public display. Why? Because by "our awe-inspired applause" we honor the artists' skillful re-creation and celebrate the mysterious forces of nature. We know that these creatures are the products of human artifice, he seems to say, but we willingly suspend disbelief in order to enhance our imaginative powers and to contemplate nature's hidden history.[17]

Diodorus's intuition that hoaxes can have potentially serious meanings is affirmed by the works of artist Beauvais Lyons, who is, like William Willers, a practitioner of the artistic-literary subculture of mock archaeology. The movement puts a self-conscious postmodern spin on the "cabinets of curiosities" of early modern Europe. But we can detect some affinities with the ancient Greco-Roman fascination with relics and marvels, both natural and man-made, too. Thought-provoking trickery offers "a timeless lesson," observes Lyons. "There is something beneficial to being duped." This kind of ironic deception, especially if at some point it reveals itself as fiction, has a didactic purpose. In Lyons's view, the "double-take" evoked by fabrications like Willers's Centaur skeleton heightens our "sensitivity to the world," a sentiment that Diodorus would have appreciated.[18]

Paleontological fictions also beckon us to confront the unknowable in nature, a realm that the ancient Greco-Roman geomyths and the later literary genre of paradoxography strove to encompass. According to some scientists, myths "meet a need in the psychological or spiritual nature of humans that has absolutely nothing to do with science." But others sense that myths express both scientific curiosity and creative speculation about the mysteries of nature. Acknowledging the concepts of unknowability and impossibility is difficult for many scientists caught up in classifying normative phenomena and rejecting popular misconceptions. That resistance may help explain why natural philosophers of antiquity ignored the "impossible" evidence of giant bones. Just as Willers's modern Centaur project breaks down the distinction between credulity and skepticism, certain ancient writers such as Diodorus, Pausanias, Philostratus, Plutarch, and Aelian seemed to intuit that challenging the imagination might be a key to coming to terms with the unknowable.[19]

Creative modern scientists harbor the same intuition. Doubt should be welcomed as "the essence of knowing," observed physicist Richard Feynman. Albert Einstein placed the experience of the mysterious or paradoxical "at the cradle of true art and true science," and he rated imagination more important than knowledge. The spirit of wonder "stands for all that cannot be understood," all "that can scarcely be believed," in the words of cultural historian Stephen Greenblatt. Never-ending wonder is the charge that energizes science, maintains philosopher Philip Fisher. Thomas Eisner, an eminent biologist, cherishes the times of "letting his mind go," as he fantasizes fanciful scenarios for imperfectly understood life-forms. That experience, he says, is "my favorite part of being a scientist." Elizabeth Fox Keller, a physicist, speaks of the difficult necessity of "reintroducing ambiguity into one's . . . world" through a "dialogue between dream and reality." Cognitive scientist Douglas R. Hofstadter hopes to recapture the "intoxicating sense of weirdness" and paradox evoked by nature's baffling enigmas, the "sense of mystery that lies at the core of science."[20]

According to these scientists, then, paradoxical evidence that questions certainty is not antiscientific. Imagination soars beyond current knowledge, and in doing so it becomes a crucial element of our potential for knowing and learning. Even missteps are valuable in creative inquiries. As Jack Horner points out, "paleontology often thrives" on "intriguing misinterpretations." If science makes major advances only when evidence is contradictory, and if flights of imagination are necessary to good science, then the creation of paradoxical, contradictory evidence is not always perverse but potentially useful. It taps into the kind of mythic imagination that, if integrated with scientific inquiry, could point to new levels of understanding the unknown. In this view, a hoax is a kind of hypothesis.[21]

The hypothetical constructions of mythical creatures may have served as ancient thought experiments about nature's hidden past. By giving tangible shape to never-seen animals, the hoaxers certainly exceeded the very real, natural evidence of remarkable skeletons of extinct mammals actually found in the earth. But their restorations satisfied people's craving to see and touch the lost world of bizarre creatures never actually glimpsed alive by human eyes.

Nostalgia for Lost Worlds

Creating models of animals that we can never see alive—either because they are long extinct or because they never existed—expresses a deep desire to bring to life either creatures that we *know* once existed or creatures that *might* have been even if they never were. Like the enduring Greek myths of the primeval world, paleontological fictions show us how the ability to suspend disbelief—if only for the duration of a mental double take—fulfills that powerful human need, a need that is crucial to all scientific endeavor, most obviously in the field of paleontology. In paleontology, "the stories will always be incomplete, ambiguous," remarks Jack Horner, "and we'll always want to fill in the missing pieces. *Because we can't help it.*" In order to create a dramatic vision of the lives of long-extinct creatures, we "develop fictional accounts of those lives." And in the process, the "interplay between fact and imagination never cease[s]." We need to visualize the strange, even the fantastic, if we are to comprehend and make real to ourselves the anomalous, tantalizing vestiges of vanished life-forms.

The ancient fascination with relics is mirrored in modern paleontology. Each and every bone that extinct creatures left behind (as well as claws, teeth, skin, fur, feathers, eggs, gastroliths, even droppings) is lovingly treasured, measured, enshrined in museums. The ancient nostalgia for the lost world of mythical creatures finds an echo in paleontologists' deep regret at the disappearance of the world of dinosaurs and mammoths. Peter Dodson ends his book about his beloved horned dinosaurs, the last dinosaurs on earth, with these words: "How keenly I regret their passing. Why did it have to end this way?" For Horner, the defining feature of our fascination for dinosaurs is the poignant realization that entire populations of glorious beasts have truly disappeared forever. Some may dream of bringing dinosaurs or mammoths back into existence through biotechnology, but the brute fact is that all those "large-scale aliens" are "long gone from this planet and will never return."[22]

That fact is simply too brutal for some to accept. The urge to resist the final loss of the prehistoric world challenges the fixity of the line between fact and fiction, science and myth. The dream of seeing a live mammoth was seriously pursued in experiments carried out in the 1960s by Soviet scientists; they implanted frozen mammoth eggs in modern female elephants in the vain hope of creating a baby woolly mammoth. Speculations about about cloning mammoths from preserved DNA material in frozen Siberian specimens continue to appear in scientific and popular media. These modern fantasies are not so far removed from the ancient notion that an ordinary mare might give birth to a Centaur foal.

In an incident that parallels the ancient discovery of a small herd of relict Centaurs in remotest Arabia, the *Explorers Journal* (1996) reported the electrifying news that mammoths still lived. Pursuing rumors of gigantic elephants in the forests skirting the Siwalik Hills of Nepal, an expedition led by John Blashford-Snell discovered a small herd of oversize elephants with domed foreheads like those of extinct mammoths. DNA tesing shows that the remarkable elephants, over 11 feet tall at the shoulder, have some genetic similarities to mammoths. They may be "throwbacks" to the prehistoric species *Elephas hysudricus* of 2 million years ago, previously known only from fossils in the Siwalik Hills.[23]

Discoveries like this fuel dreams among scientists and the general public of recovering or reanimating prehistoric beasts. In classical antiquity, people asked, What if Centaurs eluded destruction and still survived in remote mountain glens? What if one could capture a live Triton or Satyr, or at least recover a dead one as proof of their existence? Our modern questions are not so different: What if Mesozoic plesiosaurs were trapped in Loch Ness? What if a human-primate "missing link" still inhabited the slopes of the Himalayas? Could hulking Pleistocene chalicotheres somehow evade extinction in the jungles of Kenya? What if bizarre denizens of the Cretaceous seas still lurked in the deepest oceans? Can we resurrect Jurassic dinosaurs from DNA contained inside an amber-trapped mosquito? What if dinosaurs evolved into quasi-human beings?[24]

The Human-Dinosaur

Paleontologists are hardly immune to the desire to see incredible creatures that once existed or might have existed. Ivan A. Efremov, who discovered a huge new dinosaur, the *Tarbosaurus bataar* ("terrible lizard-hero," the Asian cousin of *T. rex*), on Soviet expeditions in the Gobi Desert in the 1940s, wrote a science fiction novel in which he imagined fossil hunters finding a cave wall covered in resin that had acted as a natural photographic film, capturing images of the living *Tarbosaurus.* The paleontologists gaze in wonder as the "rays of the setting sun bring out the apparition of this monstrous dinosaur . . . first in bone, and then in the flesh! The dream of every dinosaur paleontologist!" The group of eminent paleontologists who planned a memorial service to mourn the extinction of the mammoth, at the Hot Springs Mammoth Site in South Dakota (June 1999), strove to "evoke the Pleistocene" and its vanished behemoths. Their dream is to somehow, someday, restore the lost Ice Age megafaunas to the American West.[25]

Scientific reconstructions of extinct creatures known only from incomplete fossils are another form of paleontological fiction. Some innovative scientists dare to approach the vanishing point between myth and science, with results that verge on the mythical hybrid monsters of the ancients. In 1982, for example, paleontologist Dale Russell engaged in a bold thought experiment. What if dinosaurs had not all gone extinct in the Cretaceous disaster? What if one small, nimble, large-brained dinosaur survived and continued to evolve? And what if that dinosaur transmogrified into a humanoid? Russell and a taxidermist, R. Séguin, produced a museum-quality restoration of what they called the "dinosauroid hominid," based on detailed knowledge of dinosaur and human evolution.[26]

Russell and Séguin began by molding an extremely realistic, hypothetical skull and skeleton of plastic, nudging the evolution of *Stenonychosaurus*, a late Cretaceous troodont dinosaur, toward humanoid characteristics. They filled out the armature with musculature and skin, using the same latex, fiberglass, and epoxy resin that museum fabricators use to restore fossil skeletons. The four-foot-tall anthropomorphic dinosaur is a startling doppelgänger of the ancient man-made marvels, eerily reminiscent of Pausanias's Triton. Those large, staring eyes of an intelligent gecko, the wide turtle mouth, the scaly-skinned humanoid form, the three long, flexible fingers ending in formidable claws . . . the viewer experiences a disorienting jolt of *déjà* and *jamais vu*, as belief vies with incredulity. Like Willers's Centaur, Russell's uncanny "monster"

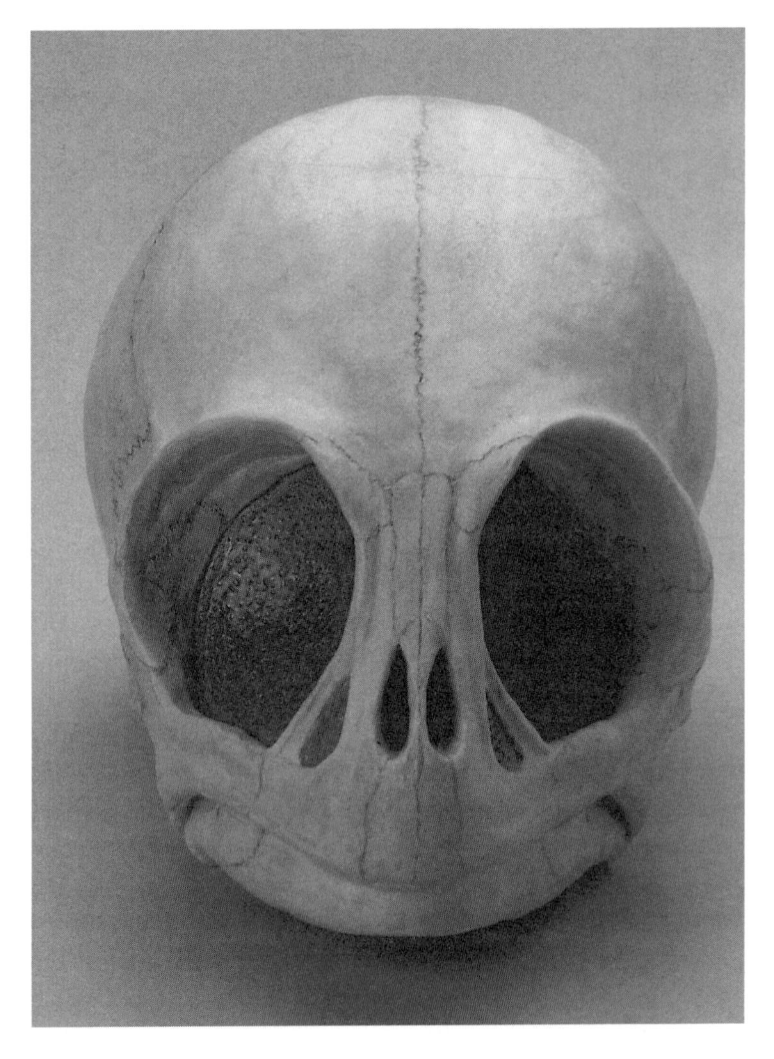

6.7. Hypothetical skull of the human-dinosaur hybrid, molded in plastic and resin by Dale Russell and R. Séguin. Dinosauroid images published with permission of the Canadian Museum of Nature, Ottawa, Canada.

allows us to feel for an instant the ancient experience of suddenly confronting impossible remains of an animal of immense magnitude or irrational form (figs. 6.7, 6.8).

If some Piltdown copycat were to plant Russell's realistic human-dinosaur skull in a fossil bed for an unsuspecting paleontolo-

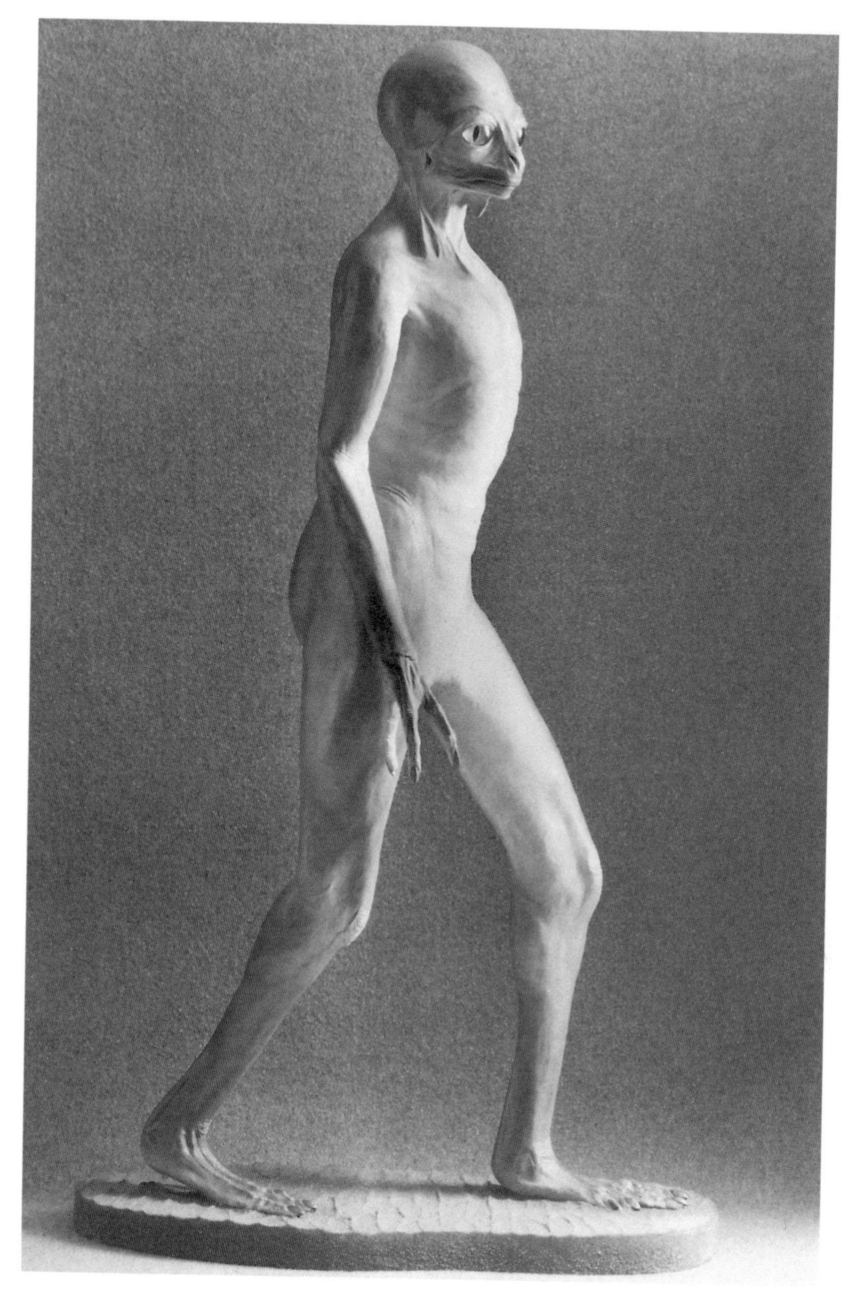

6.8. The Dinosauroid Hominid model created by Dale Russell and R. Séguin. Dinosauroid images published with permission of the Canadian Museum of Nature, Ottawa, Canada.

gist to find, it would certainly give the finder pause. In fact, Russell's hypothetical model *was* misinterpreted as a genuine dinosaur with amazingly advanced humanlike characteristics in Mary O'Neill's *Dinosaur Mysteries*, published in London in 1989. Clearly, Russell's imaginary creature had escaped the laboratory and soon began to turn up in unexpected places around the world. These developments were followed with rueful glee by a number of paleontologists attuned to popular culture and the tensions that engender hoaxing.

A gullible individual seeing the dinosauroid out of context might perceive it as some species of space alien. And sure enough, lurid photographs and artists' drawings of the creepy creature showed up in the tabloid press as evidence for the astounding discovery of "dinosaur people" and extraterrestrials. Since 1982, the image of Russell's dinosauroid has been replicated in countless popular media, from *Heavy Metal* magazine to Topps Bubble Gum collectors' cards (the *Dinosaurs Attack!* series of 1988). In at least one paleontology department corridor, a replica of Russell's original cast periodically appears in human clothing, inducing double takes in passersby. Nearly a decade after the fantasy was first publicized, a living, breathing, walking, talking dinosauroid appeared in a British television series on dinosaurs. Dinosaur specialist David Norman recounts how the museum fabricator cast a wax mold of the saurian body onto a tailor's dummy to make a full-color latex human dinosaur suit to be worn by a woman. The effect was complete when she donned three-clawed hands and a perfectly sculpted mask.[27]

This chain of alternating interpretations and reinterpretations of the dinosauroid encourages speculation that similar shifting levels of hoaxes and paleontological fictions occurred in antiquity. I think that some ancients familiar with fabricated monsters like the Triton and the Centaur of Saune might have appreciated the modern scientists' complex intentions in creating the half-human, half-dinosaur and its ironic life outside the laboratory. If the natural wonders created by William Willers and Dale Russell could be magically transported back to antiquity, we can be sure that temple curators and emperors would clamor for the exhibition rights. The natural philosophers would try to ignore such vulgar spectacles, and Diodorus, Pausanias, Aelian, and everyone else would applaud—and ponder.

Skeleton Keys

If, as Jack Horner believes, paleontology is "an attempt to bring all of natural history—that is, everything that has ever lived and died—into human awareness," then those who tried to include mythological Centaurs, Satyrs, and Tritons in the ancient paleontological record of the remains of monsters and giants believed that they were pursuing a similar concept.[28] The same mythic imagination that seeks to explain the presence of giant bones in the earth also creates paleontological fictions: these are precious keys for unlocking evidence that will never be completely understood. Just as historical narratives make the bare facts about the human past come alive, the union of mythic imagination and scientific reason gives meaning to the bare bones of prehistory.

APPENDIX

1

Large Vertebrate Fossil Species in the Ancient World

THIS APPENDIX enumerates prehistoric species (mostly Tertiary and Quaternary) reported in paleontological literature for locales where the ancients observed giant remains. This is not an exhaustive list of fossil fauna from each region, but it does indicate examples of large and unusual remains of extinct animal species that were likely to attract attention in antiquity. The survey begins in Greece and proceeds clockwise around the Mediterranean.

Greece

Attica (Pikermi): mastodons *Mammut (Z.) borsoni*, *Stegotetrabelodon*, *Choerolophodon*, *Deinotherium giganteum*, *Mastodon pentelici*; the huge herbivore *Chalicotherium*; giraffids *Palaeotragus* and giant *Helladotherium*; giant hyena; several rhinoceros species; ostrich; saber-toothed tiger; giant tortoise

Boeotia: mastodons, saber-toothed tigers, cave bears and lions, rhinoceroses

Chalkidiki, Kassandra Peninsula (ancient Pallene): mastodons, *Deinotherium giganteum*

Euboea: *Elephas* species, *Deinotherium*, *Archidiskodon* (or *M.) meridionalis*; rhinoceroses; large giraffids

Macedonia: elephants and mastodons, *Deinotherium*, *Choerolophodon*, *Palaeoloxodon antiquus*, *Zygolophodon (M.) borsoni*; cave bear *Ursus spelaeus*

Peloponnese (Arcadia, Olympia, Megalopolis, Tegea, etc.): mastodons, mammoths, *Deinotherium*(?), *Palaeoloxodon antiquus*, *Mammuthus (P.) primigenius*, *Mammuthus meridionalis*, *Anancus arvernensis*; cave bears and lions; ostriches; early horses; saber-toothed tigers; chalicotheres(?); woolly rhinoceros *Coelodonta antiquitatis* and forest rhinoceroses; giant cattle *Bos primigenius*

Thessaly: mastodons, *Anancus arvernensis*; cave bears[1]

Aegean Islands

Chios: mastodons, *Choerolophodon*, *Deinotherium*; large mammal species similar to those of Samos and the Turkish coast

Crete: giant cattle *Bos primigenius*; large, medium, and dwarf elephants *Elephas* cf. *antiquus*; *Hippopotamus amphibius*

Delos: medium-size elephant *Elephas antiquus*

Kos: *Mastodon* (or *Anancus*) *arvernensis*, *Mammuthus (E.) meridionalis*, *Palaeoloxodon antiquus*; *Hippopotamus major*; giant cattle *Bos primigenius*

Naxos: proboscideans *Apodemus* and medium-size *Palaeoloxodon antiquus*

Rhodes: medium-size (6 feet high) elephant *Elephas mnaidriensis*

Samos: mastodons *Deinotherium giganteum*, *Stegotetrabelodon*, *Mammut (Z.) borsoni*, *Choerolophodon pentelici*, *Mastodon pentelici*; giant giraffid *Samotherium*; rhinoceros *Dicerorhinus pikermiensis*; giant cattle *Bos primigenius*; huge herbivore *Chalicotherium*; large aardvark

Seriphos: pygmy hippopotamus and Pleistocene proboscideans[2]

Black Sea, Northern Turkey, Southern Russia and Ukraine, Sea of Azov, Taman Peninsula

Ancient Pontus, Moldavia (Moldava, ancient Dacia), ancient Phanagoreia (Taman Peninsula): gomphotheres, mammoths, *Mammuthus meridionalis*, *Elephas (M.) trogontherii (armeniacus)*, *Choerolophodon*, *Ambelodontinae*, *Mammuthus armeniacus*, *E.* cf. *antiquus*, *Archidiskodon me-*

ridionalis/A. planifrons, *Palaeoloxodon* cf. *antiquus*, *Deinotherium*; rhinoceroses, including the giant steppe rhino *Elasmotherium* and the huge embrithopod *Palaeoamasia kansui*; cave bears; giant deer; giant cattle *Bos primigenius*; huge herbivore *Chalicotherium*[3]

Western and Central Turkey

Ancient Troad, Chersonese (Gallipoli Peninsula), Caria, Phrygia, Ionia, Lydia, Lycia: mastodons and elephants, *Deinotherium*, *Choerolophodon*, *Anancus*, *Palaeoloxodon (E.) antiquus*; rhinoceroses; ostriches; cave bears; giant cattle *Bos primigenius*

Imroz (Imbros) island: mastodons, mammoths, *Mammuthus (E.) meridionalis*[4]

Gobi Desert, India, Pakistan

Gobi (Scythia): beaked dinosaurs *Protoceratops*, *Psittacosaurus*

Siwalik Hills: mastodons and elephants, *Mastodon sivalensis*, *Mastodon angustidens*, *Stegodon*, *Tetralophodon*, *Elephas (M.) primigenius*, *E. hysudricus*, *Deinotherium*; cave bears; rhinoceroses; anthracotheres, giant crocodile; giant tortoise; giant giraffid *Sivatherium*, *Giraffokeryx*

Baluchistan: *Indricotherium*[5]

Syria

Orontes Valley, ancient Cilicia: elephants, mastodons, gomphotheres, steppe mammoths *Elephas maximus*, *Mammuthus (E.) trogontherii*, *Mammuthus primigenius*(?), *Mammuthus armeniacus*, *Stegodon*; giant Miocene shark *Carcharodon megalodon*; giant cattle *Bos primigenius*, *Hippopotamus amphibius*[6]

Israel, Lebanon, Jordan

Ancient Palestine: mammoths, gomphotheres, *Mammuthus meridionalis*, *Stegodon*, *Deinotherium*, *Elephas planifrons*, *Mammuthus primigenius*,

Mammuthus trogontherii, Elephas maximus; rhinoceroses; *Hippopotamus amphibius*; cave bears; Eocene and modern whales; giant cattle *Bos primigenius*[7]

Egypt

Fayyum: Eocene and Oligocene mammals and whales, bizarre rhinoceros *Arsinoitherium*

Nile Valley: Pleistocene mammals, including elephants, giant cattle *Bos primigenius* and *Hippopotamus amphibius*

Wadi Natrun (ancient Nitria): Pliocene-Pleistocene mastodons, *Anancus osiris, Tetralophodon, Gomphotherium angustidens*; large giraffids *Samotherium africanum* or *Libytherium*(?), and *Sivatherium maurusium*[8]

Tunisia

Ancient Carthage, Utica, Teveste, Sbeitla: extinct proboscideans and modern elephant species, *Mammuthus africanavus, Tetralophodon, Gomphotherium angustidens, Anancus osiris*, small *Deinotherium*; Eocene whales; giant giraffids *Samotherium africanum* or *Libyatherium*(?) and *Sivatherium maurusium*[9]

Morocco

Ancient Lixus and Tingis (Larache and Tangier): Pliocene and Pleistocene elephants, mastodons, mammoths *Loxodonta atlantica, Mammuthus africanavus, Anancus osiris, Tetralophodon*; Miocene giraffid *Palaeotragus germaini*; rhinoceroses[10]

Spain

Elephants and mastodons, *Gomphotherium angustidens, Deinotherium*, and *Palaeoloxodon antiquus*; giant cattle *Bos primigenius*[11]

France

Ancient Gaul: *Deinotherium*; mammoths; Miocene dugong *Halitherium*; rhinoceroses; Jurassic dinosaurs
Jersey Island: mammoths and rhinoceroses[12]

Italy

Capri: *Mammuthus chosaricus*, rhinoceroses, cave bears
Malta: medium-size elephant *Palaeoloxodon mnaidrensis*, Miocene giant shark
Rome, Tiber Valley, Naples, Tuscany: mastodons, mammoths, *Deinotherium*, *Palaeoloxodon antiquus*, *Mammuthus meridionalis*, *Elephas primigenius*, *Anancus arvernensis*; giant cattle *Bos primigenius*; several rhinoceros species; cave bears
Sicily: dwarf and medium-size (6–7 feet at shoulder) elephants[13]

APPENDIX

2

Ancient Testimonia

THE following passages, gleaned from ancient Greek and Latin sources, present evidence of encounters with prehistoric fossils in antiquity: they refer to the remains of giants and monsters, large skeletons, footprints in stone, and petrified shells, marine organisms, and plants. This is not meant to be a comprehensive list of all classical literary allusions to fossils, but this collection from thirty-two authors (arranged alphabetically) gives a good idea of the wide-ranging chronology and geography of the ancient interest in remarkable remains that we recognize as fossils. Translations adapted from standard translations in Loeb volumes, unless otherwise noted.

1. Aelian (ca. A.D. 170–230)

On Animals 16.39. "Historians of Chios assert that near Mount Pelinnaeus in a wooded glen there was a dragon of gigantic size who made the Chians shudder. No farmer or shepherd dared approach the monster's lair. But a miraculous event allowed the discovery of how large it really was. During a violent lightning storm a forest fire destroyed the entire region of the wooded slopes. . . . After the fire, all the Chians came to see and discovered the bones of gigantic size and a terrifying skull. From these the villagers were able to imagine how large and terrible the brute was when alive."

17.28. "Euphorion says . . . that in primeval times Samos was uninhabited [except for] animals of gigantic size, which were savage and dangerous, called Neades. Now these animals with their mere roaring split the ground. So there is a proverbial saying current in Samos: 'So and so roars louder than the Neades.' And Euphorion asserts that their huge remains are displayed even to this day."

2. Antoninus Liberalis (second century A.D.)

Metamorphoses 41. The stone figure of a large wolf from mythical times was visible temporarily.

3. Apollodorus (first century A.D.)

Library 1.6. "Earth brought forth the giants, . . . who were matchless in the bulk of their bodies and invincible in their might, with terrible aspect. . . . Some say they were born at Phlegra [Italy], but according to others in Pallene [Chalkidiki, Greece]." Zeus "killed them with thunderbolts and Heracles shot them with arrows." Athena "threw Sicily on top of the giant Enceladus," while Poseidon "broke off part of Kos and heaped it on the giant Polybotes." Typhon "surpassed all the offspring of Earth. As far as the thighs he was of human shape and of prodigious bulk." Zeus fought him from Syria to Thrace and finally buried Typhon under Mount Etna, Sicily.

4. Augustine (A.D. 354–430)

City of God 15.9. "Some people refuse to believe that [in previous ages] men's bodies were of much larger size than they are now. . . . In those days the earth used to produce larger bodies. . . . As for the size of the bodies, skeptics are generally persuaded by the evidence in graves uncovered by the ravages of time, the violence of streams or various other occurrences. For incredibly large bones of the dead have been found in them or dislodged from them. On the shore of Utica [Gulf of Tunis,

Tunisia] I myself, not alone but with several others, saw a human molar so enormous that, if it were divided up into pieces the dimensions of our own teeth, it would, it seemed to us, have made a hundred of them. But that molar, I should suppose, belonged to some giant. For not only were bodies in general much larger than our own, but the giants towered far above the rest, even as in our own time, some few far surpass the size of others. Pliny the Elder, a man of great learning, declares that, as the world advances more and more in age, nature bears smaller and smaller bodies; and when he mentions that even Homer often regretted this, Pliny does not ridicule such statements as poetic fictions, but speaking as a recorder of the wonders of nature, assumes their historicity. But, as I have said, the size of ancient bodies is disclosed even to much later ages by the frequent discovery of bones, for bones are long-lasting."

5. Cicero (b. 106 B.C.)

Against Verres 2.4.46.103. "When King Masinissa [of Numidia] landed on the headland of Malta, his admiral stole the special tusks of astonishing size from the ancient Temple of Juno."

De Divinatione 13. A "figure resembling Pan" was found inside a slab of rock split open in a Chios quarry.

6. Claudian (b. ca. A.D. 370)

Rape of Persephone 1.154–59. Giants, including Enceladus, were "buried by the gods in the earth, under volcanoes," and they cause earthquakes.

3.332–43. Along the stream Acis, near Mount Etna in Sicily, is a place where Zeus set up trophies of the Gigantomachy. "Here hang the gaping jaws and monstrous hides, affixed to trees, their horrible faces still threaten, and heaped up on all sides bleach the huge bones of slaughtered dragons."

Gigantomachia 60–65. During the era of giants, "islands emerged from the deep, mountains lay hidden in the sea and rivers dried up or changed course," as mountain ranges collapsed and the earth sank.

91–103. Among the fallen giants, "Pallas was the first to change into stone."

7. Clement of Rome (ca. A.D. 96)

Recognitions 1.29. "The giants [were] men of immense bodies, whose bones of enormous size are still shown in certain places for confirmation of their existence."

8. Diodorus of Sicily (ca. 30 B.C.)

Library 1.21. Isis slew the giant monster-god Typhon at Antaeus on the Nile, a place named after the North African giant killed by Heracles.

1.26. "The Egyptian myths say that in the time of [the goddess] Isis there were beings of enormous size whom the Greeks call Giants, but in Egypt they were called [name missing]. Their colossal forms are depicted on temple walls being defeated by the allies of Osiris. Some say the giants were born of the earth when the origin of life, still rising up from the earth, was still recent. . . . They started a war against the gods . . . and were completely exterminated."

3.71–72. Dionysus fought the Amazons and the Titans in the desert of western Egypt, in the region of the oracle of Ammon (Siwa Oasis). At Zabirna, he destroyed a huge "earth-born monster called Campe," and "buried it in a huge mound, still seen in recent times."[1]

4.15. "The giants of Pallene [Chalkidiki, Greece] began a war with the gods and Heracles destroyed many there."

4.21. "According to certain writers, including Timaeus [a historian of Sicily, b. ca. 350 B.C.], the plain at Cumae, Italy, which was called Phlegraean [fiery], was inhabited by exceedingly large giants (Sons of the Earth). Here Heracles and the gods defeated the giants."

4.42. On the shore near Sigeum, in the vicinity of Troy, a monster appeared. "Those who made their living by the seashore and farmers who tilled the land near the coast were carried off." An oracle said it was sent by the god Poseidon; this Monster of Troy was slain by Heracles.

5.55. In the period "when Zeus was subduing the Titans," "giants appeared in the eastern part of Rhodes." "They were buried under the earth and men call them 'the Eastern Demons.'"

5.71. "Zeus slew the giants in Crete and Typhon in Phrygia [northern Turkey]." Giants were also slain in Pallene [Chalkidiki, Greece] and in Italy at Phlegra.

9. Dioscorides (first century A.D.)

Materia Medica 5.135. "Handsome white stones shaped like a glans are collected in Judaea for bladder medicine. They have intersecting lines as if made in a lathe [probably fossil echinoids, later known to the Crusaders as Judaean stones]."

10. *Greek Anthology* (seventh century B.C. to sixth century A.D.)

6.222 and 223. Fishermen dedicate huge ribs and bones netted in the sea to the gods.

11. Herodotus (ca. 430 B.C.)

The Histories 1.67–68. "The priestess at Delphi promised [the Spartans] victory over Tegea if they brought home the bones of Orestes." The retired cavalryman Lichas "found the body in Tegea" after a blacksmith recounted his discovery of "a huge coffin—10 feet long!" when digging a well in his yard. "'I couldn't believe that men were bigger than they are today, so I opened it—and there was the skeleton as big as the coffin! I measured it and then shoveled the earth back.' Lichas turned over in his mind the smith's account and came to the conclusion that this was the body of Orestes." He pretended to be an exile from Sparta and rented the smith's courtyard. "Then he dug up the grave, collected the bones, and took them away to Sparta."

2.12. "I have seen shells on the hills," evidence that "Egypt was originally an arm of the sea."

2.75. "In Arabia, opposite Buto [Bitter Lakes, east of Nile Delta?], I went to try to get information about flying snakes. On my arrival I saw skeletons and spines in incalculable numbers; they were piled in heaps, some were big and others smaller. . . . The place where the bones lie is a narrow mountain pass leading to a broad plain which joins the plain of Egypt."

4.82. In Scythia, "the natives show a footprint left by Heracles on a rock by the river Tyras [Dniester]. It is like a man's footprint but 3 feet long."

9.83. Long after the battle of Plataea (in Boeotia), people used to scour the fields for valuables and bones. "They found a seamless skull, and a jaw which had a continuous ridge of bone in lieu of teeth, and a 7.5 foot long skeleton."

12. Josephus (b. A.D. 37)

Jewish Antiquities 5.2.3. In the area around Hebron (Israel), the early Israelites wiped out "a race of giants, who had bodies so large and countenances so entirely different from other men, that they were amazing to the sight and terrible to the hearing. The bones of these men are still shown to this very day, unlike any credible relations of other men."

13. Lucian (ca. A.D. 180)

The Ignorant Book-Collector 13. The city of Thebes exhibits the bones of Geryon.

A True Story 1.7. A satiric account of Heracles' and Dionysus's footprints in rock: one is a hundred feet long.

14. Manilius (first century B.C.)

Astronomy 1.424–31. The giants were "broods of deformed creatures of unnatural face and shape" that appeared and were destroyed in the era "when mountains were still being formed."

15. Ovid (43 B.C.–A.D. 17)

Metamorphoses 3.397–99. When the nymph Echo died in a cave, the moisture gradually left her body and she became nothing but desiccated bones, "which they say turned to stone."

7.443–447. The giant ogre Skiron used to throw victims off a rocky cliff near Megara (Greece), until the hero Theseus threw Skiron over the precipice. After a very long period of time "his bones were tossed between sea and earth and finally hardened into rock."

9.211–229. Heracles threw Lichas over a cliff in Euboea, where he lithified. "As the moisture left his body he was transformed into hard, flinty stone." His "traces could be seen by sailors in a little reef that juts out into the sea."

15.259–67. "What was once solid earth is now changed to sea, and lands are created out of what was ocean. Seashells lie far away from the ocean and ancient anchors are found on mountaintops."

16. Palaephatus (fourth century B.C.)

On Unbelievable Tales 3. "An ancient story says that Cadmus killed a dragon and sowed the teeth like seeds [and] armed men grew from them." The truth is that Cadmus killed a king named Draco who owned some elephants' teeth among his precious possessions. Cadmus stored the teeth in a temple, but Cadmus's allies stole them and fled to various lands where they raised armies.

17. Pausanias (ca. A.D. 150)

Guide to Greece 1.17.6. "Kimon ravaged the island of Skyros and brought home the bones of [the giant hero] Theseus to Athens."

1.35.3. At Troy the sea washed out the grave of Ajax. "A Mysian told me about the size of Ajax. He said the sea had washed away a mound facing the beach. . . . He told me to judge the size of the body like this: the bones of the knees, what doctors call the millstones, are just the size of a discus for the boys' pentathlon."

1.35.5–6. "Off the city of Miletus is the island of Lade, with some baby islands broken off from it. They call one of these islets Asterios, and they say Asterios is buried on it. Asterios is the son of Anax, the son of the Earth [a giant]. That corpse is not an inch less than 15 feet tall. Another thing that surprised me was this: a mound broke open in a storm near a small city in upper Lydia called the Doors of Temenos, and some bones appeared. You would think they were human by their formation, but you would never have thought it from the huge size. At once the story got about everywhere that this was the body of Geryon. . . . They remembered how men plowing had already come across big cows' horns, and matched this to the legend that Geryon bred giant cattle. When I opposed them and said that Geryon was in Cadiz, . . . then the Lydian religious officials revealed the true story. It was the body of Hyllos the [giant] son of Earth, and the river Hyllos was named after him."

1.44.9. "Shelly marble is very white and softer than other stone; all through it are seashells."

2.10.2. "At the sanctuary of Asklepios [at Sikyon] . . . they keep an enormous sea-monster's skull."

3.3.6–7. "A Laconian named Lichas arrived in Tegea . . . when the Spartans were looking for the bones of Orestes because of an oracle. Lichas realized that they must be buried at a bronze worker's forge. . . . In later times an oracle told the Athenians to bring back Theseus's bones from Skyros. Kimon discovered the bones."

3.11.10. In Sparta "when Orestes' bones were brought from Tegea . . . this is where they buried him."

3.22.9. "At the sanctuary of Asklepios [at Asopos, southern Peloponnese] . . . the bones they worship at the training ground are enormous but human."

3.26.10. The bones of Asklepios's son Machaon, a hero killed at Troy, "were believed to have been rescued by Nestor and brought back from Troy to be buried at Gerenia" near Sparta.

4.32.3. In Messenia, "there is a monument to [the hero] Aristomenes, which they say is no empty memorial. When I asked them how they brought home his bones and from where, they claimed to have fetched them from Rhodes, on the orders of the Delphic oracle."

5.12.3. "I am not writing from hearsay: I have seen an elephant's skull [probably the skull of a living elephant] at the sanctuary of Artemis, near Capua [Italy]. . . . Elephants' size and shape are nothing like any other beast."

5.13.1–7. "There is this story that when the Trojan War was dragging on, prophets announced that Greece would never capture Troy . . . until they brought a bone of Pelops to Troy. So they sent for . . . a shoulder blade from the bones of Pelops to be fetched from Pisa [in Elis, at or near Olympia]. As they were returning to Greece from victory at Troy, the ship carrying Pelops's bone was wrecked by a storm off the coast of Euboea. Many years after the fall of Troy, a fisherman of Eretria called Damarmenos let down his net at sea and hauled up this bone. He was staggered by its size and hid it in the sand. But in the end, he went to Delphi to ask whose bone it was and what he should do with it. By coincidence, an Elean embassy was there asking for a cure for the plague. The Pythian priestess told the Eleans they needed to recover Pelops's bones and told Damarmenos to give them his discovery. He did so . . . and the Eleans made him and his descendants the guardians of the bone. In my time Pelops's shoulder blade had already crumbled away, in my opinion because it was hidden so long on the sea-floor, and the sea and other processes of time wore it away."

6.20.7. The bones of the wife of the giant hero Pelops, Hippodameia, "were brought from Midea [Dendra] to Olympia by command of an oracle."

7.1.3. The hero Tisamenos, Orestes' son, "was originally buried at Helike, but the Spartans brought the bones back to Sparta."

7.2.4. Ephesus was founded by the giant "Koresos, a son of the Earth."

8.29.1–4. Near the Alpheios River, "the Arcadians say the legendary battle of gods and giants took place here . . . and they sacrifice here to lightning storms and thunder. . . . Giants were mortal and not a divine race. . . . The story of giants having serpents instead of feet is ridiculous. . . . The Romans diverted . . . the Syrian river Orontes. . . . When the old riverbed drained they found a coffin more than 10 cubits [16 feet] long and the corpse was the size of the coffin, and human in every detail. When the Syrians consulted the oracle at Claros, it said that the giant was Orontes, an Indian. . . . India has beasts monstrous in size and appearance today."

8.32.5. At Megalopolis "there are some bones dedicated at the sanctuary of the Child Asklepios that seem too vast for a human being; there was a story about them that they belonged to one of the giants that Hopladamos recruited to fight for Rhea when she had the baby Zeus."

8.46.1, 5. "The tusks of the Calydonian Boar were taken [from Tegea] by the Roman emperor Augustus after the victory over Antony. . . . The keepers of the wonders say one of the boar's tusks is broken, but the surviving one is in the Emperor's Gardens in a Sanctuary of Dionysus [beyond the Tiber in Rome], and it measures 3 feet long."

8.54.4. "The Tegeans say that it was in Thyrea [east of Tegea] where the Spartans found the bones of Orestes."

9.18.1–4. On the road to Chalkis, "The Thebans point out the grave of the [giant] heroes Melanippos. . . . and Tydeus. . . . They have a grave of Hector [hero of the Trojan War]. . . . They say his bones were brought here from Troy, because of an oracle."

10.4.4. "In Phokis . . . beside a torrent is a mound where [the giant] Tityos is buried; the circumference is about 75 yards. They say that the verse in the *Odyssey* that Tityos sprawled 9 acres on the ground refers not to the actual size of Tityos but to the placename '9 Acres.' . . . Kleon, one of the Magnesians who sailed on the *Hermos*, used to say that people who have not encountered indescribable marvels in their own lives are skeptical about anything marvelous, but he was convinced that Tityos and other giants existed, because when he was at Cadiz. . . . he and his companions found a sea-giant washed up on the beach [beached whale carcass?]. It covered 5 acres." It was still burning, for "it had been struck by lightning by the gods."

18. Philostratus (ca. A.D. 200–230)

Life of Apollonius of Tyana 3.6–9. Northern "India is girt with dragons of enormous size; not only are the marshes full of them but the mountains as well and not a single ridge is without one. . . . The dragons of the foothills have crests, of moderate height when young but they grow with them and extend to a great height when they reach full size." The bodies of plains dragons are sometimes found with elephants, a great reward for hunters. Their tusks resemble those of swine, but more twisted and sharp. "They say that in the skulls of the mountain dragons are stored stones of flowery colors that flash out all kinds of hues." They tell us that "a great many dragons' skulls are enshrined" in the center of the great city of Paraka (Peshawar?) close by the mountain.

5.16. Apollonius states, "I agree that giants once existed" because "gigantic bodies are revealed all over earth when mounds are broken open," but "it is mad to believe that they fought the gods."

On Heroes 7.9. When the Phoenician merchant doubts that ancient heroes "were 15 feet tall," the grape-farmer of the Gallipoli Peninsula replies, "My grandfather said that the grave of Ajax was destroyed by the sea [near Rhoeteum] and a skeleton came to light about 16 feet tall. He said that the emperor Hadrian laid it out for burial, embraced and kissed some of the bones, and built a tomb for it at Troy."

8.3–14. "Orestes' body was discovered by the Spartans in Tegea—it was 10 feet tall." The great skeleton "in the bronze horse, buried in Lydia and miraculously revealed to shepherds by an earthquake, was larger than a man could imagine. But even if you doubt these stories because they happened so long ago, it's hard to contradict events of our own day. Not long ago, an excavation on the banks of the Orontes River brought to light Aryades, a giant 45 feet tall."

"And less than fifty years ago, Sigeum, across the Hellespont, disclosed on one of its headlands the body of a giant. . . . I myself sailed to Sigeum to witness exactly what had happened to the land as well as the giant's size. Many others sailed there too, from the Hellespont, Ionia, and all the islands and Aeolia. The huge body was visible on the cape for two months. Until the oracle cleared things up, everyone proposed different explanations." It "measured 33 feet, and lay in a rocky cave, with the head inland and the body extending out to the end of the cape. . . . The skeleton was human."

"About four years ago, Hymnaios [of Skopelos], a friend of mine," and his son found "a similar wonder on the island of Ikos [Alonnisos]. While digging up some vines, the earth rang under the shovel. When they cleared away the dirt, there lay exposed a skeleton 18 feet tall." An oracle said it was a fallen giant, so they reburied it.

"The largest giant of all was the one on Lemnos, discovered by Menecrates of Steira. I myself sailed over last year from Imbros to see it. It was no longer possible to see the bones in their proper position because the backbone lay in pieces—separated by earthquakes, I imagine—and the ribs had been wrenched away from the vertebrae. But as I examined the bones, all together and one by one, I got an impression of terrifying size, impossible to describe. The skull alone held more than two Cretan amphoras!"

"There is also a cape on southwest Imbros called Naulochos . . . where a piece of land broke off and carried with it the skeleton of a huge giant. If you don't believe me, we can sail there. The body is still exposed there and it is a short trip."

"But you must not believe all this, until you sail to Kos, where the first earth-born Meropes [daughters of giants] lie, until you see the bones of . . . Hyllos in Phrygia, or the Aloadae [young giants slain by the gods] in Thessaly—they were actually 54 feet long." At Phlegra in Italy "they have made wonders of the bones of Alkyoneus and other giants laid low there, buried under Mount Vesuvius."

"And at Pallene, which the poets call Phlegra [Chalkidiki], the earth still contains the skeletons of many giants, since that was their camp. Thunderstorms and earthquakes still bring the bones to the surface. . . . After Heracles killed Geryon, the largest creature he ever encountered, in Erythia [Tartessus, Spain], he dedicated the bones at Olympia so that his feat would not be dismissed as incredible."[2]

19. Phlegon of Tralles (ca. A.D. 130)

Book of Marvels 11–19 (finds of giant bones). "In Messene not many years ago . . . it happened that a storage jar made of stone broke apart in a powerful storm when it was pounded by much water." Inside were large bones with three skulls and two jawbones with teeth. "Idas" was inscribed on the jar. "The Messenians prepared another storage jar at public expense, placed the hero in it," and carefully tended the relics. "They perceived that he was the man that Homer" described as "the strongest of all men on earth at that time."

"In Dalmatia [former Yugoslavia] in the so-called Cave of Artemis one can see many bodies whose rib-bones exceed 11 cubits [ca. 15 feet; 4.5 m]."

"In the time of Tiberius there was an earthquake in which many notable cities of Asia Minor utterly disappeared, which Tiberius rebuilt at his own expense." At the same time, quakes struck Sicily and the toe of Italy and "affected numerous peoples in Pontus," south of the Black Sea. "In the cracks of the earth huge skeletons appeared. The local people were reluctant to move them, but as a sample they sent to Rome a tooth from one of the bodies. It was greater than a foot long. The delegates showed it to Tiberius and asked if he wished the hero to be brought to him. Tiberius devised a shrewd plan, such that, while not depriving himself of a knowledge of the hero's size he avoided the sacrilege of robbing of the dead. He summoned a certain geometer, Pulcher by name, and bade him fashion a face in proportion to the size of the tooth. The geometer estimated how large the entire body as well as the face would be by means of the weight of the tooth, hastily made a construction, and brought it to the emperor. Tiberius, saying that the sight of this model was sufficient, sent the tooth back to where it had come from."

"One should not disbelieve the foregoing narrative, since in Nitriai [Nitria, Wadi Natrun] in Egypt" huge skeletons are exhibited that "are not concealed in the earth but are unencumbered and plain to see. The bones do not lie mixed together in disorder but are arranged in such a manner that a person viewing them recognizes some as thigh bones, others as shin bones, and so on with the other limbs. One should not disbelieve in these bones, either, considering that in the beginning when nature was in her prime" she produced substantial creatures but "just as time is running down so also the sizes of creatures have been shrinking."

"I have also heard reports of bones in Rhodes that are so huge that in comparison the human beings of the present day are greatly inferior in size."

The Athenians fortified "a certain island near Athens. As they were digging foundations for the walls, they found a coffin 100 cubits long." It contained a body in proportion to the coffin's size. The inscription read "I, Makroseiris, am buried on a small isle after living a life of 5,000 years."

"Eumachos says in his *Geographical Description* that when the Carthaginians were surrounding their territory with trenches they found in the course of their digging" two giant bodies, one 24 cubits, the other 23.

"Theopompus of Sinope says in his work *On Earthquakes* that in the Cimmerian Bosporus [Taman Peninsula] there was a sudden earthquake" that "tore open a ridge and discharged huge bones. The skeletal structure was 24 cubits." The local people "cast the bones into the Maiotis [Azov] Sea."[3]

20. Plato (ca. 429–347 B.C.)

Republic 2.378d–e. "The story goes . . . there was a violent thunderstorm, and an earthquake broke open the ground and created a chasm where a shepherd was tending his flock. . . . He went down into it and in addition to many wonders of which we are told, he saw a hollow bronze horse with little window-like openings. Inside he saw a body of larger than human size."

21. Pliny the Elder (ca. A.D. 77)

Natural History 2.226. Wood is petrified ("covered with a film of stone") in the River of the Cicones in Thrace and in the Sabine region of Italy at the Veline Lake at Picenum. Petrified logs with bark are seen in the River Surius in Colchis (east of the Black Sea). In "the Sele River beyond Sorrento [Italy] twigs and leaves are petrified."

7.73–75. "It is a matter of observation that the stature of the entire human race is becoming smaller. . . . When a mountain in Crete was cleft by an earthquake, a skeleton 46 cubits long was found, which some people thought must be that of [the giant hunter] Orion and others of [the young giant] Otus. The records attest that the skeleton of Orestes dug up at the command of an oracle measured 7 cubits. Moreover, the famous bard Homer nearly 1,000 years ago never ceased to lament that mortals were of smaller stature than in the old days." Augustus preserved the bodies of two giants (Secundilla and Pusio) over 10 feet tall at Sallust's Gardens in Rome.

8.31. "Exceptionally large specimens of tusks can be seen in temples."

9.7. "Alexander the Great's admirals stated that the Gedrosi who live by the river Arabis [Baluchistan, Pakistan] make the doorways of their houses out of monsters' jaws and use their bones for roofbeams, many of them 40 cubits long."[4]

9.9. "An embassy from Lisbon reported to Tiberius that a Triton had been seen playing a conch shell in a certain cave. . . . A Nereid was seen on the same coast. . . . The governor of Gaul wrote to Augustus that a large number of dead Nereids were seen on the shore."

9.10–11. "During the rule of Tiberius, in an island off the coast of the province of Lyons [northern France] the ocean revealed more than 300 monsters of marvelous variety and size." Among them were elephants, strange rams, and Nereids. "The skeleton of the monster to which Andromeda in the story was exposed was brought by Marcus Scaurus [ca. 58 B.C.] from the town of Joppa in Judaea and shown at Rome among the rest of the marvels of his aedileship. It was 40 feet long, the height of the ribs exceeding the elephants of India, and the spine 1.5 inches thick."

28.34. "At Elis there used to be shown a shoulderblade of Pelops, which was stated to be of ivory" and had healing powers.

31.29–30. Petrifying springs exist in Euboea, Colossae (south of the Meander River), and Eurymenae (northern Greece). "In the mine at Skyros, there are trees turned to rock, branches and all."

35.36. A white stone called "paraetonium" from Paraetonium (on the coast of Libya), "is said to be sea foam hardened with mud and that is why tiny shells are found in it. The same substance occurs at Cyrene [coast of Libya] and on Crete."

36.14. In Paros "there is an extraordinary tradition that once, when stone-breakers split a block with their wedges, a likeness of Silenus was found inside the rock."

36.81: "All around [the Egyptian pyramids] far and wide the sand is shaped like lentils [nummulite fossils]; the same sand is found on most of the African side of the Nile."

36.134: "Theophrastus states that fossil ivory mottled black and white is found, that bones are produced from the earth, and that stones resembling bones come to light."

36.134–35: "Around Munda" in southern Spain "there are stones that contain the likeness of a palm branch, which appears when they are split open."

36.161: In Segobriga, Spain, "wild animals are petrified in deep shafts where selenite is mined. Over a single winter, their bone marrow is replaced by sparkling selenite crystals." Selenite (a form of gypsum) "is formed by petrifying exhalations of the earth."

37.42–46: "Amber is formed of resin seeping from the interior of pine trees. . . . The exudation was hardened by cold or heat or by the sea, then washes up on shores of the North Sea. . . . That amber originated as sap is proved by the visible presence of gnats, ants, and lizards trapped inside as it hardened."

37.150: "Bucardia stones resemble the heart of an ox and are found only at Babylon [*Protocardia* mollusk fossils resemble large stone hearts, known as bull's hearts in later European fossil lore]."

37.167: "Hammonis cornu or horn of Ammon [iridescent ammonite fossils], among the most sacred stones of Ethiopia, has a golden yellow color and is shaped like a ram's horn."

37.170: "Idaian Dactyloi or Fingers of Ida [mythical iron-working gnomes of Mount Ida in the Troad] are stones the color of iron that reproduce the shape of a human thumb [probably belemnites, the fossilized guards of extinct cuttlefish, which are solid cylinders with pointed ends]."

37.175. A stone called "nipparene, which gets its name from a city in Persia, is like the tooth of a hippopotamus [fossil hippopotamus tooth?]."

37.177: A stone "called 'ostritis' or 'oyster stone' owes its name to its resemblance to an oyster shell."

37.182. A stone called " 'spongitis' or 'sponge stone' is absolutely true to its name. . . . 'Syringitis' or 'pipe stone' is a hollow tube resembling the length of a stalk between two of its joints [probably crinoid stem fossils]."[5]

37.187–93. Various stones resemble the trunk of an oak, or look as though they contain millet grains, fishbones, ivy leaves, scorpions, tree roots.

22. Plutarch (ca. A.D. 50–120)

Greek Questions 56. Why is Panaima (Samos) called the Blood-Red Field? "Because the Amazons in flight from Dionysus crossed from Ephesus to Samos. But he crossed and joined battle with them and slew many of them at that place, which owing to the amount of blood that flowed, those who saw it marvelled and called it the Blood-Red Field. Now some of Dionysus's elephants are said to have died near Phloion [Hard Crust of the Earth] and their bones are shown there. But some say also that Phloion was rent [and collapsed] on some great creatures [Neades?] in their time as they uttered great and piercing cries."

Isis and Osiris 40. "Egypt was once all sea, which is why to this day people find an abundance of mollusc shells in its mines and on its mountains."

The Names of Rivers and Mountains and the Things Therein (Pseudo-Plutarch). "On a mountain near the river Meander [Turkey] are stones that resemble cylinders [crinoid stems], which children . . . gather and dedicate to the Temple of the Mother of the Gods." "In the river Erotas [Sparta] is a stone shaped like a helmet [helmet echinoids? *Gryphaea* bivalves?] . . . many of these are dedicated in the bronze Temple of Athena."

Kimon 8. "Kimon learned that the ancient hero Theseus . . . had been treacherously murdered in Skyros." The "Athenians had been given an oracle commanding them to bring back the bones of Theseus," but the Skyrians "denied the story and forbade any search. Kimon attacked the task with great enthusiasm and after some difficulty located the sacred spot. He had the bones placed in his trireme and brought them with great pomp and ceremony to Athens."

Theseus 6. The distant era of heroes "produced a race of humans who for sheer strength . . . were indefatigable and far surpassed our human scale. . . . Some of these creatures were destroyed by Heracles." But others were killed by Theseus, such as the huge Sow of Crommyon and the giant ogre Skiron.

36. In about 475 B.C., "the Athenians consulted the Delphic Oracle and were instructed to bring home the bones of Theseus, give them an honorable burial in Athens, and guard them as sacred relics. . . . Kimon captured Skyros and made it a point of honor to find the spot where Theseus was buried. He caught sight of an eagle pecking and scratching at a mound of earth." He ordered his men to dig there, and they found "a skeleton of a man of gigantic size and lying beside it a bronze spear and sword. When Kimon brought these relics home on

his trireme, the Athenians were overjoyed and welcomed them with magnificent processions and sacrifices. . . . The hero now lies buried in the heart of the city."

Sertorius 9. The Roman commander Sertorius captured Tingis [Tangier, Morocco] and learned that "the city was the burial place of the giant Antaeus. Skeptical because of the enormous size of the burial mound, Sertorius had his men dig it up. It is said that the skeleton was 60 cubits long. Sertorius was dumbfounded and, after offering a sacrifice, he reburied the skeleton. He personally confirmed the local story and paid honors to Antaeus."

23. Procopius (ca. A.D. 540)

De Bello Gothico 5.15.8. The tusks of the Calydonian Boar at Beneventum, Italy, are "well worth seeing, measuring not less than three hand spans around and having the form of a crescent."

24. Pseudo-Aristotle (date contested)

On Marvelous Things Heard 834a23–32. In mines near Pergamum (Turkey), bones are turned to stone.

838a11–14, 29–35. "In Cumae, Italy, bones are turned to stone. At Iapygia [heel of Italy], legends relate that Heracles fought the giants, and ichor [burning naphtha or petroleum] flows there in great abundance. . . . Near Pandosia, footprints of Heracles are shown."

846b3–6, 21–25. "On Mount Sipylos they say there is a stone like a cylinder [crinoid stems], which children place in the temple of the Mother of the Gods." "In the river Nile, they say that a stone like a bean [nummulites] is found," which has medical and magical uses.

25. Quintus Smyrnaeus (third century A.D.)

The Fall of Troy 3.724–25. The hero Achilles' "bones were like an ancient Giant's."

12.444–97. During the Trojan War, a pair of "fearsome monsters, the brood of Typhon" emerged from the rocky cliffs of Tenedos, a tiny island across from Troy (Bozcaada, Turkey), and ravaged the city until they "vanished beneath the earth."

26. Solinus (ca. a.d. 200)

Collectanea rerum memorabilium 1.90–91: "As for the hugeness of men in olden times, the relics of Orestes are proof. His bones were found at Tegea by the Spartans on the information from the Oracle and we are assured that they were 7 cubits long. Also, writings register events in ancient times. [For example, during the Roman war against the pirates in Crete, ca. 106–70 b.c.], rivers flooded outrageously and broke up the ground. After the water receded, among the many clefts in the ground was found a skeleton of 33 cubits. [The Roman commanders] Lucius Flaccus and Metellus were amazed at what they thought was a fable, so they went to see for themselves the wonder revealed by the torrent."

9.6: "At Phlegra [ancient Pallene, Kassandra Peninsula, Chalkidiki], before there were any men there, the story goes that a battle was fought between the gods and the giants. . . . Great proofs and tokens of that war have and continue to appear to this day. Whenever the streams rise with rainstorms, the waters overflow their banks and flood the fields, they say that through the action of the water are discovered bones like men's carcasses but far bigger. Due to the immeasurable hugeness of the bones they are reported to have been the monstrous bodies of the army of giants." In Thessaly, there is evidence of Deucalion's Flood: "in dark caves in the hollowed out hills, [the receding waters] left behind shells and fishes and many other things which are cast up by the rough sea. Although these places are inland they resemble the seashore."

27. Strabo (b. 64 b.c.)

Geography 1.3.4. "Eratosthenes [of Cyrene, Libya] points out that far inland, 2,000–3,000 *stadia* from the sea, are vast numbers of cockle, oyster, and scallop shells. . . . By the Temple of Ammon [oracle of Siwa, southwest of Nitria, Egypt] and along the road to it for 3,000 *stadia* there are huge masses of oyster shells [Ammon was associated with ammonites]. . . . Xanthos of Lydia [Turkey] also saw seashells, cockles, and scallops a long way from the sea. . . . leading him to conclude that the land had once been sea."

5.4.6. At Puteoli, Baiae, and Cumae (Italy), the water "has a foul smell, because it is full of sulphur and fire. And some believe that it is for this reason that the Cumaean countryside was called 'Phlegra' [burning fields] and that it is the wounds of the fallen giants, inflicted by Zeus' thunderbolts, that pour forth those streams of fire and hot water."

6.3.5. In Italy, near Leuca is a fountain of "malodorous water; the mythical story is told that those of the Giants who survived at the Campanian Phlegra and are called the Leuternian Giants were driven out by Heracles, and on fleeing hither for refuge were hidden in the earth, and the fountain gets its malodorous stream from the ichor of their bodies [These regions have sulphurous volcanic phenomena and large fossil remains]."

7.25, 27. In ancient Pallene, Chalkidiki, later called Cassandreia (Kassandra Peninsula), "writers say that in earlier times the giants lived here and that the country was named Phlegra. The stories of some are mythical but the account of others is more plausible, for they tell of a certain barbarous tribe that occupied the place but was destroyed by Heracles."

11.2.10. At Phanagoreia, on the Taman Peninsula, between the eastern Black Sea and the Sea of Azov, "there is a temple of Aphrodite," where according to a certain myth the giants attacked the goddess. But she called on Heracles for help. He hid in a cave there and she drove the giants into the cave where they were slaughtered by Heracles.

13.1.30–32. Ajax's grave is at Rhoeteum, "near a low-lying shore," by Sigeum where rivers silt up the coast, and across from Elaeus, where the hero Protesilaus's shrine lies. There was a sea monster (the Monster of Troy) at the promontory of Agameias (near Sigeum).

16.2.7. The river Orontes was "formerly called Typhon, a dragon of myth. Struck by bolts of lightning, Typhon fled underground and formed the bed of the river."

16.2.17. "Posidonius tells of the fallen dragon [stranded whale?] seen on the plain at Macras [near the river Lycus, Syria]. It was nearly a *plethrum* [100 feet] long and so thick that men on horseback on either side could not see across to one another. The open jaws could contain a man on a horse and the scales were larger than a shield."

17.1.34. "At the base of the Pyramids in Egypt are found heaps of stones like lentils in form and size [nummulites]. They say that what was left of the workers' food has petrified. . . . I have seen similar lentil-shaped pebbles of porous stone in my own country [Pontus by the Black Sea]."

17.3.8: "Gabinius, the Roman historian, . . . tells marvelous stories of Maurusia [Morocco], for example, he tells a story of a tomb of Antaeus near Lynx [Lixus, modern Larache, south of Tangier], and a skeleton 60 feet in length, which, he says, Sertorius exposed to view, and then covered again with earth."

17.3.11. In Libya, "there are deserts of oyster shells and mussel shells in great quantities."

28. *Suda* (ancient scholarship compiled ca. tenth century A.D.)

S.v. *Menas.* A "great quantity of giant bones was found under the foundations of the Church of Saint Menas in Constantinople [Istanbul]." The Byzantine emperor Anastasius (A.D. 491–518) placed them in his palace.[6]

29. Suetonius (b. A.D. 70)

Augustus 72. "At Capri, Augustus collected the huge skeletons of land and sea monsters, popularly known as 'Giants' Bones,' and the weapons of ancient heroes."

30. Tacitus (b. A.D. 56)

Germania 45. "The Germans gather amber but they have not investigated the natural cause or process. . . . It is sap from trees, which can be inferred from the fact that you can see creeping things and winged insects which were trapped when the substance gradually hardened."

31. VIRGIL (b. 70 B.C.)

Aeneid 12.899–900. On giants buried in the earth: "Scarce could that stone [a huge boundary stone] twice six picked men upraise / With bodies such as now the earth displays."

Georgics 1.494–97. "As the farmer toils at the soil with curved plow, he shall find rusty javelins, or with heavy hoes shall strike empty helmets and marvel at the giant bones in the plowed-up graves."

32. XENOPHANES (sixth century B.C.)

Fragment in Hippolytus. To demonstrate that the earth was once covered by sea, Xenophanes says "shells are found inland and in the mountains, and in the quarries of Syracuse an impression of a fish and seaweed has been found, and impressions of fish were found in Paros in the depth of the rock and in Malta impressions of many marine creatures. These, he says, were produced when everything was long ago covered with mud and the impressions were dried in the mud."[7]

NOTES

Introduction

1. See Greene 1992, xvii–xviii, on classicists' traditional blind spot for natural knowledge embedded in ancient texts. For an emerging scientific appreciation of accurate natural knowledge in prescientific, preliterate cultures, see "Digging into Natural World Insights" 1996. A handful of classicists have noted a relationship between large fossil bones and mythical giants and monsters, e.g., Frazer 1898, commentary at Pausanias 1.35.7; Pfister 1909–12; H. Rackham (Loeb), n. *b* at Pliny *Natural History* 7.73; Levi 1979, n. 241 at Pausanias 8.32.5; Huxley 1979; and Hansen 1996, 137–38. One reason the large, extinct mammal fossils of the Mediterranean are so little known is that since the mid–twentieth century, paleontologists have concentrated on studying the fossil remains of the tiniest mammal species. The fossils of small species reveal more clues to evolution than do large species (Bernor, Fahlbusch, and Mittman 1996, 135). Similarly, Buffetaut, Cuny, and Le Loeff (1995) found that the rich dinosaur remains of France are poorly known because modern evolutionists focus on small mammal fossils.

2. William A. S. Sarjeant in Currie and Padian 1997, 340. "The bones which we would now find most impressive—the great vertebrae, the ribs, and the limb bones—were essentially too big to be noticed—or if noticed, to be taken seriously as bones of animals": Sarjeant in Farlow and Brett-Surman 1997, 4. In *A Short History of Vertebrate Paleontology* (1987, 3–5), Eric Buffetaut acknowledges that big fossil bones attracted attention in antiquity and cites examples from ancient sources. Authorities on science in Greco-Roman antiquity generally assume that large fossil bones were ignored, and that only small marine fossils attracted notice: see, e.g., Sarton 1964, 180, 560; Kirk, Raven, and Schofield 1983, 168–79.

3. The first section of Cuvier's three-part monograph of 1806 is a history of vertebrate paleontology from the fourth century B.C. to 1802, with a discussion of the ancient knowledge of living elephants. On 4–5, 14, and 54, Cuvier cites discoveries recorded by Theophrastus, Pliny, Herodotus, Suetonius, Strabo, the *Suda*, and Saint Augustine, noting that fossil elephant bones were often mistaken for remains of giants in antiquity. Henry Fairfield Osborn (1936–42, 2:1147)

briefly noted Cuvier's interest in ancient fossil finds. In 1997, Martin Rudwick published the first modern translations of Cuvier's writings, with commentary. Only the geological conclusions (the final five pages) of the 1806 monograph are translated. There is no mention of Cuvier's historical research, which Rudwick calls a section "on the geographical distribution of finds of fossil elephants" (91–94, 92 n1). See 46, 216 for Cuvier's other allusions to finds of giant bones in antiquity.

4. I owe the phrase "institutional myth of modern paletontology" to Peter Dodson. Rudwick 1985 devotes less than one page to classical antiquity, with no mention of vertebrate fossil discoveries: 24–25, 39.

5. Russell 1981, 80. I myself once repeated this false "fact" in print (Mayor and Heaney 1993, 7), along with the following authorities. Thenius and Vávra 1996, 15, 20: "Empedocles had already reported the existence of an extinct race of giants in Sicily." Novacek 1996, 141: "Empedocles, writing in 400 B.C., noted that fossil elephant skulls common in the Mediterranean region could be associated with the Homeric legend of the Cyclops." Dominique Lecourt, introduction to Tassy 1993, 10: Empedocles "describes the discovery of huge bones in Sicily, and sees in them the vestiges of the mythical race of 'giants' whose features had first been outlined by Homer." Burgio 1989, 72: "Empedocle . . . affermò che nelle caverne presso le coste della Sicilia esistevano le testimonianze sicure di una stirpe estina di giganti." Buffetaut 1987, 5, and 1991, 19: Empedocles was familiar with "finds of large bones in Sicily and considered them the remains of a race of giants." Reese 1976, 93: "The first notice of fossils in Sicily—in this instance of dwarfed elephants—was by . . . Empedocles [who] thought the bones were the skeleton of Homer's Polyphemus." Wendt 1968, 18–19: "Empedocles thought he recognized the skeleton of Homer's [Cyclops] Polyphemus in the fossil bones of dwarf elephants. [Boccaccio] described similar finds in the fourteenth century. In the fourth book of his *Genealogy of the Gods*, he refers to these 'bones of the cyclops' and cites Empedocles as his authority." Swinton 1966, 21: Empedocles "nearly two thousand years before [Boccaccio] had called such things [elephant fossils] 'the bones of Polyphemus.'" Matthews 1962, 145: "Empedocles . . . noticed an abundance of fossil shells and bones in Sicily and concluded that they had originally been animal forms. Empedocles' studies of these fossils led him to postulate that plants appeared before animals." See also Ley 1948, cited below.

6. Boccaccio *Genealogy of the Gods* 4.68. Boccaccio does not mention elephants. Local mathematicians calculated the giant's length at 200 cubits (nearly 300 feet or 86 m). I thank Boccaccio scholar Robert Hollander, Princeton University, for translation and analysis of Boccaccio's narrative. In the seventeenth century, the learned German scholar Athanasius Kircher came to Sicily to study

fossils from caves, including Boccaccio's giant. He corrected the length to 30 feet (9 m) and suggested that some bones in caves may have been elephants brought by ancient invaders from Africa or Asia, a common explanation for mammoth remains in Europe from the 1600s on. Kircher's measurements were probably exaggerated too, since the "giant" skeleton was almost certainly that of a Pleistocene dwarf elephant, whose skeletons are now known to be extremely common in coastal caves all around the Mediterranean, especially in Sicily (see chapter 2). It's possible that Empedocles saw such remains in Sicily, but there is no evidence in his extant writings. Inwood 1992; Wright 1995. Brad Inwood, professor of classics (University of Toronto) and translator of Empedocles, states, "There is certainly, to my knowledge, no evidence that Empedocles was familiar with prehistoric faunal remains in Sicily or that he proposed that such remains might be evidence for giant creatures." Personal communication, October 16, 1997. The theories of Empedocles and other ancient Greek and Roman natural philosophers are discussed in chapter 5, the Greek knowledge of elephants in chapter 2.

7. Abel 1914, 32; 1939, 40. Abel cited no sources for his claim. Ley 1948, 47–51. According to Ley, Boccaccio said his find of Cyclops bones in the cave "vindicated Empedocles, who, in 440 B.C., had claimed that Sicily had once been the dwelling place of ferocious giants" (49). I thank David Large for translating Abel's passages about Empedocles. Eric Buffetaut (Centre National de la Recherche Scientifique, Paris) helped me track the origin of the Empedocles myth to Abel. "Abel had a lot of interesting insights about the folklore of fossils and the early history of paleontology, but he was prone to ingenious explanations that went beyond what the facts really suggested." Personal communication, December 1997–January 1998.

8. E.g., Wendt 1968, 76: "Aristotle delivered the deathblow to evolutionary ideas in natural philosophy. Instead the dogma of the immutability of species became established. All living beings, it maintained, had been given their present form by a single act of creation." Horner and Dobb 1997, 24: "From the dawn of humanity . . . the standard belief was that all the plants and animals on Earth had been created at the same time and that all of them had continued to exist in the same way from that day onward." I thank Dale Russell for suggesting an inquiry into this question, which is discussed in chapter 5.

9. On the perceived estrangement between popular belief and science, see, e.g., Douglas Hofstadter's essay "Popular Culture and the Threat to Rational Inquiry" (1998). "To try to turn a myth into a science, or a science into a myth, is an insult to myths, . . . and an insult to science," says Michael Shermer in *Why People Believe Weird Things* (1997, 130).

10. Tassy 1993, 30–31. Cuvier also compared paleontology to archaeology: Rudwick 1997, 34–35, 174.

Chapter 1

The Gold-Guarding Griffin: A Paleontological Legend

1. The "scientific" movement to discredit old travelers' tales began with Sir Thomas Browne's declaration that the griffin was a mystical symbol (*Vulgar Errors*, 1646). His notion was rejected by Andrew Ross in 1652 (*Arcana Microcosmi*), who believed that the ancients were describing an unusual, real animal, but it is Browne's pronouncement that has dominated. For the development of my theory that griffins were inspired by observations of prehistoric fossils, see Mayor 1989, 1990, 1991, 1994. See Mayor 1991 and Mayor and Heaney 1993 for full ancient and modern references.

2. The terms *myth*, *legend*, *folklore*, and *popular belief* are often used interchangeably. Their features frequently overlap, but a culture's *myths* usually describe actions of gods or divine beings to explain the origins of the natural world in the primeval past. *Legends* are folklore narratives that typically feature animals or human beings engaged in unusual exploits, often set in historical times and places. The story purports to be true, although it may not always be believed every time it is told or heard. Legends and *popular beliefs* often coalesce around actual events or natural facts and, as they circulate by word of mouth or in writing, they often accumulate details. For standard definitions of folklore genres, see Bascom 1965. Traditional lore is often transmitted in mythical and legendary narratives, but it is also embedded in belief statements that do not take the form of developed stories. What we have of ancient griffin lore survives in the form of belief statements about the animals' appearance and habits.

3. Ancient discoveries of the Samos fossils are discussed in chapter 2. The archaeological museum of Samos has been expanded, but many of the bronze griffins I saw in the old museum in 1978 are now in storage. The bone room above the Mytilini post office was transformed in the 1980s into the Samos Paleontological Museum, selling postcards and souvenir key chains of the *Samotherium*. In 1988, I noticed a very large skull, still caked with dirt, on the floor. The curator explained that the skull had just been dug up on Ikaria, a nearby island, and brought by boat to Samos. In 1992, the new Natural History Museum of the Aegean was established on Samos.

4. Erman 1848, 1:87–89, 380–82, 163–64, 250–51, 2:377–82; Erman 1834, 9–11. In the 1840s, Erman published paleontological studies of northern Asia, Spain, and Germany. The Siberians also referred to the giant limb bones as the birds' "quills." In 1865, Edward Burnet Tylor suggested that the fossil rhinoceros's "paired tusks [*sic*] united by part of the skull might well be taken" by the Siberians as two talons of a bird's foot, 1964, 179.

5. Horner and Dobb 1997, 18, 26, 165. Cf. Lanham 1973, 168–67.

6. Erman 1848, 1:88. Bolton 1962, 84, 93, 101, 176.

7. Aristeas and his epic poem *The Arimaspea*: Bolton 1962 and Phillips 1955.

8. Scythian art and archaeology: Rolle 1989; Mayor and Heaney 1993. Recently discovered Altai tattooed mummies: Bahn 1996.

9. Aeschylus *Prometheus Bound* 790–805. Aeschylus (or his son, according to some scholars) may also have borrowed from other folklore sources about Scythia. Griffith 1983, 230, 266. Aeschylus painted a poetic picture of the geological processes of fossilization in the same play; see chapter 2.

10. Grene 1987, 1–32. Ascherson 1996, 78.

11. Herodotus *The Histories*, 2.44, 3.116, and bk. 4, esp. 4.13–27. The mounted nomads who fought griffins beyond Issedonia were called Arimaspeans, a Scythian word that the ancient Greeks mistakenly translated to mean "one-eyed." It probably means "owner of desert horses." Mayor and Heaney 1993, 10–16; Mayor 1994, 55. Vases: Bolton 1962, 37; Mayor and Heaney 1993, 47, fig. 6. Translations adapted from standard Loeb translations, unless otherwise noted.

12. Photius *Bibliotheca* 72.46b30; Ctesias *De Rebus Indicis* 12, my emphasis.

13. Pomponius Mela 2.1. Pliny *Natural History* 10.136, 7.10. Philostratus *Life of Apollonius of Tyana* 3.48.

14. Pausanias 1.24.6, 8.2.7. Aelian *On Animals* 4.27; see also Bolton 1962, 65–72. In the Middle Ages, the griffin crystallized into a symbolic creature defined by Christian allegorical attributes. History of the griffin: Armour 1995.

15. Medieval giantology and legends of dragons and monsters related to fossil fauna are discussed in, e.g., Stephens 1989; Sutcliffe 1985, chap. 3; Buffetaut 1987, chap. 1; Wendt 1968, 18–25; Thenius and Vávra 1996. Klagenfurt dragon: Abel 1939, 82–83; Buffetaut 1987, 13; Thenius 1973, 37–38; Thenius and Vávra 1996, 23–24. Cyclops: Abel 1914 and 1939; Thenius and Vávra 1996, 19–21; Homer *Odyssey* 9; Hesiod *Theogony* 139–46, 501–6. Dwarf elephants (actually miniature mammoths): Shoshani and Tassy 1996, chap. 22. Theodorou 1990 suggests that the dwarf elephants of Tilos died seeking shelter from the Thera volcanic eruption (see chapter 2). Davis 1987 compares the sizes of the pygmy elephants *E. mnaidrensis* (about 6 feet at shoulder) and *E. falconeri* (3 feet at shoulder) with the huge extinct elephant *Palaeoloxodon antiquus* (13 feet), 119. Dwarf elephants weighed about 1.5 tons: Lister and Bahn 1994, 149. For a history of interpretations of the Cyclops legend, see Glenn 1978.

16. Gold-miner's skeleton: Rolle 1989, 52. Romans in Central Asia and China: Dubs 1941.

17. Turfan and Lop Nur: Mayor and Heaney 1993, 25 n. 39. Dragon bones in canal: Joseph Needham cited in Oakley 1975, 40; Lanham 1973, 6.

18. *I Ching*: Crump and Crump 1963, 14–17; Bassett 1982, 25; Canby 1995,

22–28. Provenience: Osborn 1921, 332–34. Buffetaut 1987, 17. See also Andrews 1926, chap. 2; Sutcliffe 1985, 30–32. Chinese farmers in fossiliferous regions still find bone hunting more profitable than agriculture. Use of fossils for medicine is widespread and goes back to classical antiquity: Kennedy 1976. E.g., nummulites were collected as medicine in ancient Egypt: see Pseudo-Aristotle in appendix 2.

19. Paleontology in China: Spalding 1993, chaps. 13–15; Crump and Crump 1963; Oakley 1975, 40–42; Oakley 1965, 123–24. For the collection of marine fossils in ancient China, see Rudkin and Barnett 1979, 16.

20. Andrews 1926, 180, 222. See also Dodson 1996, 206–9; Spalding 1993, chap. 13.

21. Andrews 1926. Dodson 1996, 206–39, 260. Norman 1985, 128–33. Novacek 1996, 10.

22. Dodson 1996, 180, 217, 232–40; Spalding 1993, 211; Norman 1985, 13–14. Uzbekistan: Dodson 1996, 239; Dale Russell, personal communication, April 1991; Philip Currie, personal communication, May 1991.

23. Russell, personal communications, April–May 1991, March 1992. Currie, personal communication, May 1991. Taquet 1998, 142–46. Glut 1997, 741. Dodson 1996, 209, and personal communication, July 1995.

24. Theophrastus *On Stones* 6.35. Pliny *Natural History* 37.65, 37.112, 37.146. Michael Novacek discounts the chance of gold dust's blowing over fossil exposures (1996, 141), but see Monastersky 1998 and Wilford 1998 on recent sedimentologists' studies of aeolian and colluvial deposits at Gobi fossil sites. The nests may have served as riffle boxes to trap gold flakes. "Fossil" gold may have been deposited over Cretaceous sediments by paleo–stream channels: Higgins and Higgins 1996, 108; Currie, personal communication, May 1991. Ancient riverbeds in Central Asia deserts: Brice 1978, 319–34.

25. Horner and Dobb 1997, 21, 28, 37, 164, 176.

26. Aristotle *Parts of Animals* 697b14–26. Lloyd 1996, 80 n. 13. Other flightless, ground-dwelling birds were well known in antiquity; e.g., Herodotus 8.250–60. Ostrich: Bodenheimer 1960, 132. Aristotle discussed the appearance of various "unclassifiable" sea monsters in *History of Animals* 532b19–29; see chapter 5.

27. Wendt 1968, 230–31. Hecht 1997; Browne 1997. Ostrom, personal communication, August 1997. Currie quoted in Spalding 1993, 207. Debate over the origin of flight and "transitional" dinosaurs: Norman 1985, 191–94; Shipman 1998; Horner and Dobb 1997, 27–28, 165.

28. Standing positions: Glut 1997, 741. Dodson 1996, 225–26, 221, 214, 215, 209. Horner and Dobb 1997, 21–24; Taquet 1998, 143–44 (standing positions, clavicles). Elongated shoulder bones are typical of birds: Spalding 1993, 182.

29. Claws: Russell, personal communication, May 1991; Mayor and Heaney 1993, 63 n. 34. Norman 1985, 48–49. There are many examples of mistakes and accidental composites in modern dinosaur restorations.

30. Horner and Dobb 1997, 160–62, 7–8.

31. Horner and Dobb 1997, 24. According to historian of dinosaur discoveries William Sarjeant, heretofore only one legend could be directly traced to dinosaur bones—a nineteenth-century Piegan Indian tradition. Most other legendary monsters have been based on observations of very large mammal fossils. Sarjeant in Currie and Padian 1997, 340; and in Farlow and Brett-Surman 1997, 4.

32. Ostrom, personal communication, August 1997. Horner and Dobb 1997, 12, 232, 101–2, 224–25. On 226–27, we catch a glimpse of Horner as a kind of reincarnated nomad: "I have wandered all of my life, . . . through barren hills and rocky washes, . . . letting my imagination run free" and testing ideas against ever-shifting evidence. Taquet 1998, ix, also compares paleontologists to nomads. Meaning of fossils for humans: Horner and Dobb, 101–2, 203, 218, 225; Spalding 1993, 294–97; Norman 1985, 8–9.

Chapter 2

Earthquakes and Elephants: Prehistoric Remains in Mediterranean Lands

1. Plutarch *Greek Questions* 56. Plutarch's account conflates two separate legends, a Hellenistic story (ca. 325–30 B.C.) about Amazons and Dionysus's Indian elephants and another, earlier story (fifth century B.C.) about monsters called Neades (below). The various strands are untangled by Halliday 1928, 210–11. Dionysus also battled Amazons, Titans, and a huge monster in North Africa—he buried the monster under an enormous mound that could be viewed in Roman times, according to Diodorus of Sicily 3.71–72. Dionysus versus giants: Rose 1959, 180 n. 36. "Hard Crust" for Phloion: Solounias 1981a, 18. The so-called graves of Amazons along the coast of Turkey may have been exposures of oversize prehistoric skeletons similar to those of Samos. Other Amazon graves were pointed out around Megara, Chaeronea, and Thessaly in Greece. A legendary battle between Amazons and war horses was set on Leuke in the Black Sea: see chapter 3. Elephants were unknown in Greece until Asian war elephants were first encountered by Alexander the Great in India in 331 B.C. Scullard 1974, 32–33, 37–52, chap. 2. Tourist attractions and guides in antiquity: Friedländer 1979, 373–74; Pausanias 5.20.4. The fossil beds of Samos contain giant giraffes, mas-

todons, and other large mammals of the Miocene epoch, discussed later in this chapter.

2. Plutarch's syntax relates the big bones to a seismic event that caused the animals' demise "in their time," that is, in the distant past. It's not clear whether the Neades' bellowing caused the ground to crack or whether they uttered shrieks during the upheaval. See Halliday 1928, 207 and 208. Euagon in *Fragmente der grieschischen Historiker* (*FGrHist*) 535 F 1. Huxley 1973, 273. The fourth-century B.C. philosopher Heraclides Ponticus also referred to the Neades: frag. X, *Fragmenta Historicorum Graecorum* (*FHG*) Müller, ii, p. 215, cited in Halliday 1928, 208.

3. Aristotle (or one of his students) *Constitutions*, in Heraclides Lembos *Excerpta Politiarum* 30 (p. 24 Dilts *GRBS* monograph 5; Müller iii, p. 167) cited by Huxley 1973, 273.

4. Aelian (ca. A.D. 170–230) *On Animals* 17.28. On Euphorion and others who refer to Neades, see Huxley 1973, 273 n. 6; fragment in *FHG* Müller, iii, p. 72, cited by Halliday 1928, 208; Shipley 1987, 281. See chapter 4 for the large fossil femur found at the Heraion in 1988 and other examples of fossils in archaeological sites. In 1996, the Samos bone deposits were proclaimed "the oldest recognized fossil beds in Western Eurasia and possibly the world" by paleontologists Raymond Bernor and colleagues, 138.

5. Erol 1985, 491. Ager 1980, 441, 437, 447. Tziavos and Kraft 1985, 445. Brinkmann 1976 for geology of Turkey, and Brice 1978 for the Pleistocene environment of Greece, Turkey, the Near and Middle East. Miocene-Pliocene geology and faunas: Bernor, Fahlbusch, and Mittman 1996.

6. Coastline changes: Vitaliano 1973, 32–34; Erol 1985; Higgins and Higgins 1996, 116, 125–27 (Gallipoli and Troy); 142–43 (Ephesus), 140–41 (Izmir and Sardis), 147–50 (Priene, Miletus, Herakleia); Brinkmann 1976, 82–83; Brice 1978, chap. 6; Tziavos and Kraft 1985; Gore 1982, 727; Attenborough 1987, 118. Pliny *Natural History* 2.87–96; 2.191–210. Ancient writers' knowledge of geology: Bromehead 1945. Herodotus: Lloyd 1976, 2:36–35, 66–67.

7. Philostratus *Life of Apollonius of Tyana* 4.36. Pausanias 7.24–25, with Levi's nn. 129–33. Severe earthquakes and coastal changes in Crete in the first century A.D. and the disasters at Helike and Sipylos: Higgins and Higgins 1996, 69–70, 199, 205–6, 213. Black Sea: Ascherson 1996, 79. Dr. Steve Soter, American Museum of Natural History, is using geoarchaeological techniques to search for Helike.

8. Pliny *Natural History* 2.200–206. Pindar, Pliny, Plato, Aristotle, Strabo, Ovid, and Philo discussed in Bromehead 1945, 93–97, 98–102; Brinkmann 1976, 2. Diodorus of Sicily on Rhodes's emergence: 5.56. Ovid on geomorphic change: *Metamorphoses* 15.252–306. Geological history of Greece and Aegean:

Higgins and Higgins 1996. See also Huxley 1973, 273–74; Vitaliano 1973; Greene 1992, 46–88.

9. Older vertebrate fossils exist in some parts of the old Greco-Roman world; for example, there are Jurassic deposits in Portugal and France and Eocene-Oligocene remains in Egypt. In 1983, an amateur fossil hunter found a 9-inch long (23-cm), exquisitely preserved baby dinosaur in the Pietraroia Cretaceous limestone formation thirty miles northeast of Naples (previously known for its finely preserved fish remains). The fossil, the only dinosaur known in Italy, was rediscovered fifteen years later in a dusty drawer in Salerno in 1998 and named *Scipionyx samniticus*. It is now fueling the debate over dinosaur warm-bloodedness and relationship to birds. *New York Times*, March 26, 1998, A24; January 26, 1999, F5.

10. The well-preserved buildings at ancient Olympia in the Peloponnese are good examples of shelly limestone: Ager 1980, 445. Higgins and Higgins 1996, 68. Pausanias 1.44.9; cf. Pliny *Natural History* 35.36. The paleogeology of the circum-Mediterranean is controversial. This chapter relies on Melentis 1974, 18–24; Ager 1980, 354, 435–517; Higgins and Higgins 1996, esp. 1–25 (see 210–14 for their predictions of future severe geologic upheavals in the Aegean); Bernor, Fahlbusch, and Mittman 1996; Jameson, Runnels, and van Andel 1994, 154–57; Embleton 1984, 341, 380–81; Bird and Schwartz 1985, 445; Maglio 1973, 111–19; Vitaliano 1987; Bodenheimer 1960, 13–21; Gore 1982; Stanley 1989, esp. 551–608, with maps of Tethys; Attenborough 1987, 10–15; and Carrington 1971, chap. 2.

11. Exactly which islands were connected when during the Pleistocene is unknown, but differences in faunal remains are valuable evidence for determining paleogeography. Swimming by prehistoric species is also controversial. Dermatzakis and Sondaar 1978; Theodorou 1990. The petrified forest of Lesbos is a Greek national monument: Higgins and Higgins 1996, 135, 133, 187–95. The Thera eruption is thought to have been ten times that of Krakatoa (Indonesia) in 1883. Thera's ashes rained down on Sicily and Crete, the Nile Delta, and western Turkey. Melentis 1974, 21–22. See Vitaliano 1973, and Greene 1992, 46–88, for vivid descriptions of Mediterranean volcanoes.

12. Higgins and Higgins 1996, 213; Gore 1982, 713–15. Vitaliano 1987, 13–14, 16; Phillips 1964, 177; Sondaar 1971, 419; Klein and Cruz-Uribe 1984. Elephant skulls break up owing to their porous structure: Shoshani and Tassy 1996, 9, 68; Lister and Bahn 1994, 60; Harris 1978, 338. Survival of limb bones, scapulas, kneecaps: Attenborough 1987, 29. Italy: Lister and Bahn 1994, 24, 58. Cf. crushed bones at Samos: Nikos Solounias, personal communication, September 23, 1998; George Koufos, personal communication, October 1997. Bernor, Fahlbusch, and Mittman 1996, quote 416.

13. *Fossil*: Bromehead 1945, 104; Eichholz 1965, 4, 17, 39, 42–47, 71, 86, 113 (noting that fossils in Greece are indeed mottled brown and white), 121, 71. *Elephas* in Homeric times, fossil ivory, and ancient knowledge of elephants: Scullard 1974, chap. 2 and 260–61. Mammoth ivory in the Mediterranean: Barnett 1982, 8, 70 n. 68, 74 n. 46; Krzyszkowska 1990, 12, 22, 37–38; Lapatin 1997 and 2001, chap. 2. Theophrastus *On Stones* 37; Philostratus *Life of Apollonius of Tyana* 2.13; Pliny *Natural History* 8.7, 36.134. Pliny was speaking of buried tusks in Africa and Asia.

14. Pliny *Natural History* 36.161. I thank members of the Vertebrate Paleontology Internet discussion group who clarified the formation of selenite. Kenneth Carpenter, Denver Museum of Natural History, wrote (January 5–7, 1999), "Selenite is a long crystal precipitate of calcium sulfate (gypsum). It's water soluble, so its presence in solution in a selenite mine is highly probable assuming that the mine was wet. In solution, it could precipitate out within bones of animals that fell into the shafts." But for marrow to be mineralized over a single winter (as Pliny assumed) "the rate of decay would have to be rapid enough to decompose all of the soft tissue so that the solution could get to the bone marrow." Carpenter suspects that instead "the bones were those of Pleistocene animals, primarily rodents and carnivores." It is "a common mistaken belief that fossil bones at the bottom of sink holes belong to living animals that fell in." Gypsum can coat the surface of objects rapidly, but it is not clear how long it would take for bone marrow to be completely replaced by crystals. Della Collins Cook, Anthropology, Indiana University, January 18, 1999, notes that "deep mines would exclude microorganisms and insects that cause decomposition of corpses," so it seems likely that the marrow cavities of animal bones would have mineralized over centuries. The selenite mines were in the region of ancient Segobriga, central Spain, east of modern Toledo. Pleistocene faunal deposits do occur around Toledo, according to Shoshani and Tassy 1996, chap. 14, 138 map; Lister and Bahn 1994, 160 and map; Bromehead 1943, 326. These include mastodons and mammoths, but fossil bones seen in mines probably were smaller, familiar-looking animals. Fossils often occur in gypsum quarries: e.g., Rudwick 1997, 133, 146–47; Wendt 1968, 24.

15. Footprints: Diodorus of Sicily 4.24.1–4; Philostratus *On Heroes* 13.3. Heracles' 3-foot (1-m) print at the Tyras (Dniester) River in Scythia: Herodotus 4.82. Heracles' footprint was supposedly 12.6 inches long; as a boy he was 4.5 cubits tall (6–7 feet, over 2 m). Apollodorus 2.4.9 with Frazer's n. 3. Giant footprints in rock were common enough to mock during the Roman Empire: see Lucian *A True Story* 1.7. Modern hoofprint/bivalve fossil folklore: Ager 1980, 524; Thenius 1973, 32–34, figs. 22–23; Thenius and Vávra 1996, 35–36. It is not clear what kind of impressions were taken as Heracles' footprints: the Dniester area contains late Neogene and Pleistocene deposits, but fossil footprints are

not known there, according to paleontologist Roman Croitor, personal communication, January 19, 1999. Human footprints embedded in rock figure in folklore around the globe; see, e.g., Krishtalka 1989, 230, 239, 242, 247–48. Footprint lore in Malta: Zammit-Maempel 1989, 12, 22. In Algeria: Buffetaut 1987, 180–81. In the Americas: Spalding 1993, 5, 6, 79–83. Human footprints from the Pleistocene in Turkey: Brinkmann 1976, 81. Historian of paleontology William Sarjeant believes that fossil footprints were noticed by humans long before fossil bones: see Farlow and Brett-Surman 1997, 3–4.

16. See appendix 2 for ancient texts describing amber, ammonites, nummulites, bones, teeth, shells, plants, and other fossils. Amber: Aristotle *Meteorology* 388b19–21; Pliny *Natural History* 37.42–46 (cf. Hesiod frag. 311 MW). Strabo 17.1. On accounts of tongue-stones (fossil sharks' teeth) in Pliny 37.164, see Bromehead 1945, 105. Eocene nummulitic limestone in Egypt: Stanley 1989, 551. Neogene plants: Brice 1978, 40–44 (eastern Mediterranean); Bernor, Fahlbusch, and Mittman 1996, chap. 30 (Europe).

17. Fossilization processes and taphonomy: Savage and Long 1986, 1–7; Lister and Bahn 1994, 141, 142, 144, 145–46; Higgins and Higgins 1996, 8. Aeschylus *Prometheus Bound* 1015–25. Griffith 1983, 265–66.

18. Pausanias 8.29.2–4, 6.5.1. Cubit = about 17 inches. Philostratos *On Heroes* 8.3 related that the Orontes giant was 45 feet long, and identified it as Aryades of Assyria (see chapter 3). Typhon: Strabo 16.2.5. Tiberius (b. 42 B.C.) was probably the emperor who diverted the Orontes: see Levi 1979, n. 217 at Pausanias 8.29; Bodenheimer 1960, 140. English translations say the Orontes skeleton was found in a "sarcophagus," but Pausanias's wording, *keramea soros*, "burial place in clay," could mean that the bones lay in clay sediment, not a man-made coffin. Cf. fossil skeletons found in "bubble" vaults in rock, Carrington 1971, 78. Mastodon, hippo, and steppe mammoth remains are found along the Orontes River: Adam 1988, 2. Femur size: Osborn 1936–42, 2:1251. Early European travelers in the Levant brought home fossil curios, such as "the huge molar of a petrified giant" collected by Paul Lucas in 1714 in Aleppo, northwest Syria: Halliday 1928, 208. See chapters 3 and 4 for examples of gigantic bones buried in coffins as early as the seventh century B.C.: in later times these heroic burials may have been rediscovered. Cf. burials of giant aurochs in sarcophagi in ancient Egypt, Attenborough 1987, 77–80. Oversize coffins in antiquity: Wood 1868, 22–23. On the mythical war between the gods and the giants (the Gigantomachy) and various battlegrounds, see Vian 1952.

19. Asterios, Geryon, Hyllos, and Koresos: Pausanias 1.35.4–6; 7.2.4, with Frazer's notes. Plato referred to another huge skeleton exposed in Lydia: *Republic* 2.359c–e (see chapter 5). Geryon's bones were also displayed in Olympia and Thebes. The port of Cadiz, west of Gibraltar, was once an island. Pausanias knew

of a fellow Magnesian named Kleon who had sailed to Cadiz and saw a giant skeleton on the beach: it "covered five acres and was burning, because it had been struck by lightning." Pausanias 10.4.4. Pliny mentions a giant sea-monster skeleton on the shore at Cadiz, *Natural History* 9.11. Were these beached living whales, or Eocene fossil whale exposures? Gomphothere, *P. antiquus*, and mastodon remains exist in Spain, and Tartessus, around Cadiz, has burning lignite soils associated with other burial grounds of giants (chapter 5). Shoshani and Tassy 1996, 138, 141–42.

20. Sevket Sen, personal communications, October–November 1997. William Sanders, Museum of Paleontology, University of Michigan, personal communications, October–November 1997. Harris 1978. See also Christopoulos and Bastias 1974, 26–35. Western Turkey has the same fossils as Samos, Kos, Chios, and Lesbos: Brinkmann 1976, esp. 60–73; Gates 1996, 281 (Ozluce). For proboscidean species and maps, see Savage and Russell 1983, chaps. 6–8; Shoshani and Tassy 1996; Bernor, Fahlbusch, and Mittman 1996.

21. Lloyd 1979, 127. See also Greene 1992, 86.

22. Imaginative and mythohistorical scientific nomenclature: Krishtalka 1989, chap. 2. *Brontotherium* (*Titanotherium*): Wendt 1968, 277–81; Kindle 1935, 451. *Indricotherium* (also called *Baluchitherium*): Andrews 1926, 192–93. *Quetzalcoatlus*: Norman 1985, 171. *Achelousaurus*: Dodson 1996, 197. Examples could be multiplied.

23. Saints' relics: Bentley 1985; Ley 1948, 45; Sutcliffe 1985, chap. 3. Buffetaut 1987, 5–11, citing Ginsburg 1984. By the seventeenth and eighteenth centuries, some observers with a historical bent concluded that mammoth bones were those of African elephants taken by Hannibal over the Alps during the Second Punic War between Rome and Carthage (late third century B.C.). The first mammoth remains discovered in Britain were believed to be elephants brought by the Roman emperor Claudius, and Peter the Great identified mammoth skeletons in Russia as Alexander the Great's lost war elephants. The ancient theory that the big bones of Samos belonged to Dionysus's war elephants, discussed at the beginning of this chapter, is the earliest example of similar historical thinking.

24. Ancient ideas about the stature of heroes and giants are discussed in chapter 5. I have heard modern Greeks in rural areas refer to ancient Greeks as physical giants; cf. Hansen 1996, 138. The Luka bone, the end of a femur (about 6 inches or 15 cm wide), was discovered by a cousin of Chrysanthi Gallou, archaeology graduate student at the University of Nottingham and a native of Luka, personal communications, November 1997–February 1998. Villagers refused to allow Gallou to photograph this fossil and others (perhaps not surprisingly, given the long history of appropriated fossil finds in Greece). "Skeletons of all mammal species tend to comprise roughly the same variety of skeletal parts arranged in the

same anatomical order." Klein and Cruz-Uribe 1984, 11. Ubelaker and Scammell 1992, 76–85; thigh bone can be used to calculate total stature, 88, 45.

25. Thanks to William Hansen for recommending Guthrie 1993: see 90, chap. 4, "Anthropomorphism as Perception," and 63 (Xenophanes). The higher the level of schemata perceived, the more efficient the cognitive process: 101–3. Even Darwin's theory of natural selection anthropomorphizes: 173–74. Xenophanes' anthropomorphic theory: Kirk, Raven, and Schofield 1983, 168–69. The anthropomorphizing drive also helps account for today's popular images of extraterrestrial aliens and explains why the earliest dinosaur found in North America (1820) was thought to be human (Spalding 1993, 80), and why a fossil whale rib was taken for a hominid clavicle as recently as 1982 (Krishtalka 1989, 66). Myriad medieval and modern examples of mammoth and rhino bones mistaken for human exist in paleontological literature: see, e.g., Ley 1948, chap. 2; Buffetaut 1987. Krishtalka 1989, 64 (quote), 63–71. Stephens 1989 argues that Renaissance nationalism was fueled by the "science" of giantology and the search for giant ancestors. The modern search for missing human-primate links: Carrington 1971, 65–66. Paleontological hoaxes often involve anthropomorphic monsters: chapter 6.

26. Pikermi bone rush: Buffetaut 1987, 114–17; Wendt 1968, 237–44; Woodward 1901; Gaudry 1862–67; Osborn 1921, 267–71; Melentis 1974, 22–23, 392; Solounias 1981a, 1981b, 232–33; Rudwick 1985, 244–46; Bernor, Fahlbusch, and Mittman 1996, 138–39. Excavations at Pikermi were undertaken by Wagner, Othenio Abel, Albert Gaudry, Arthur Smith Woodward, Edouard Lartet, Hercules Mitsopoulos, and Theodore Skoufos, among others. Some Pikermi fossils are displayed at the University of Athens Paleontological Museum and at the Goulandris Natural History Museum in Kifissia, Greece. The spectacularly cluttered Paleontological Gallery of the old Natural History Museum in Paris displays dusty Pikermi specimens collected by Gaudry.

27. Gigantic limb bones predominated at Pikermi: Osborn 1921, 268; Osborn 1936–42, 1:93. Gaudry saw many *Deinotherium* limb bones. Woodward 1901, 484, noted numerous isolated rhinoceros bones, the large "limb-bones of Giraffidae . . . and several very large limb-bones" of mastodons. The bones were very "rotten" and eroded, broken up by roots, etc. (483). Buffetaut 1987, 114–17. Gaudry 1862–67. Pikermi theories summarized in Melentis 1974, 22–23; Bernor, Fahlbusch, and Mittman 1996, 138; Solounias 1981a, 242–45.

28. Buffetaut 1987, 114–17; Wendt 1968, 241; Brown 1927, 32. Osborn 1921, esp. 270–71, 323–336; Van Couvering and Miller 1971, 560; Lister and Bahn 1994, 140; Symeonidis and Tataris 1982, 148, 182; Bernor, Fahlbusch, and Mittman 1996, esp. chap. 6. Proboscideans at Pikermi and Siwaliks: Shoshani and Tassy 1996, 340. Miocene fauna: Savage and Russell 1983, chap. 6. Excellent

maps of Miocene and other geological sediments of Greece, the Aegean islands, and western Turkey are shown in Higgins and Higgins 1996.

29. These late Miocene fossils of Samos are important in new understandings of the prehistoric zoogeography and paleochronology of the eastern Mediterranean and Asia Minor. Major 1887, 1891, 1894; Solounias 1981b and 1981a, 19, 210; Melentis and Psilovikos 1982; Brown 1927; Higgins and Higgins 1996, 144–46 and n. 286; Weidmann et al. 1984, 488; Sondaar 1971, 419; Buffetaut 1987, 4. Major's collections went to Lausanne, Geneva, and Basel, and to the British Museum (Natural History). Lausanne boasts the second largest Samos collection after that of Brown at the American Museum of Natural History. Dr. Aymon Baud, director, Musée Géologique Cantonal, Lausanne, personal communications, August 1997.

30. Panaima and Phloion: Shipley 1987, 281 nn. 24, 29. Phloion thought to be a cult name for Dionysus, or a place with earthquake-caused features: Halliday 1928, 209. Brown 1927, 25, thinks Phloion was a collapsed mountainside. Solounias, personal communications, September–November 1998; Solounias and Mayor, in preparation.

31. Native American prospectors: Kindle 1935, 449–51. Theophrastus *On Stones* 63–64, and commentary by Eichholz 1965, 129–32. Theophrastus does not mention the fossils of Samos in his discussion of "earths," but he may have done so in another, lost, work, *On Petrifactions* (discussed in chapter 5). Pliny *Natural History* 35.19. Cf. Buffetaut 1987, 181: mineral prospecting in North Africa led to fossil discoveries. Stonecutters may also have come across fossils in Samos. In 575 B.C., workers constructed the great Temple of Hera (to replace the older temple) with blocks quarried from Neogene sediments in the south of the island, the same type of rock that occurs around Mytilini. Higgins and Higgins 1996, 146–47, 186.

32. Dr. Elmar P. J. Heizmann of the Staatliches Museum für Naturkunde of Stuttgart told me that fossils from Samos, Euboea, Kos, and Pikermi were acquired in the early twentieth century, but cannot explain how so many bones from Greece "found their way into our collection." Personal communication, July 28, 1997. A prominent fossil dealer in Bonn still offers a "substantial stock" of bones "from the Isle of Samos" but how the fossils were obtained is "an enigma." Renate Krantz, for Dr. F. Krantz, fossil dealer, Bonn, to David S. Reese, July 7, 1992. In Greece, Samos specimens are exhibited in the Goulandris Natural History Museum, Kifissia; the Museum of the University of Athens; Aristotle University of Thessaloniki; and the Museum of the Aegean at Mytilini, Samos. George Koufos of Aristotle University, Thessaloniki, is the current excavator of Samos. For a list of the museum collections of Samos fossils, see Solounias

1981b, 234–35. Karl Acker was also the German consul. Solounias 1981a, 19–26; Bernor, Fahlbusch, and Mittman 1996, 138. Nikos Solounias began excavating in Samos in 1976. I thank him for showing me Brown's trove of Samos fossils in the AMNH.

33. Simpson 1942, 133. Disputed ownership of fossils: Horner and Dobb 1997, 27, 233–44. *Mosasaurus*: Simons 1996; Taquet 1998, 170–74.

34. Brown 1927, 19–20. Solounias 1981a, 211 ("warped to include fossils"), 24. Brown remarked that Skoufos wanted to "loot" his AMNH collection. In his negotiations, Brown offered Greek officials free cameras, lifetime AMNH memberships, and some "common American fossils." Brown Papers, Osborn Library, AMNH, cited in Solounias 1981a, 211–18. As a curator of the AMNH, Brown represented the finest scientific establishment of his day, and he complained about the incompetence of his English, German, and Greek predecessors in Samos. Yet today his methods would be deemed "grubbing," and his trophies have little scientific value. Brown (like his namesake, P. T. Barnum) was obsessed with acquiring spectacular skeletons for public display, not systematic research. Scientific revision: Horner and Dobb 1997, 18–19, 32, 163.

35. Major fossil deposits in Greece and the Aegean receive little notice, even where one might expect coverage. For example, *A Geological Companion to Greece and the Aegean* (1996), by Higgins and Higgins, a geologist and a classical archaeologist, gives detailed descriptions and maps of Neogene and Pleistocene sediments without mentioning any vertebrate remains except for one fleeting reference to Samos. *Mammoths* (1994), by Adrian Lister, a leading authority on mammoths, and Paul Bahn, a zooarchaeologist, omits both Greece and Turkey from the maps showing mammoth distribution in Europe and Asia (152–55). Lister acknowledges that fossil elephant remains are very common in the eastern Mediterranean lands, but the material is "little-known and poorly stratified": personal communications, July 22–25, 1997. Henry Fairfield Osborn's exhaustive *Proboscidea* (1936–42) made no mention of the Megalopolis mammoths. The most comprehensive study of prehistoric elephants since Osborn, Shoshani and Tassy's *Proboscidea* (1996), also omits Megalopolis and gives the impression that dwarf elephants are the only species of fossil elephant found in Greece. Despite the abundant remains of a succession of evolving elephants in the circum-Mediterranean, proboscideans are excluded from analysis in Bernor, Fahlbusch, and Mittman, *The Evolution of Western Eurasian Neogene Mammal Faunas* 1996, 418, 431, 463. Gantz, *Early Greek Myth* 1993, the most up-to-date handbook on the development of Greek myth, makes no mention of ancient writers who related mythical giants and monsters to finds of remarkable bones in specific locales. The useful footnotes of early commentators, such as Bather and Yorke,

Frazer, and Levi, who did connect those ancient accounts to known paleontological discoveries, have been discarded in the effort to modernize classical scholarship. Levi 1979, n. 241 at Pausanias 8.32.5. The new Paleontological Museum of the University of Athens may be visited by appointment only.

36. Melentis 1974, 23; Huxley 1979, 147. In 1890–93, British archaeologist William Loring excavated at the fourth-century B.C. ruins of Megalopolis. The curator at the Dimitsana museum, an old teacher named Hieronymus, showed Loring a "large semi-fossilized bone which he calls the shoulder-blade of an elephant." Loring mocked the teacher in his excavation report. "His explanation would hardly satisfy a geologist," sniped Loring, but it "is at least as near the truth as that given by Pausanias." According to Pausanias and Herodotus, the bones of giants had been unearthed in the area since the days of the Trojan War (see chapters 3 and 5). Loring in Gardner et al. 1892, 121. Pausanias 5.13.4–7, 8.32.5, 8.36.2–3. Herodotus 1.68. Actually, a geologist would probably have *confirmed* the teacher's identification. In fact, Loring could have consulted with the eminent German geologist Alfred Philippson, who was exploring the Megalopolis basin at the same time and reported "many fossil bones" in the "tectonically active region northwest of Megalopolis." Pritchett 1982, 46 (quote), 45, n. 49. Frazer 1898, 4:352: Hieronymus of Dimitsana "may be more nearly right than Mr. Loring seems to think." Cf. Bather and Yorke 1892–93. I thank George Huxley for references to Megalopolis fossils in classical studies, personal communication, May 9, 1997.

37. *M. primigenius*, *M. meridionalis*, and *P. antiquus* at Megalopolis: Melentis 1974, 23–25; Symeonidis and Tataris 1982, 182; Dermitzakis and Sondaar 1978, 815–16. Large bones are abundant along the Alpheios River and around Megalopolis, Trapezus, Bathos, Olympia, Tegea, Gortyn, Mantinea, Apidhitsa, and Kalamata. Pritchett 1982, 45–46. Peloponnesian geology: Brice 1978, 54–66; Higgins and Higgins 1996, chaps. 5–7. Kenneth Lapatin, classical archaeologist, Boston University, observed the two huge tusks found in road operations at Varda and Simiza in Elis. Personal communication, December 5, 1997; Lapatin 2001, chap. 2. For information about modern lignite mining uncovering fossils and classical wells in Arcadia, I thank Dr. James Roy, Department of Classics, University of Nottingham, personal communications, October 10–16, 1997. Chapter 5 discusses burning lignite deposits associated with the Gigantomachy. See Higgins and Higgins 1996, chap. 7, for Neogene-Pleistocene sediments in Elis and Arcadia; and 84 on lignite formation. Pausanias's accounts: chapter 3.

38. Prehistoric elephant remains have been studied in more than seventy localities in Greece: Symeonidis and Theodorou 1990. Femur lengths: Osborn 1936–42, 2:1251. Macedonian proboscidean remains: Solounias 1981a, 218.

Evangelia Tsoukala, personal communications, August–November 1998. The broad term "mastodon" here includes mammutids, gomphotheriids, and stegodons that appeared in the Miocene. Mastodons have cusped teeth, whereas mammoths and elephants have flat, plated teeth (see fig. 3.12). Elephants and mammoths evolved separately over millions of years; reasons for the extinction of the mammoths (about 20,000 to 10,000 years ago) are unknown. Carrington 1958, chaps. 6–8; Savage and Long 1986, 143–59; Maglio 1973; Lister and Bahn 1994, 19–21 and glossary.

39. Proboscideans in the circum-Mediterranean and Eurasia: Savage and Long 1986, chap. 10; Shoshani and Tassy 1996, 44, 43, 58–74, 84, 90, 122, 136–42, 184, 188, 196, 205–11, 215, 219, 225–39, 259, 274, 340, 361, 364, 383, passim; Maglio 1973; Dermitzakis and Theodorou 1980; Melentis 1974, 25; Lister and Bahn 1994, 152, 157, 160, passim; Lister, personal communications, July 22–25, 1997. Ancestral mammoths were first recognized in Italy by F. Nesti in 1825; more recently an extra-large ancestral mammoth was found at Scoppito (central Italy), and numerous mammoth skeletons have turned up near Rome, Florence, and in the Arno Valley. In the 1980s, several mammoth skeletons were found in a lignite mine at Pietrafitta, Italy. Lister and Bahn 1994, 24, 159, 160, 161; Osborn 1921, 311, 320, 398; Osborn 1936–42, 1:1240–51. A large Pleistocene elephant tusk recently unearthed at Rebibbia, Aniene Valley, is in the National Museum at Terme: Lapatin, personal communication, December 5, 1997. African prehistoric elephants: Maglio 1973, and Shoshani and Tassy 1996; for shoulder heights, 208 and s.v. "measurements." Sizes of femurs, scapulas, and tusks: Haynes 1991, chap. 1; Osborn 1936–42, 2:1251 with photo, 1602–5. The exact species and dates of the large elephant remains in the Peloponnese and Eurasia are disputed, and nomenclature for extinct proboscideans has changed to reflect new taxonomies. For example, *M. meridionalis* is also called *Archidiskodon meridionalis*; *M. primigenius* is also called *Elephas primigenius*. *M. trogontherii* remains from Erzurum, Turkey, have been called *Mammuthus armeniacus*, a middle Pleistocene precursor of the woolly mammoth, although some assign them to *Elephas*, the modern Asian elephant genus. I thank Adrian Lister for discussing nomenclature and the likelihood of *E. antiquus* and *M. meridionalis* in western Turkey, personal communication, July 22, 1997. Taxonomies and controversy: Lister and Bahn 1994; Shoshani and Tassy 1996.

40. Rhinos, hippos, cave bears: Osborn 1921, 272, 309, 313, 390, 505. *Bos*: Attenborough 1987, 64; Bodenheimer 1960, 51; Rapp and Aschenbrenner 1978, 64. Giant rhino species: Savage and Long 1986, 139–40, 194–98. Range of elephants and *Bos*: Clutton-Brock 1987, chaps. 6, 11. Ice Ages: Brice 1978. Higgins and Higgins 1996, 65–73. Dermitzakis and Sondaar 1978, 815–16. Melentis 1974, 23–25. Aelian *On Animals* 12.11.

Chapter 3

Ancient Discoveries of Giant Bones

1. Pelops's bone: Pindar *Olympian* 1.90–93 (fifth century B.C.); Ps.-Lykophron 52–55 (third century B.C.?); Apollodorus 5.10–11 (first century A.D.); Pausanias 5.13.1–7, 6.22.1 (second century A.D.). In the myth, the goddess Demeter inadvertently devoured the shoulder. Gantz 1993, 2:532–34, 646; Huxley 1975, 15–16, 45; Burkert 1983, 99–101; Schmidt 1951.

2. Burkert 1983, 6–7, 13–14, 98–101. Scapulas are still used for fortune-telling in rural Greece: Nikos Solounias, personal communication, September 1998. Pelops's bone came from Pisa, the district in which Olympia lies. See chapter 2 for fossils of the Peloponnese, and mammal bones mistaken as human. Bone was substituted for ivory in antiquity (e.g., Pliny *Natural History* 8.7) and often fools archaeologists: Kenneth Lapatin, personal communications, December 2 and 5, 1997. Old bone resembles ivory: Scullard 1974, 260–61; Barnett 1982, 74 n. 45. William Loring saw a semifossilized mammoth scapula in Arcadia in 1890: Gardner et al. 1892, 121. Fossils from the Peloponnese are often stained dark brown; I am assuming that Pelops's bone resembled antiquated ivory. But a fossil scapula could have been covered with an ivory veneer, as were other sacred items in Olympia. Lapatin 1997. The use of mammoth ivory in antiquity is controversial; Lapatin 2001, chap. 2; Krzyszkowska 1990, 22, 37–38.

3. Measurements and weights: Aymon Baud, Musée Géologique Cantonal, Lausanne, personal communication, August 12, 1997. Adrian Lister, personal communication, November 20, 1997. I thank Sue Frary, Natural History Exhibit Hall (NHEH), Livingston, Montana, for helping me measure stegodon and mammoth femurs and scapulas. Scapula sizes compared: Osborn 1936–42, 1:1225, 1249. Sailing around Euboea was notoriously perilous: e.g., Strabo 10; Herodotus 8.13; Philostratos *Life of Apollonius* 4.15. Sailing the Aegean: Casson 1974, esp. chap. 9; Herodotus 4.86–87. Barry Strauss, Cornell University, helped me visualize how heroic relics might have been transported by sea. Fossils and large tusks have been found in ancient shipwrecks: chapter 4.

4. Pausanias 5.13. The netted bone was apparently a scapula, to match the bone lost at sea off Euboea. Huxley 1975, 45, and 1979, 147, suggested it may have been a whale bone. A huge shoulder blade, recently identified as a cetacean's, was stored in Athens in the ninth century B.C.: see chapter 4. Hellenistic epigrams describe fishermen dedicating huge bones netted at sea to sanctuaries (third and second centuries B.C.): *Greek Anthology* 206, 243 (6.222, 223). On the motif of fishermen netting sacred or unusual objects from the sea, Buxton 1994,

101. Neogene grabens off Euboea depressed by tectonics: Higgins and Higgins 1996, 84. Trawlers in the Channel and North Sea routinely net mammoth and Cretaceous dinosaur remains: "Fishing for Mammoths," Lister and Bahn 1994, 60–61; and Buffetaut, personal communications, March 10 and 30, 1997.

5. Heroic burial traditions were in place by the tenth century B.C., but widespread "ancestral yearning" for great heroes' relics emerged by the late eighth century B.C. Hero cults: Antonaccio 1995, 5, 247, 250–54.

6. The seventh-century Pelopion was built over an archaic cult site dating back to the tenth century (the shrine of the original Pelops bone?), with rebuilding in the fourth century B.C. and in Roman times. Levi 1979, n. 116 at Pausanias 5.13, and n. 192 at Pausanias 6.22. See Antonaccio 1995, 170–76. A century before Pausanias, Pliny 28.34 indicated that the ivory shoulder blade of Pelops was no longer displayed. Clement of Alexandria *Exhortation to the Greeks* 4.

7. Herodotus 1.66–68; Pausanias 3.3.5–6, 3.11.10, 8.54.4. Pliny *Natural History* 7.74: "records attest that Orestes' body was 7 cubits." Huxley 1979, 147–48. See Pritchett 1982, 45–46; Parke and Wormell 1956, 1:96; Boedeker 1993. Hero cult: Antonaccio 1995. Well digging bores through deep columns of strata and often reveals fossils. In the Peloponnese, the wells that people dug in the early Bronze Age and the classical and Hellenistic periods tapped Pleistocene deposits. Jameson, Runnels, and van Andel 1994, 157, 171. See chapter 2 for classical wells and Pleistocene fossils recently uncovered by lignite miners near Megalopolis, and for tusks recently found by well-diggers in Luka, near Tegea. Tegea geology: Higgins and Higgins 1996, 70–72.

8. "Hominid fever" (chapter 2); the possession of fossil ancestors makes city-states "feel somehow anointed": Krishtalka 1989, 64.

9. Plutarch recorded the taphonomic details, interference by an animal (eagle), human digging, and the presence of man-made artifacts in *Theseus* 36 and *Kimon* 8; see also Pausanias 3.3.7, 1.17. Skyros archaeology and geology: Higgins and Higgins 1996, 93–95. Cf. Wood 1868 for numerous European examples of oversize bones buried in coffins with armor, weapons, and inscriptions.

10. Oracles and bones: Parke and Wormell 1956, vol. 1. Huxley 1979; Pritchett 1982, 45–46; Rose 1959. Hero cults involving bones: Pfister 1909–12, 1:196–208, 223–38. Hesiod: Antonaccio 1995, 128–30. Pausanias 1.28.7 (Oedipus), 9.29.8 (Linus), 9.18.2–5 (Melanippos, Hector, and the giant Tydeus), 7.1.3 (Tisamenus), 6.20.7 (Hippodamia), 6.21.3 (Oenomaus, Saurus), 9.38.3 (Hesiod), 8.9.3–4 and Levi n. 67, 8.36.8 (Arcas), 4.32.3 (Aristomenes), 2.22.4 (Tantalus). Herodotus 5.66 (Melanippos). The acquisition of heroes' bones has a "rich afterlife well into our own times," notes Boedeker 1993, 171; see Bentley 1985 on saints' relics. Cf. the modern mania for acquiring spectacular dinosaur and mammal skeletons and ancestral primate/human remains.

11. Pausanias 1.35.3 (Ajax), 6.19.4 (discus). Ajax: Homer *Iliad* 3.226–32 (giant heroes), Pliny *Natural History* 5.125, Strabo 13.1.30–32 (the low-lying shore of Rhoeteum, site of Ajax's grave, is near Sigeum, site of Achilles' grave, at a place where rivers "carry down great quantities of alluvium and silt up the coast"), Quintus Smyrnaeus *Fall of Troy* 5.650–56. Philostratus *On Heroes* 8.1 says the skeleton was about 16 feet long. Alternative legends place a second Ajax's corpse in an eroding rock just offshore, or at the tip of Euboea, or on the islands of Tenos or Mykonos. *Lemprière's Classical Dictionary* (1978 ed.), s.v. Ajax; Apollodorus Epitome 6.6.

12. Pausanias 3.22.9 (Asopos), 8.32.5 (Megalopolis), 8.36.2–3 (Kronos), 5.12.3 (elephant skull). Pausanias's use of technical terms of anatomy and his devotion to Asklepios leads to speculation that he was a medical man: Levi 1979, 1:2. Pausanias also saw a monstrous skull at the Asklepios sanctuary at Sikyon (2.10.2). Barnum Brown (1926) found evidence (in the form of a *P. antiquus* molar) that fossils were kept at the Asklepios temple at Kos (see chapter 4). Temple "explainers" and curators: Friedländer 1979, 373–74. See Lloyd 1983, 153–56, on ancient medical terminology for bones.

13. Harris 1964, 86. Comparative measurements abound in modern fossil history. For example, cowboys who lassoed a huge *Triceratops* skull poking out of a gully in eastern Wyoming in 1888 described "horns as long as a hoe handle and eye holes as big as your hat." Spalding 1993, 121. In antiquity, Aelian (*On Animals* 16.14) compared immense turtle shells of India to "small skiffs," while a 1993 Athenian newspaper likened the gigantic Miocene tortoise fossil at Pikermi to a compact car. See Higgins and Higgins 1996 for severe shoreline changes around Troy in antiquity (cf. Pliny *Natural History* 5.23) and Troy's Neogene-Pleistocene sediments, 114–15, 127. Large extinct mammal remains, including mastodon, rhino, and giraffe, along the Dardanelles: Sevket Sen, personal communications, October–November 1997. See chapter 4 for Heinrich Schliemann's discovery of a Miocene fossil bone in his excavation of ancient Troy: Schliemann 1880, 323.

14. The erudite *On Heroes* and *Life of Apollonius of Tyana* are examples of the Second Sophistic, which produced the cultivated salon literature in A.D. 60–230. These two works, long assumed to be unoriginal and derivative romances, are coming to be appreciated on their own merits. Drew 1987, chap. 3; Flinterman 1995. Giants: *Life of Apollonius* 5.16; see below for Apollonius's account of dragons in the fossiliferous hills of northern India. His views on griffins are discussed in chapter 1. *On Heroes* synthesizes ancient disputes and legends about the Trojan War in a fictionalized setting, but the material on heroes' bones appears to record authentic new fossil finds in the second century A.D. Philostratus's details conform to the way fossils are exposed in the eastern Mediterranean. Translation is based

(with alterations) on that of Jeffrey Rusten, Cornell University, who shared his unpublished translation and notes on this long-neglected work.

15. Philostratus *On Heroes* 8.1–8. Philostratus gives certain figures consistent with other texts (e.g., the Orestes skeleton) but exaggerates some other dimensions attested by previous writers: e.g., according to Pausanias 8.29.3 the Orontes skeleton was only about 16 feet long, not 45.

16. The oracle claimed that Apollo killed the Sigeum giant while defending Troy; some traditions held that Apollo guided the fatal arrow shot by Paris into Achilles' heel. Achilles buried at Sigeum: Apollodorus 5.5–6 with Frazer's n. 1. Quintus Smyrnaeus *Fall of Troy* 3.724–26. Strabo 13.1.33–31. Achilles' height is 12 cubits in Philostratus *Life of Apollonius* 4.16. Alexander the Great visited the grave of Achilles in 334 B.C.; the emperor Caracalla came in about A.D. 200; and even Sultan Mehmed the Conqueror paid his respects after 1453. Troy's geology: Higgins and Higgins 1996, 115, 126–27. Sigeum was also the site where the Homeric Monster of Troy appeared: chapter 4.

17. Philostratus *On Heroes* 8.8. Geology: Higgins and Higgins 1996, 93–95. An extensive ancient city lies submerged beyond the rocky islets north of Alonnisos. Gigantomachy: chapter 5.

18. Philostratus *On Heroes* 8.8. Philostratus a native of Lemnos: Flinterman 1995, 15. Intact proboscidean skulls are rare but they have been discovered in the Mediterranean region: see, e.g., Osborn 1936–42, 2:1240–41, for photos of a complete elephant cranium of imposing volume found by an Italian farmer in 1926. There are many ancient references to measuring the volumes of animal remains. For example, Aelian *On Animals* 16.14 and 17.3 calculates the capacity of giant tortoise shells as "10 *medimnoi*" and "6 Attic *medimnoi*" (ca. 120 and 72 gallons, respectively). Pliny *Natural History* 9.93 says a sea monster's skull (captured off Spain ca. 151 B.C.) held 15 Roman amphoras, about 90 gallons. Ps.-Aristotle *On Marvelous Things Heard* 842b (129) discusses gigantic horns that held the equivalents of 3 pints and 2 gallons. Strabo 16.2.13, 15.2.13, records that giant sea urchins held the equivalent of a pint. An interesting parallel occurred in colonial America in 1705 when Dutch farmers found a gigantic prehistoric skeleton at Claverack, New York: one of the "fangs" held "Half a pint of Liquor!" Levin 1988, 767. See Marangou-Lerat's 1995 study of Cretan amphoras, 25, 68, 77, 82, 158. William Sanders, personal communication, March 10, 1998. Adrian Lister, personal communication, March 18, 1998. It takes four men to lift a mammoth skull weighing about 260 pounds: I thank Libby Caldwell for his descriptions of assembling mammoth and mastodon skeletons at the NHEH, Livingston, Montana. Geology of Lemnos: Higgins and Higgins 1996, 115, 123–25.

19. Philostratus *On Heroes* 8.8. Neogene sediments on Imroz: Higgins and

Higgins 1996, 115. Sevket Sen, personal communication, October 22, 1997. The tiny island of Tenedos (ancient Calydna, now Bozcaada, Turkey) has similar sediments, and an ancient legend spoke of monsters that emerged from its rocky clefts. See chapter 4's discussion of the Monster of Troy.

20. Philostratus *On Heroes* 8.14, 9.1, 13.3, 54–57. The farmer also claims to have footprints proportionate to a 15-foot man on his land, and the grave of the hero Protesilaus. Strabo 13.1.32 also located the shrine of Protesilaus at Elaeus. The Gallipoli Peninsula has Neogene sediments: Ager 1980, 438. Higgins and Higgins 1996, 107, 112 (Pallene's sediments); 125–29 (Gallipoli and Troy). Prehistoric mammal remains in Thessaly, Macedonia, and the Kassandra Peninsula (discussed below) include deinotheres and mastodons, cave bears, and woolly rhinos: Melentis 1974; Tsoukala and Melentis 1994; Tsoukala, personal communication, January 4, 1999. Philostratus describes the formation of Leuke island from river-borne silt. Leuke (Insula Sharpelor in Romanian): Roman Croitor, Department of Paleontology, Kishinau, Moldova, personal communication, January 20, 1999.

21. Lloyd, personal communication, January 21, 1997. "Holy Dacian" 1979; see also Buffetaut 1987, 16.

22. It should also be noted that the perceived length of a skeleton's scattered bones is exaggerated. See Dodson 1996, 33–55, on the difficulty of arranging a heap of bones into the anatomy of an extinct animal, even for a professional. See Burkert 1983, 6, 15, 99 (Pelops), for evidence that the ancients ritually arranged animal bones in "the proper order" after sacrifices and hunting. Sizes of living animals in exotic locales were often exaggerated: e.g., Pliny *Natural History* 8.35 gives 20 cubits (28 feet; 8.5 m) for Ethiopian elephants.

23. Plutarch *Sertorius* 9. Strabo 17.3.8. Antaeus: Pindar *Isthmian* 4; Apollodorus 2.5.11; Diodorus of Sicily 4.17.4 and see 1.21.4, for an ancient city on the Nile named after Antaeus; Pliny *Natural History* 5.1–4. Konrad 1994, 113–14. Cuvier 1806, 5, suggests that the skeleton might have been a cetacean's. Fossil mammal remains of Morocco: Lister and Bahn 1994, 23–24, 142; Maglio and Cooke 1978, 345, 353–54, 358, 516, 522; Shoshani and Tassy 1996, 83, 196, 233, 338n; Osborn 1936–42, 1:232. There are truly stupendous dinosaur remains in the Atlas Mountains, about 150 miles (250 km) southeast of Lixus, but it seems unlikely that specimens would have been transported to the coast in antiquity. Taquet 1998, 95–121.

24. Solinus 1.90–91. Solinus is assumed to copy Pliny *Natural History* 7.73 (see, e.g., Hansen 1996, 142), but Pliny described a different skeleton some 69 feet long, revealed by an earthquake cleaving a mountain, not a flood. That skeleton was identified as either the giant Orion or Otus (chapter 5). Philodemos of Gadara *On Signs* 4 De Lacy, cited in Hansen 1996, 142.

25. If we apply Cuvier's formula (above) of dividing exaggerated ancient figures by 8 or 10, the Cretan giants would measure 5–8 feet. Early travelers such as Richard Pococke saw "petrified bones of unusual size" in grottoes in 1745, and V. Simonelli found the remains of a large elephant and extinct deer in caves near Rethymno in 1883. Human artifacts (some of great antiquity) often accompany the remains of the extinct elephants, hippos, deer, and giant oxen on Crete. Bate 1913, 241, 240–49. Bate reported *E. (P.) antiquus* in 1905; large and medium elephant remains were also reported in the 1960s and 1970s, but some question the existence of large elephants in Crete. Shoshani and Tassy 1996, 237–39. Thomas Strasser, a zooarchaeologist who works in Crete, comments that large elephantid remains "would be surprising because the large terrestrial fauna on Crete nannized [became dwarfed]." The remains described by Pliny, Philodemos, and Solinus "could be specimens of elephants that date to their earliest arrival on the island, before they nannized. That would be *very* interesting." Personal communications, November 14–17, 1997. Dwarf elephants evolved from mammoths; Lister and Bahn 1994, 34–35. Pleistocene remains of Crete, including *Elephas* cf. *antiquus*: Reese 1996.

26. Solinus 9.6–7. Pausanias (1.25.2) saw a statuary group on the Athenian Acropolis representing the "legendary war of the giants who used to live around Thrace and Pallene"; cf. 8.29.1 and Apollodorus 1.6. For other ancient references, see chapter 5 on the Gigantomachy. Pallene fossils: Tsoukala and Melentis 1994. The species of mastodons are not yet determined; the material was still being cleared in fall 1998: Evangelia Tsoukala, personal communication, January 4, 1999.

27. Philostratus *Life of Apollonius* 3.6–9 (route from Taxila, 2.42–3.9). Eusebius *Treatise against Apollonius of Tyana* 17. Philostratus may have added contemporary knowledge about India to the Apollonius manuscripts. Drew 1987, chap. 3; Flinterman 1995, 83–86 (he translates *drakontes* as "snakes," but the word also indicated anomalous monsters). Shrine of the Thousand Heads (one for each of Buddha's incarnations): Cunningham 1963, 91–94; Peshawar known as Parasha or Parashawar, 40, 66; Taxila (Takshasila, Takshasira) means "cut rock" or "severed head," 91–92. Thanks to Michelle Maskiell, Montana State University, for referring me to Cunningham, the discoverer of ancient Taxila, and to Joshua Katz, Princeton University, for linguistic advice.

28. Murchison 1868 reprints Falconer's complete notes with illustrations of Siwalik topography and Pliocene-Pleistocene fossils. For selenite, mica, quartz, crystalline carbonate of lime and sulphate of lime, iron pyrites (fool's gold), and jet, in the Siwalik deposits, see 15, 17–18, 33–34, 36, 191; black fossils impregnated with "pyrites," 34; "lightning bones," 4. Falconer related the Siwalik fossil species to fantastic animals in Indian mythology: 43, 367–69, 376–77. Kenneth

Carpenter (personal communication, January 7, 1999) and John C. Barry, Harvard (personal communications, November 12, 1998; January 27, 1999) note that iron pyrites might form on the fossils deposited in boggy, marshy areas of the Siwaliks, but the "gems" were most likely calcite and selenite. Barry has observed "large calcite crystals in the marrow cavities" and Anna K. Behrensmeyer (personal communication, January 11, 1999) has noted fossils around Taxila with "large crystals of sparry calcite, which has an iridescent gleam." She suggests that the "dragon skulls" on display were "most likely anthracothere, rhino, suid, giant crocodiles, and *Sivatherium* skulls." On dragons fighting with elephants, cf. Pliny *Natural History* 8.32; see 36.161 (selenite replaces bone marrow of petrified animals, discussed in chapters 2 and 5). Siwalik fossils: Ghosh 1989, 164, 205–13, 307–14. Buffetaut 1987, 169–70, and personal communication, September 12–15, 1998. Savage and Long 1986, 210–11, 226–29; Maglio and Cooke 1978, 509, 524, also s.v. "Siwalik Series"; Osborn 1921, 323–32; Osborn 1936–42, s.v.; and Shoshani and Tassy 1996, 113, 117, 120, 125, 188, 340. Giant turtles of India described by Aelian (*On Animals* 16.14) were said to have shells as "large as a full-sized skiff" with a capacity equal to 120 gallons—could this be a reference to the immense tortoise fossils of the Siwaliks? See chapter 6 for discussion of giant living elephants recently discovered in the Siwalik Hills of Nepal, apparently "throwbacks" to the extinct *Elephas hysudricus.*

29. Herodotus 2.75. How and Wells 1967, 1:203–4, 246, locate Buto near the Bitter Lakes (the Suez area). The Sinai has Cretaceous formations: Norman 1985, 62–67; Jurassic and Cretaceous fossil beds in Egypt, 199. David Weishampel, Johns Hopkins University, personal communication, March 22, 1998. Bodenheimer 1960, 17, discusses remains of Cretaceous reptiles and ratite birds in Egypt. The two identifications most often proposed are locusts (but those were well known to Herodotus) and parachuting agamid lizards (unknown in Egypt). Herodotus does not tell us whether the bones he saw actually matched the image of flying reptiles. Oakley 1975, 42–44, discusses tons of fossil bones stored at ancient Set shrines near Asyut; see chapter 4.

30. Aelian *On Animals* 16.39. Chios and Paros geology and ancient mines: Higgins and Higgins 1996, 136–38, 180–82. The northern part of Chios is mostly Paleozoic shales and sandstones and Triassic-Jurassic limestones; fossil deposits are known in the southern part of the island. Chios fossils: Tobien 1980, and see chapter 5. Cicero (d. A.D. 43) reported the Chios quarry monster but doubted that it really resembled Pan in *De Divinatione* 13. Paros: Pliny *Natural History* 36.14. Bromehead 1945, 105–6, claims that *Prolebias* fish fossils, observed by Xenophanes on Paros, resemble a "hairy Silenus." See chapter 6 for Satyr sightings.

31. Pliny *Natural History* 9.8–15 (Joppa monster and Ostia whale spectacle);

perhaps the sea monster was displayed in Scaurus's Theater, the largest building ever built in Rome (it held 80,000 people), 36.114–17. Friedländer 1979, 1:373, 4:6–8 (whale model). Andromeda: Apollodorus 2.4; Ovid *Metamorphoses* 4.660–736; Strabo 16.2.28, 16.3.7 (100-foot whale beached in Persia), 4.4 (zoological hoaxes in theaters); Pausanias 4.35.6 with Levi's n. 190. See fig. 4.4 for an ancient illustration of the Joppa monster. Folklore about "bloody" sea and earth: Halliday 1928, 208. Chains shown at Joppa: Josephus *Bellum Judaicum* 2.9.2, 3.9.3. Shipping: Casson 1974, chap. 9. Whale strandings in eastern Mediterranean: Posidonius quoted by Strabo 16.2.17; Arrian *Indica* 30.8–10 (76-foot beached whale); Reese, personal communications and unpublished bibliography on whales in the Mediterranean, September 1989, December 27, 1997. Koch: Ley 1948, 105–6; Carrington 1971, 62. Large elephant bones have been found at ancient Ostia: de Angelis d'Ossat 1942, 6–7. More on Roman hoaxes: chapter 6.

32. Fossil remains of the Levant: Bodenheimer 1960, 13–21, 52; Shoshani and Tassy 1996, 225–33, 340; Reese 1985a, 393–96. Josephus *Jewish Antiquities* 5.2.3.

33. Relics displayed in antiquity: Friedländer 1979, 1:367–94; "We face the dangers," 1:378. Rouse 1902, 318–21; Pfister 1909–12. "People were expected to gape in wonder [at relics] without making any particular sense of them," Casson 1974, 251. On European cabinets of curiosities, see Weschler 1995, 75–90, esp. 82–83; Purcell 1997, 21–40. Humphreys 1997, 216–17, 220. Diodorus of Sicily 4.8.2–5. Kuhn 1970. Fisher 1998; see 38–39 for his argument that myth historicizes and gives meaning to wonders.

34. As Pliny (*Natural History* 8.31) observed, "exceptionally large specimens of tusks can indeed be seen in temples." Malta: Cicero *Against Verres* 4.46. See chapter 5 for the myth of the Calydonian Boar. Pausanias 8.46.1–5, 8.47.2, 2.7.9 (the spear that killed the Calydonian Boar was kept in the Shrine to Persuasion at Corinth). Procopius *De Bello Gothico* 5.15.8. A span (*spithame*) was the space between extended thumb and little finger, ca. 23 cm. I thank Kenneth Lapatin for valuable comments on the various Calydonian trophies, personal communications, August 17–18, 1998.

35. Suetonius *Augustus* 72.3. The inclusion of ancient heroes' weapons with the giant bones has puzzled many commentators from 1889 on; some believe that Suetonius meant to say "bones" of heroes, not weapons. See Leighton 1989, 184–85. But the Latin clearly specifies weapons: Robert Kaster, Princeton University, helped clarify the translation. The mystery is solved by the modern discovery of Capri's special fossil beds, discussed in chapter 4. De Angelis d'Ossat 1942, 7, suggested that the large elephant bones found by paleontologists at ancient Ostia, Rome's port, had been transported from Africa or the Levant because of the emperor Augustus's interest in giant bones.

36. Virgil *Georgics* 1.494–97. See Osborn 1936–42, 2:1240–51, for an Italian farmer who unearthed the colossal skull and tusks of *P. antiquus* on his land in 1926. De Angelis d'Ossat 1942, 3–6, documents several fossil elephant remains excavated in the ruins of ancient Rome. See Osborn 1936–42, e.g., 1:263, 618–19, 2:941, 1137, 1240–51, for Italian fossils. Capri fossils: Maiuri 1987, 9–10. Giant pair: Pliny *Natural History* 7.74–76. Manilius cited in Friedländer 1979, 4:9. Wood 1868, 22–23, first suggested that the two giants shown in Rome were counterfeits. Aelian *On Animals* 17.9. Diodorus of Sicily 4.8.4–5. Zoological fabrications are discussed in chapter 6.

37. Suetonius *Tiberius*: see 70 on his interest in Greek mythology. "Demons of the East": Diodorus of Sicily 5.55.5–6. Pliny *Natural History* 9.9–10. Jersey was called Caesarea, which suggests that the island enjoyed special ties with the emperor. Mammoths at La Cotte de St. Brelade, Jersey: Lister and Bahn 1994, 129, 158. Mesozoic and Tertiary remains in France: Buffetaut, personal communications, March 8–11, 1997; August–September 1998; Norman 1985, 198–99; Oakley 1975, 42. Triton and Nereid sightings are discussed in chapter 6.

38. Translation adapted from Hansen's translation of Phlegon of Tralles' *Book of Marvels* 13–14. Sicily suffered earthquakes at the same time as the Pontus quakes. Hansen 1996, 43–44, 141–43. Pliny *Natural History* 2.200; and see 34.46, for Pliny's visit to an artist's studio to admire a clay model and wooden armature for a colossal statue commissioned by the emperor Nero. Phlegon's story of a commissioned bust from fossil remains has an interesting parallel in the 1996 discovery of Kennewick Man, a complete 9,300-year-old skeleton exposed by the Columbia River in Washington. The Umatilla Indians claimed their "ancestor" for reburial, while academics and the federal government insisted on keeping the bones for study. The anthropologist who first analyzed the skeleton created a controversial clay model of the man's head showing Caucasian features. Timothy Egan, "Old Skull Gets White Looks, Stirring Dispute," *New York Times*, April 2, 1998, A12.

39. Theopompus says that, after the earthquake revealed huge bones on the Taman Peninsula, "the local barbarians cast the bones into the Maiotis [Asov] Sea." Phlegon *Book of Marvels* 19. This site is thought to be the extremely productive Blue Ravine fossil exposure on the Taman's Asovian coast, first scientifically studied in 1910. Alexey Tesakov, Moscow, personal communication, January 12, 1999 (see chapter 5). Large elephant remains from the Miocene to Pleistocene around the Black Sea and in northern Turkey include the steppe mammoth *M. trogontherii*, and *M. meridionalis*, *Choerolophodon*, *Deinotherium*, *Archidiskodon*, *Ambelodon*, and the *Elasmotherium*, an immense rhinoceros with skull about 30 inches long (75 cm) topped by a horn over 6 feet long (2 m). Shoshani and Tassy 1996, 209–13, 340; Ann Forsten, Finnish Museum of Natu-

ral History, Helsinki, personal communication, January 8, 1999; Tesakov, personal communication, January 12, 1999. Mammoth molar dimensions: Osborn 1936–42, 2:1061; Lister and Bahn 1994, 78–79. Arthur H. Harris, director of the Laboratory for Environmental Biology, University of Texas at El Paso, kindly sent me life-size scanned images of mastodon, mammoth, and human molars. Suetonius *Tiberius* 8, 48. Hansen 1996, 9, 43–44, 45. Colossal statues were extremely popular in Rome: Pliny *Natural History* 34.38–47. Phlegon's source for the Tiberius story may have been Apollonius Anteros, who wrote during the reign of Claudius (A.D. 41–54). Robert Kaster, personal communication, April 2, 1998.

40. Phlegon 11; Hansen 1996, 43, 139–41. Geryon was sometimes said to have had three heads. Multiple limbs, heads, and jaws as folk attributes of traditional strongmen like Idas: see chapter 5. Geology: Higgins and Higgins 1996, 52, 63–64. Pausanias doubted Sparta's claim to own the bones of Idas, pointing out that the Spartans had tried to obliterate Messenia's knowledge of its own antiquities when they conquered it. He felt it was more likely that Idas was buried somewhere in Messenia. Pausanias 3.13.1–2; cf. 4.32.3 (Aristomenes). The Messenians would have revered local heroes during their wars with Sparta in the eighth and seventh centuries B.C., and a craze for heroic tombs swept Messenia again in the Hellenistic period after they were freed from Spartan rule. Archaeological evidence of a heroic burial in a large jar and a large fossil femur placed on the acropolis of a Messenian town: chapter 4.

41. Phlegon 15; Hansen 1996, 143–45. Wadi Natrun fossils: personal communications, Sevket Sen, November 4, 1997; Eric Buffetaut, April 1998; William Sanders, April 7 and 13, 1998; and Elwyn Simons, Duke University, February 17, 1999. Simons describes a fossil exposure he saw at Wadi Natrun in the 1960s, called Garmaluk ("Hill of Kings"). It was about the size of a city block, but the fossils had been removed to the Cairo Geological Museum. Bodenheimer 1960, 17–18; Maglio and Cooke 1978, 35–36, 345–48, 517, 525; Savage and Long 1986, 27, 226–29. Shoshani and Tassy 1996, 62 (sweeping technique in Fayyum), 70, 89, 233.

42. Diodorus of Sicily 1.26, 3.71–72 (giants and mound at Zabirna), 1.21–24 (Set dismembered Osiris at a place named after the giant Antaeus). The desert as domain of "typhonic" monsters: Fischer 1987, esp. 16–17. Cult of fossils: Ray 1998, 17. Set shrines: Oakley 1975, 42–44; see chapter 5. Wendt 1959, 416–18. Hansen 1996, 146–47. Strabo 17.1.23, Plutarch *Isis and Osiris* 14–18. Kees 1961, 131, 224, 328. The Fayyum Basin, about sixty miles south of Nitria, has abundant Oligocene mammal remains, some quite bizarre, such as the demonic-looking horned rhinoceros-like creature *Arsinoitherium*. Eldredge 1991, 164–69. "Wadi Natrun was considered a secret mound of Osiris," according to *Lexikon*

Ägyptologie (1986), s.v. "Wadi 'n-Natrun." Joseph Manning, Stanford University, helped me with the sacred geography of Osiris.

43. Phlegon 17; Hansen 1996, 145–47. Higgins and Higgins 1996, 27. The Hellenistic era was notorious for gargantuan monuments. The hyperbolic round figure of 100 cubits and the epigraph composed in traditional funeral verse seem to have been inspired by an actual discovery of extraordinary remains.

44. Phlegon 18; Hansen 1996, 147. Warmington 1964, 63, 140, 233, 235. Shoshani and Tassy 1996, 117, 121, 196, 233 (Neogene mammals on Tunisian coast, including area of Utica). Sufetula (Sbeitla, Tunisia) and Theveste (Tebessa, Algeria): John Harris, personal communication, May 12, 1998; Harris 1978, 315–16.

45. Augustine *City of God* 15.9, citing Virgil *Aeneid* 12.899–900; Pliny *Natural History* 7.73–75; and Homer *Iliad* 5.302–4. Ancient Utica is now 7 miles (11 km) inland owing to silting. Pliocene large mammal remains are abundant at Lake Ichkeul, near ancient Utica: Shoshani and Tassy 1996, 233. The tooth may have belonged to a mastodon, but Swiny 1995, 1–2 and plate a, suggests that Augustine found a molar of *Hippopotamus amphibius*, which resembles a gigantic human tooth. Augustine was defending the historicity of Noah's Flood, which had apparently destroyed giants that were not saved in the Ark, creating a logical lapse in the biblical dogma about unchanging, persevering species. Some 1,500 years after Augustine, Barnum Brown would feel the same thrill of touching time in the fossil elephant tooth he found in the temple ruins of Kos (chapter 5).

46. *Oxford Classical Dictionary* (1996), s.v. "Augustine" and "Claudian." Claudian *Rape of Persephone* 3.331–56, ca. A.D. 402–8. Augustine's *City of God* was composed in A.D. 413–426.

Chapter 4

Artistic and Archaeological Evidence for Fossil Discoveries

1. Monster of Troy: Homer *Iliad* 20.146; Apollodorus 2.5.9; Diodorus of Sicily 4.42. Gantz 1993, 1:400–402. Typical sea monsters in ancient art: Boardman 1987, 77 (of the Monster of Troy he notes that "in literature the beast is amphibious" and attacks "on dry land"); quote 79; see plates 21–28 (note that in the plates, figure numbers 10 and 11 are transposed). Schefold 1992, 150, 314. The Monster of Troy in ancient art: Oakley 1997, quote 624; see also 628. By the sixth century B.C., Corinthian vases were exported all over the Greek world. Amyx 1988. This column-krater (sometimes called the Hesione vase), 13 inches high, was

acquired by the Boston Museum in 1963. According to the then-curator, it was found "years ago at Caere," an Etruscan site north of Rome, and shows a sea monster "in his cave under the sea." Vermeule 1963, 162. See Mayor 2000.

2. For the legend of the pair of dragons that emerged from Tenedos (Bozcaada, Turkey), see *Oxford Classical Dictionary*, 3d ed., s.v. "Laocoön," and Quintus Smyrnaeus *The Fall of Troy* 12.444–97. Tenedos has Neogene sediments just like Imroz, an island a few miles north: Higgins and Higgins 1996, 115. Personal communications May–September 1998 with paleontologists Eric Buffetaut (Centre National de la Recherche Scientifique, Paris); Christine Janis (Brown University); George Koufos (Aristotle University, Thessaloniki); Dale Russell (Museum of Natural Sciences, Raleigh, North Carolina); Sevket Sen (Laboratoire de Paléontologie du Muséum, Paris); Matt Smith (Natural History Exhibit Hall, Livingston, Montana); and Nikos Solounias (New York College of Osteopathic Medicine). The artist could have observed Neogene fossil skulls near Corinth or Athens, or in Arcadia, an Aegean island, or the Troad. Neogene sediments in the Corinthia: Higgins and Higgins 1996, 40–44. Miocene ostrich remains occur at Pikermi, Attica, and in Turkey, and ostrich eggshells have been found in a seventh–sixth century B.C. context in ancient Corinth: Reese, personal communication, October 24, 1998. Fossil ostrich remains: Bodenheimer 1960, 17; Melentis 1974, 22. Ostrich eggshells in ancient sites: Poplin 1995, 130–34. For illustrations of *Samotherium*, *Indricotherium*, *Palaeotragus*, horse, giraffe-camels, and other possible skull models, see Savage and Long 1986. Skeletons are rare in Greek art, except for small animal knucklebones used in gaming and *bucrania*, decorative skulls of horned bulls: e.g., Boardman 1989a, fig. 350. John Boardman, Ashmolean Museum, Oxford, personal communication, June 30, 1998. On hybrid monsters on Corinthian vases, see Amyx 1988, 2:660–62.

3. Black-figure amphora showing the monster at Joppa, Corinthian, dated to the second quarter of the sixth century B.C., Berlin, #1652; see Boardman 1987, 79 and plate 24, fig. 10 (mislabeled as 11); and Amyx 1988, 2:392–93, plate 123, fig. 2a. Black-figure skyphos showing Heracles dragging a giant out of a cave, ca. 500 B.C., Danish National Museum #834; see Buxton 1994, 104–5, fig. 10b.

4. Andrews 1926, chap. 15. William Sanders, personal communication, November 13–14, 1998. Reese in Rothenberg 1988, 267. David Reese, an archaeologist and paleontologist who specializes in invertebrates and ivory, has long encouraged the collection and study of fossils from archaeological excavations, e.g., Reese 1985b. Reese's extensive publications and bibliographies of published and unpublished material on fossils in Mediterranean archaeology were indispensable in the writing of this chapter. Oakley 1965, 9–11 (quote 9). Oakley 1975, 15–19, 37. Kennedy 1976. Rudkin and Barnett 1979. *Gryphaea* and camel

bones: Jeannine Davis-Kimball, personal communications, March 9, 1996; October 21, 1997.

5. Antonaccio 1995, 75–98, 128, 246–47. Contrast Rackham 1994, chap. 2, on the way bones from northern European human occupation sites were carefully described by nineteenth- and early-twentieth-century paleozoologists and anthropologists. History of zooarchaeology: Davis 1987, esp. 20–21. Large, rare, exotic, and fossil antlers in graves and sanctuaries: Reese 1992, 775–76; Rothenberg 1988, 267 (fossil antlers at Minoan shrine on Crete); Rouse 1902 (dedications of large antlers in antiquity).

6. Reese 1985b. Kos: Brown 1926. Buffetaut 1987, 4.

7. David Jordan, American School of Classical Studies in Athens, told me about the whale bone, to be published by John Papadopoulos and Deborah Ruscillo: "A Ketos in Early Athens," in preparation.

8. David Reese learned about the find from personnel at the U.S. Embassy. Cape Kormakiti has rich pygmy hippo and elephant deposits. Reese, unpublished memoirs of work in Cyprus, 1971–1973, Field Museum of Natural History, Chicago; personal communications, November 29, 1990; September 21, 1998. Emily Vermeule received the tooth (now lost) from Christakis Loizides (now dead), personal communications, October 31, 1978; October 9, 1985; October 25, 1991. Dr. Vassos Karageorghis, director of antiquities on Cyprus at the time, does not recall hearing of a classical shipwreck discovered in 1973. Personal communication, August 24, 1998. Authorities of the Turkish Republic of Northern Cyprus know nothing of the wreck; personal communications, September–October 1998.

9. Cargoes of tusks in eastern Mediterranean: Krzyszkowska 1990, 20, 28–29 nn. 13, 22. Tusks shipped from Syria to Greece via Cyprus: Bass 1986; Reese and Krzyszkowska 1996, 325, 326. Heavy tusks as "profitable ballast": Gill 1992, 235. Kas: Cemal Pulak, Texas A&M University, personal communications, October 14 and 18, 1998. David Reese plans to identify the several dozen fossil shells recovered from the Kas shipwreck. Thanks to Brendan McDermott for telling me of the fossil shells.

10. Oakley 1975, 7, 18, and plates 1.c and 3.c. Cycad fossil: Edwards 1967, 1, fig. 1. Pliny *Natural History* 36.29; Theophrastus *On Stones* 7.38.

11. Replicas: Zammit-Maempel 1989, 2; Oakley 1975 and 1965, 17; Reese 1984, 189; Buffetaut 1987, 16. Zammit-Maempel 1989 discusses the abundance of Tertiary fossils on Malta, their discovery in several archaeological sites, and lore attached to the fossils. See also Leighton 1989.

12. The Quisisana Hotel is in the Tragara Valley, the southwest part of the island; the location of Augustus's museum is unknown. The bones and weapons found in 1905 lay beneath a layer of volcanic material in Quaternary clay, so this

particular deposit was not disturbed in antiquity, but similar finds were likely to have occurred in antiquity. The fossils and weapons are kept in the Centro Caprense Ignazio Cerio, Capri Town. The link to Augustus's museum has been published in local Italian histories, according to Filippo Barattolo, paleontologist, University Federico II, Naples, and Centro Cerio, personal communication, October 14, 1998. For example, Maiuri's 1956 guidebook to Capri (1987, 9–10), and, more recently, Federico's 1993 article in *Civiltà del Mediterraneo* related the fossils and weapons to Augustus's museum. In his 1989 paper on antiquarianism, archaeologist Robert Leighton (184) concluded that "heroes' bones," not weapons, were shown in Augustus's museum. Leighton cited an Italian paleontological study of 1906 about the Capri fossils, but he was unaware of the associated weapons. I thank Kenneth Lapatin for bringing this important find to my attention and Robert Kaster for discussing its meaning. It's interesting to note that Virgil, a poet of Augustus's time, mentioned that Italian farmers plowed up huge bones along with "rusty spears," *Georgics* 1.494–97.

13. Rothenberg 1988, 266–68. Oakley 1975, 8, 18, and plate 3.b. The ancient Egyptians used petrified logs to make corduroy roads over sand, but we don't know whether they speculated on the origin of the fossil wood. Edwards 1967, 1. Plutarch *Isis and Osiris*. Fischer 1987, 25–26. Ray 1998, 16–17, citing Scamuzzi 1947. The Timna sea urchin is in the Turin Museum, Italy. Other fossil sea urchins, some drilled for suspension, were found in Neolithic and Iron Age sites in Jordan: Reese 1985b. Pliny called the bladder-shaped fossil sea urchins common in Palestine *tecolithos*; according to Pliny and other ancient writers, these were collected as medicine for gall- and kidney stones, a remedy that continued in the Middle Ages: Kennedy 1976, 51.

14. Brunton 1927, 1–3, 12; 1930, 15, 18, 20. Petrie 1925, 130. Sandford 1929, 541. Oakley 1975, 42–44. The boxes at Wandsworth are "extremely heavy," and unpacking them would be "quite a big job": Andrew Currant, personal communication, November 6, 1998. Angela P. Thomas to Reese, January 6, 1999. Thanks to David Reese for sharing his correspondence with the Bolton Museum. Angela P. Thomas, personal communication, March 2, 1999. David Reese hopes to study the rediscovered Qau fossils. See chapter 3 for literary evidence of Osiris relics in various Egyptian shrines.

15. Schliemann 1880, 323. I thank David Reese for calling Schliemann's find to my attention. Andrew Currant, personal communication, November 6, 1998.

16. Boessneck and von den Driesch 1981 and 1983; Marinatos and Hägg 1993, 138 (*Crocodylus niloticus*). The tyrant Polycrates of Samos (sixth century B.C.) enlarged the sanctuary; his strong ties with the Egyptian king Amasis may account for the remains of crocodiles, antelope, and other North African species at the Heraion. Ostrich eggshells, some fossilized, have been found in numerous

ancient Greek (Crete, Corinth, the Argolid, Delphi, Rhodes, and Chios, etc.) and North African sites: Poplin 1995; Reese 1985a, 403.

17. Kyrieleis 1988, 215. The formation of stalagtites was understood in antiquity: Pausanias 10.32, 4.36 and Frazer's notes, cited in Bromehead 1945, 95–96. Crinoids and helmet stones: Bromehead 1945, 106; Ps.-Aristotle *On Marvelous Things Heard* 846b (162). The bronze-plated temple on Sparta's acropolis was excavated in 1907. Fishermen's netted bones: *Greek Anthology* (ed. Beckby, 1957) 6.222, 223.

18. Reese 1985a, 403 (quoting Riis); 391–409; hippos, 392. Hippo molars look like giant human molars: Swiny 1995, 2. Experts disagree on the species of elephant teeth in many ancient sites, and the distribution and extinction of elephants in these areas are also controversial. Reese and Krzyszkowska 1996, 324–26.

19. Kyrieleis 1988, 220–21; and personal communications, May 28, 1998, and March 2, 1999. Professor Zafiratos, Athens University, made the preliminary identification of the fossil. The bone was misplaced after it was sent to Athens in 1988; a decade later, in February 1999, Kyrieleis recovered it. As noted in chapter 2, fossils from Samos are chalky white. Solounias, personal communication, September 23, 1998. Hippo tusk at the temple: Boessneck and von den Driesch 1983.

20. The fossil femur was found in a context of mixed age excavated on the acropolis of Nichoria; it is impossible to know exactly when in antiquity the bone was collected. Robert E. Sloan and Mary Ann Duncan's faunal analysis in Rapp and Aschenbrenner 1978, 17, 65–73, 287, plates 6–2, 6–3, 6–8; and Reese 1992. That the fossil femur belongs to a chalicothere has not been ruled out, but rhino seems most likely. Fossil deposits similar to those of Megalopolis also exist near Kalamata, not far from Nichoria. On heroes' tombs at Nichoria, see Antonaccio 1995, 87–94; and personal communication, September 27, 1998: heroic burials can include "tombs securely identified as belonging to a hero by epigraphic or other textual evidence," tombs reused for ancestor/hero cult activity, and rich "warrior tombs." Ivory: Reese and Krzyszkowska 1996, 325. Fossil shark teeth at Nichoria and other ancient sites: Reese 1984, 190. Smithsonian zooarchaeologist Lynn Snyder is reevaluating the Nichoria vertebrate remains (personal communication, September 30, 1998). She notes discrepancies between the published find-spot of the Great Horse tooth and its trench label. In 1999, the tooth could not be located. The Great Horse arrived in Greece in the third century B.C., possibly from Gaul, Libya, or Scythia: Clutton-Brock 1987, 88. Pleistocene large-bodied horse at Megalopolis: Melentis 1974, 25. The fossil femur relic would never have been recovered for evaluation without the efforts of Rip Rapp, personal communications, August–September 1998.

21. The Asklepieion is located midway between two Pliocene fossil beds containing *E. (P.) antiquus* and other remains: Brown 1926 (with photo of the tooth); Symeonidis and Tataris 1982; Buffetaut 1987, 4. Other Asklepieions displayed fossils in antiquity (perhaps for anatomy studies). Pausanias visited several in the Peloponnese where he examined colossal remains (chapter 2). David Reese attempted to find Brown's Kos specimen in the AMNH, without success.

Chapter 5

Mythology, Natural Philosophy, and Fossils

1. Plato *Phaedrus* 229c–230a (Socrates has neither time nor inclination to rationalize monster myths); *Republic* 2.378a–379b (Gigantomachy stories are untrue and dangerous); *Laws* 663e–664a (the general belief in thousands of fairy tales, such as sowing dragon's teeth, demonstrates popular credulity). Lydian anecdote: Plato *Republic* 2.359d–e. Yet Plato also has Socrates say that "philosophy begins in wonder": *Theaetetus* 155c. See Fisher 1998, 10–11, 41, 61.

2. Vitaliano 1973, 3, 1–7. Greene 1992, xi–xvii, 85–88. Symbolic perception of myths in antiquity: Huxley 1975. Rationalizing movement in antiquity: Sarton 1964, 310, 587–88; Stern 1996, 7–16. For example, Virgil and Pliny saw the Cyclopes as personifications of volcanoes: Ley 1948, 48. Greene 1992 finds a similar idea in Hesiod. Claudian (fourth century A.D.) saw the Gigantomachy as a metaphor for geomorphic changes, in his unfinished poem *Gigantomachia*, discussed below. Pausanias 8.8.3. Strabo 10.3.23. Cf. Aristotle *Poetics* 25 on the credibility of mythohistorical sources.

3. Hesiod's *Theogony*; Apollodorus 1–2; quote 1.6. Diodorus of Sicily 1.26 says the giants were enormous beings that appeared when the "origin of life was still recent." Greene 1992, chaps. 3 and 4; Gantz 1993, 1:1–56, 445–54. *Terata*: Hansen 1996, 148–50. In an interesting fragment, Aristotle (F172 Rose, Schol. to Homer *Odyssey* 9.106) questions the pure pedigree of the Cyclops Polyphemus. One modern definition of a species is that "any two organisms capable of begetting viable, fertile offspring belong to the same species." Horner and Dobb 1997, 168.

4. Greene 1992 proposes that the events described by Hesiod form a chronological record of the volcanic eruptions of prehistory. Claudian *Gigantomachia* 60–65; and see 93–103 for fallen giants transformed into stone. Cretan giants: Diodorus of Sicily 5.71. Rhodes: Strabo 5.55.4–7. Giants buried on Kos: Gantz 1993, 1:446, 453. Skiron: Pausanias 1.44.8; Ovid *Metamorphoses* 7.432–38; Apollodorus Epitome 1.1; Gantz 1993, 1:252. Geology of Corinthia: Higgins

and Higgins 1996, 40–43. According to Apollodorus 1.6, Zeus killed giants with thunderbolts, and Heracles shot them with arrows; Athena threw Sicily on top of the giant Enceladus, while Poseidon broke off part of Kos and heaped it on the giant Polybotes. Zeus fought Typhon from Syria to Thrace and, according to one version of the myth, finally buried Typhon under Mount Etna, Sicily. Vian 1952 surveys the Gigantomachy in ancient art and literature; see 223–35 for naturalistic interpretations of the giants as symbols of the violent forces of nature.

5. Phanagoreia: Strabo 11.2.10. *M. meridionalis* and *Elasmotherium* remains of the Taman Peninsula: Lister and Bahn 1994, 160; and Tesakov, personal communication, January 12, 1999. For a restoration of *Elasmotherium*, see Savage and Long 1986, 195–96. Pleistocene cave bear remains also occur in that area, according to Ann Forsten, personal communications, January 1999. *Choerolophodon* (mastodon) and deinothere remains occur across southern Ukraine: Shoshani and Tassy 1996, 340.

6. Rose 1959 makes the important point that giants were "not exactly human," 73 n. 720. Quadruped giants: Snodgrass 1998, 83–86, 154–55. Pausanias dismissed as "ridiculous" the artistic cliché that giants had coiled serpents for feet: Levi 1979, 2:446 n. 216. Manilius *Astronomy* 1.424–31. Claudian *Gigantomachia* 60–65 associated the giants with vast geophysical change, above.

7. The original Greek terminology for extra limbs or heads of giants and monsters seems to indicate the appendages of grotesque mixed animal and human creatures, according to Guthrie 1965, 203 and n. 4. Multiplicity of limbs and rows of teeth, etc., as folklore motif expressing superhuman capabilities: Solmsen 1949, 23; Hansen 1996, 140. But as early as the fourth century B.C., this motif was not taken literally: Stern 1996, 50, 54–55. A set of multiple bones was identified as the heroic strongman Idas in Messenia, chapter 3. Typhon's human thighs: Apollodorus 1.6.3, Typhon "surpassed all the offspring of Earth. As far as the thighs he was of human shape and of prodigious bulk." See Gantz 1993, 1:50, for human-legged Typhon(?) in early art. Anthropomorphism: chapter 2.

8. Geryon and oxen: chapter 2 and Herodotus 4.9; Livy 1.6; Philostratus *On Heroes* 8.14; Lucian *The Ignorant Book Collector* 13; Stern 1996, 55, 58–59 (Chimera). Flames streamed from clefts in the ground where Typhon (or Typhoeus) fell, as the soil burned and melted, according to Hesiod *Theogony* 820–1022. Apollodorus 1.6.3; Strabo 13.4.6 and 16; Diodorus of Sicily 5.71; Homer *Iliad* 2.783; Ovid *Metamorphoses* 1.1–140; Rose 1959, 58–60; Bodenheimer 1960, 140. Greene 1992, chap. 3, interprets Typhon as a volcano.

9. Classical scholar Gantz 1993, 1:419, 445, 449, wonders why battles with giants and monsters were located in specific places, such as Megalopolis, Pallene, and Kos. The presence of colossal bones and/or burning soils at those sites surely explains their ancient association with fallen giants and monsters, an idea first

expressed by the classical archaeologists Bather and Yorke in 1892–93, 231, 227 (Bathos/Trapezus) and reiterated by J. G. Frazer in his 1898 commentary on Pausanias 8.29.1. Burning soils: Forbes 1936, 19. Burning giant remains at Cadiz, Tartessus: Pausanias 10.4.4. Lignite is now mined extensively in Greece and Turkey to feed power stations. The digging frequently turns up large, dark-colored fossils: see chapter 2. Higgins and Higgins 1996, 72, 84, 108, 132, 151. Cilicia: Brinkmann 1976, 2. See Frazer n. 2 at Apollodorus 1.6 on the coexistence of fossils and volcanic phenomena at Gigantomachy battlefields. Pallene, also known as Phlegra ("Burning Fields"): Pliny *Natural History* 2.110–11, 2.237; Strabo 5.4.6, 6.3.5, 7.25, 27; Diodorus of Sicily 5.71. Flammable natural bitumen and naphtha were said to be the spilled blood of fallen giants and monsters: Ley 1948, 54; e.g., Pseudo-Aristotle *On Marvelous Things Heard* 838a; Strabo 5.4.6, 6.3.5.

10. Skoufos: chapter 2. Pausanias 8.29.1. Appian *Syr.* 58. Pallene (Kassandra Peninsula, Chalkidiki) is almost entirely "made up of Neogene sedimentary rocks": Higgins and Higgins 1996, 112, map 107. Fossil faunal remains include woolly rhinoceros (*Rhinoceros antiquitatis*: Melentis 1974, 25) and mastodons (Tsoukala and Melentis 1994; cf. Shoshani and Tassy 1996, 340). Siwalik Hills: chapter 3. Native American lightning scenarios: Wendt 1968, 277–78; Kindle 1935, 451. Combustible lignite is also common in the Dakota and Siwalik beds. Note that the white fossil bones of ancient Samos and Kos, for example, were associated with destruction by seismic events rather than lightning.

11. Herodotus 3.13, 9.83. See Burkert 1983. Until the fourth century B.C., the largest living land animal known to the Greeks was the horse.

12. Humans shrinking as symptom of diminishing energy of the universe: Hansen 1996, 137–39, 143–45, 147–48 (on traditional stature of heroes). Homer *Iliad* 5.302–4; Herodotus 1.68, 2.91; Pliny *Natural History* 7.73–75; Virgil *Aeneid* 12.899–900; Lucretius *On the Nature of Things* 5.26–30; Solinus 1.90–91. Pausanias 1.27.9, 6.5.1 (quote). Quintus Smyrnaeus *Fall of Troy* 3.24–26. Women of the mythic past were sizable too: e.g., Pyrra the Titan (below), Amazons (chapter 2), Antaeus's wife Tinga (chapter 3), and Atalanta in Aelian *Historical Miscellany* 13.1.

13. Aristophanes *Frogs* 1014; Herodotus 1.60.4; How and Wells 1964–67, 1:83, 2:170. Pliny *Natural History* 7.73. I thank Rosalind Helfand and Katura Reynolds, who drew on their classwork in forensic anthropology to help me imagine how the ancient Greeks estimated the stature of giant heroes by comparing thigh bones of mammoths and humans. Personal communications, November 1998. They applied a formula for determining human stature from long bones found in Douglas Ubelaker, *Human Skeletal Remains* (Washington, DC: Taraxacum, 1989), 61.

14. Giantism tends to die out, owing to environmental and other factors. Larger animals maintain more body heat; warming climates tend to favor reduced body size. See Swinton 1966, 116–17. "The larger the species, the more vulnerable it was to extinction": Lister and Bahn 1994, 124–25; cf. "megaherbivore" extinctions: Krishtalka 1989, 207–9, 165–67. Island fauna: Davis 1987, 118–19. *Bos*: Clutton-Brock 1987, chap. 6. Dwarf species: Shoshani and Tassy 1996, chap. 22. Similar beliefs arose among Native American tribes whose members encountered immense mammoth remains. A Shawnee legend recorded in 1762 in Kentucky identified the skeletons as herds of giant animals hunted in the deep past by heroic "Great and Strong Men" of a size proportionate to that of the animals. After those supermen died out, the god killed the mammoths with lightning bolts. The Delaware had a similar scenario. Simpson 1942, 140; Carrington 1958, 236–37.

15. Giants: Hansen 1996, 137–39; Herodotus 9.83 (giant skeleton at Plataea). Skepticism over giants' stature, not existence per se: e.g., Herodotus 1.68, Pausanias 1.35.3, 10.4.4, Philostratus *On Heroes* 7.9, Augustine *City of God* 15.9. Skeletons in the ground often appear to be larger than expected: Leighton 1989, 185. Giants in myth: Gantz 1993, 1:419–20, 445–53. Cyclopean walls were known in Italy, Sicily, and Greece: Leighton 1989, 194–95. Thucydides *History of the Peloponnesian War* 6.2. Plato *Republic* 2.359; see note 1 above. Living giants: e.g., Pliny 7.73–75. Births of human and animal monsters resembling atavisms of mythological creatures were also recorded: see chapter 6.

16. Pausanias 1.38.7.2. Emil Dubois-Reymond quoted by Wendt 1968, 237.

17. Rudwick 1985, 186. Creation science quotes: Goodstein 1997. See also Krishtalka 1989, chap. 22. Horner and Dobb 1997, 24. Cf. Wendt 1968, 76. Some scholars find evidence for the idea of protracted creation and evolving species of giants and humans through reproduction in the Old Testament: DeLoach 1995, 106–8. See Josephus *Jewish Antiquities* 5.2.3, on Hebrew traditions of giants wiped out by the early Israelites.

18. Modern evolution concepts require the ability to observe minute differences, similarities, and changes in fossil remains of contemporary organisms and those that precede and follow. Horner and Dobb 1997, 49, 168–72, 197. Definitions of species and "apomorphies": Krishtalka 1989, 191, 268–69, 279–81. Mythical monsters: Lloyd 1996, 113. Ongoing generations of ancient monsters in myth: Solmsen 1949, 71, "Monsters are not all confined to one or two early generations." Centaur families in Greek art: Gantz 1993, 1:143–47; Blanckenhagen 1987, 87–90. The young giants Otus and Ephialtes were nine years old: Pausanias 9.22 and 29, with Levi's n. 118. Shoshani and Tassy 1996, 238.

19. Might the Gigantomachy represent ancient folk memories of hunting to

extinction the large mammals of the late Ice Age? This idea was proposed in the nineteenth century by Tylor 1964, 172–75. Now, new theories of human roles in mammoth extinction are reviving that old notion. Some paleontologists propose a *Blitzkrieg* scenario (in the sense of lightning-fast overkill by early humans) to solve the mystery of the sudden extinction of mammoths. Lister and Bahn 1994, 129: "The idea of a 'Blitzkrieg' has been put forward: the mammoths were killed off so quickly that little evidence [of human hunting] has been preserved. The idea is ingenious, but difficult to prove." See also Krishtalka 1989, 205–9. Cf. Cuvier's Theory of Cataclysms in Ley 1948, 9–11; Rudwick 1997, ix–x, 261–62.

20. Heracles eliminated the predators of Crete and Libya: Diodorus of Sicily 4.17.3. Heracles cleared the seas of monsters: Boardman 1987, 77 and n. 26; Boardman 1989b. Centaurs: see chapter 6. Orion: *Lemprière's Dictionary*, s.v.; Gantz 1993, 1:271–73; Hughes 1996, 91 and n. 1; Bergman 1997, 73–79. Chios: Sen, personal communication, November 4, 1997; Higgins and Higgins 1996, 137. Shoshani and Tassy 1996, 340. Pliny *Natural History* 7.73. The skeleton's size, 46 cubits, could mean that the remains of several animals were taken for one, or that people laid bones end to end in an attempt to reconstruct a skeleton. On the other hand, the length matches that of a 70-foot Eocene whale skeleton, a fossil that occurs in the Levant and North Africa (chapter 2). Other Cretan giants, chapter 3. Geology of Crete: Higgins and Higgins 1996, chap. 16.

21. Nemean Lion, Erymanthian and Calydonian Boars, Crommyon Sow, Cretan Bull, Fox, Stymphalian Birds and Wolves, Geryon: Apollodorus 2.4.5, 2.5.4, 1.8.2, 2.4.6–7, 2.5.6–7, 2.5.10; Epitome 1.1. Pausanias 1.27.8–9, 8.22. Erymanthian and Calydonian boar tusks at Cumae, Tegea, and Rome: Pausanias 8.24.5, 8.46.1–5; Ovid *Metamorphoses* 8.287–89; and see chapter 3. Albert Gaudry named the giant Miocene ancestral pig found at Pikermi *Sus erymanthius* after the mythical Erymanthian Boar.

22. Hughes 1996, 91, 105; historical extinctions, 27, 91, 93, 104–7, 192. Pleistocene mammoth/mastodon extinctions: Maglio 1973, 111, 117. Ostriches: Xenophon *Anabasis* 1.5; Ley 1968, 15–17. Elephants: Scullard 1974, 29–31. Pliny *Natural History* 10.37 (extinct birds), 9.1–11 (monstrous whales, squid, octopus, sea turtles). Bodenheimer 1960; Clutton-Brock 1987, 62–65, 114; Brice 1978, chap. 9. Some Pleistocene-Holocene animals overlapped with early humans: Davis 1987; Vigne 1996; Bodenheimer 1960; Haynes 1991, 196, 263.

23. Pindar *Olympian* 9, Ovid *Metamorphoses* 1.140–443, Apollodorus 1.7, Pausanias 1.10. Gantz 1993, 1:165–67. According to Pindar, some current humans descended from the race of heroes, but others descended from the stones cast by Deucalion: Huxley 1975, 23.

24. Horner and Dobb 1997, 146. Comparison of flesh/bone to earth/stone a "naive popular" notion: Kirk, Raven, and Schofield 1983, 176, 178. Ovid's

Metamorphoses alludes to scientific petrifaction theories in 3.397–99, 4; 9.226–29. Seriphos: Gantz 1993, 1:309–10. Seriphos has Pleistocene hippo and elephant remains (Dermitzakis and Sondaar 1978, 827), and archaeologists have found evidence of ancient petrifaction rituals in caves there, Higgins and Higgins 1996, 175. Ancient petrifaction folklore detailed in Felton 1990; see also Forbes Irving 1990.

25. Eichholz 1965, 114, 38–40 on Aristotle's *Meteorologica*. Pseudo-Aristotle *Problems* 24.11; *On Marvelous Things Heard* 834a27–28, 838a14. Rudwick 1985, 24–25. Aristotle *Parts of Animals* 641a19–21. Pliny *Natural History* discussed petrifactions at 2.226, 31.29–30, 36.131, 36.161 (selenite crystals replace marrow in petrified animal bones). Claudian described the lithification of the giants in his unfinished *Gigantomachia* 91–128.

26. Petrified trees around ancient Erasos (chapter 2): Higgins and Higgins 1996, 133, 135. Lesbos also has fossil elephant deposits, according to Sevket Sen. Theophrastus *On Stones* 1.4, 6.37–38. Eichholz 1965, 14–15, 91, 113–14. Theophrastus's lost treatises: Diogenes Laertius 5.36–57, esp. 42–43; Edwards 1967, 2. Fossil fish: Sarton 1964, 559–61. Mammoth ivory in classical antiquity: Lapatin 1997, 664 n. 9, and 2001, chap. 2. Eichholz, 113, thinks that "mottled fossil ivory" referred to Miocene remains with brown and white markings like those found at Pikermi (see chapter 2); tusks could be found wherever the remains of prehistoric elephants repose. Pliny *Natural History* 36.134 (fossil ivory and Theophrastus). Cuvier credited Theophrastus with the earliest knowledge of fossil elephants in "Sur les éléphans vivans et fossiles," 1806, 4.

27. Pre-Socratic philosophers wrote between the sixth century B.C. and the death of Socrates in 399 B.C. Myth and philosophy: Guthrie 1962, 140–42; Sarton 1964, 587–88; Lloyd 1996, 111–12. Aristotle and myth: Huxley 1973.

28. Solinus 9.6. Bromehead 1945, 102–4; Rudwick 1985, 36–38; Burnet 1948, 2, 26. Plato *Critias* 110e–112c. Guthrie 1962, 386–89; Sarton 1964, 179–80; Kirk, Raven, and Schofield 1983, 168–79; Lloyd 1979, 143; Osborn 1894, 36. Coastal changes: chapter 2.

29. Bromehead 1945, 104–5; Phillips 1964, 177; Herodotus 2.12; Strabo 1.3.4. Pliny *Natural History* 35.36. The western desert of Egypt is "riddled with marine fossils": Lloyd 1976, 2:36–37, 67. In eastern Libya, echinoid and bivalve fossils cover the ground for hundreds of miles: Ager 1980, 445. Kirk, Raven, and Schofield 1983, 177; Guthrie 1962, 388; Sarton 1964, 180, 558–59.

30. Rudwick 1985, 36–41, esp. 39. Buffetaut 1987, 1, 19. Some ancient theories for marine fossils seem to attribute them to mysterious molding forces within the earth, an idea that resurfaced in the Middle Ages. Moreover, confusion arises over modern misunderstandings of the word *fossil* in ancient discussions of "fossil fish." Since living fish were occasionally found in mud, Theophrastus and Aris-

totle speculated that some fish spawn might have been trapped underground and petrified. Bromehead 1945, 104; Ley 1968, 191–98.

31. Brad Inwood, personal communications, October 16, 20, 1997. Lloyd 1979, 142–43. Burnet 1948, 25–26. Empirical research: Guthrie 1962, 292; Lloyd 1979, 126–43. For a biologist's view of the way orthodox science legitimates social constructions of reality, see Ruth Hubbard in Harding and Hintikka 1979, 45–52.

32. Plato *Phaedrus* 229c–230a. Osborn 1921, 1, citing Cuvier 1826. Cuvier's remarks, first made in 1812, referred to his immediate predecessors, but they could just as well apply to the ancient natural historians. Rudwick 1997, 216–17. On philosophers' distaste for wonders appreciated for shock value, see Fisher 1998, 46–48. On philosophers and the unknowable, 9, 67.

33. Establishing that "nature has a history" is the essence of paleontology: Horner and Dobb 1997, 21; Rudwick 1985, 69. Philostratus, for example, tried to mediate the gap between philosophy and physical evidence of big bones: see below and *On Heroes* discussed in chapter 3. Gradual change in environment resulted in "evolution" of life-forms in Plato *Laws* 6.782. Aristotle *Meteorology* 1.14, 351a–b, discusses vast cycles of geomorphological change over extended periods of time. Catastrophic versus gradual demise of dinosaurs: Dodson 1996, 279–81; Farlow and Brett-Surman 1997, 662–72.

34. Guthrie 1962, 72–104; Sarton 1964, 176; Kirk, Raven, and Schofield 1983, 177–78; Osborn 1894, 33–35; Phillips 1964, 172; Loenen 1954. Time scales, definition of evolution: Horner and Dobb 1997, 33–38, 49, 168–72, 197. Pausanias 8.29.4: "If the first men were created by the heating of the earth when it was still wet, what other country is likely to have produced earlier or bigger men than India [a hot and wet country]—which even today breeds wild beasts of extraordinary size and peculiar appearance?"

35. Some sources say Empedocles died in the Peloponnese. Inwood 1992; Guthrie 1965, 122–265; Sarton 1964, 246–50. New fragments of an Empedocles poem were discovered in the papyrus collection of the Strasbourg Library in 1999. The Pelops shrine at Olympia probably established by the seventh or sixth century B.C.: Pausanias 5.13.1–7. Orestes: Herodotus 1.67–68; see chapter 3.

36. A few of these products of chance became self-sustaining animals and survived. It is not clear whether Empedocles' theory embraced gradual development of higher forms of life through natural selection, by successive minute genetic modifications of animals over generations. Blundell 1986, 40–41, 86–90. Artistic evidence suggests that Centaur-type creatures (half-human, half-animal) were meant to represent all manner of primal monsters, according to Gantz 1993, 1:50, 144. This imagery may reflect an anthropomorphic way of making sense of puzzling fossil evidence: see chapters 2 and 6. Empedocles' concept of Love and Strife in his

"survival of the fittest" doctrine intended to counter faith in myths finds a striking echo in the writings of Albert Gaudry, the nineteenth-century investigator of Pikermi. Gaudry struggled to reconcile Darwinian evolution with Christian doctrine. He used very similar terminology, "Love" and "Union," in advancing his mystical version of evolution. Buffetaut 1987, 117; Gaudry 1862–67.

37. Blundell 1986, 41. Guthrie 1965, 205.

38. Lucretius *On the Nature of Things* 4.726–43, 5.93–292, 5.787–933. Blundell 1986, 89–93. Since Lucretius states that "at the same time, viable creatures were being born," and that both fit and unfit animals still appear on earth, both the "negative and positive sides of survival of the fittest" are present, but he also stated his opinion that species never changed. Blundell, 92. Osborn 1894, 61–64. Horner and Dobb 1997, 198, 203; cf. 218. A similar meaning of fossils is expressed by Krishtalka: extinct humanoids are "harbingers of our own primeval biological fate": 1989, 64.

39. Blundell 1986, 40, 86–88. Aristotle *Physics* 198b17–32. After the classical era, Aristotle's view dominated Christian and Muslim "science" from the Middle Ages to the early modern period. Scientifically oriented thinkers were constrained to make tortuous arguments around scriptural dogmas of creationism; Osborn 1894, 39–57, 111–250. Rudwick 1985 discusses Aristotelianism's influences on European paleontology.

40. Aristotle lived in Assos and Lesbos in 347–344 B.C. Sarton 1964, 472–76, 529–45. Pliny *Natural History* 8.44. Sarton 1964, 534–35, notes that Aristotle's scale "might be conceived as static and the idea of the fixity of species is not incompatible with it." Lloyd 1996, 184.

41. "Deathblow": Wendt 1968, 76; cf. Ronan 1973, 77. Geoffrey Lloyd, personal communication, January 21, 1997. Lloyd 1996, 113, 122, 125, 164–65 and n. 4.

42. Brad Inwood, personal communication, October 20, 1997. Huxley 1973 shows that Aristotle often cited archaeological discoveries of prehistoric artifacts in his arguments—this is another context where finds of giant bones might have been relevant. Horner remarks on the serious shortcomings of Linnaean-type systems for modern paleontology: Horner and Dobb 1997, 166–65. Hubbard in Harding and Hintikka 1979, 48–49. Mistakes and monstrosities: Aristotle *Physics* 198b17–199b32; *History of Animals* 496b18, 507a23, 544b21, 562b2, 575b13, 576a2; *Generation of Animals* 769b11–30; and see Lloyd 1996, esp. chap. 3. Aristotle's discussion of sightings and capture of sea monsters, *History of Animals* 532b19–29, draws attention to his silence on discoveries of unusual skeletons. Another lost work that might have shed light on fossil finds was a treatise on sea monsters by Demostratus (second century A.D.?) cited by Aelian *On Animals* 13.21, 15.19; Levi 1979 n. 105 at Pausanias 9.4; and see chapter 6.

43. Kuhn 1970, ix, 52–76 (quote 52). See also Eldredge 1991, 210; Vitaliano 1973, 7; Hubbard in Harding and Hintikka 1979, 47–49. Paleontological advances from 1565 to Darwin are summarized in Rudwick 1985: see 36–38 on the problems of reconciling biblical Diluvianism with natural law *after* the era of classical antiquity. Diluvianism, which in antiquity needed to account only for Mediterranean lands, became more and more unwieldy as the known world expanded. Species diversity severely strained the capacity of the Ark. As more and more "giants' bones" were discovered on new continents, the Flood theory was forced to inundate the entire globe to account for giants and unknown animals in the Western Hemisphere. Cohn 1996, chap. 7.

44. Diogenes Laertius 5.22–27. There are a few other "gaps" Aristotle's writings; for example, he left no treatises on mathematics, medicine, or mineralogy. Sarton 1964, 478, 559.

45. Pseudo-Aristotle *On Marvelous Things Heard* 839a5–11; 834a23–32; 838a11–14, 29–35; 842b26–27; 846b3–6, 21–25. Bromehead 1945, 106. On the relationship between curiosity about the rare and uncanny and philosophical inquiry, see Fisher 1998.

46. Stern 1996, 16–17, 22, 29–31, 33–35, 58–59. Centaur sightings are discussed in chapter 6. For elephant molars and hippo incisors in kept in Minoan, Mycenaean, and classical sanctuaries in Greece and Aegean islands, see chapter 4 and Reese 1985a, Reese and Krzyszkowska 1996. The rationalizing movement was one aspect "of the eternal war between reason and superstition," according to historian of science Sarton 1964, 588. Rose 1959, 4–5. Cf. Plato on rationalizing myths, *Phaedrus* 229c–230a; on Centaurs, *Axiochus* 369c; on Cadmus, *Laws* 663e–664a.

47. Philostratus *Life of Apollonius of Tyana* 5.14, 5.16. Historicity of the Neoplatonic Apollonius, problems of authorship and sources: Drew 1987, 83–113; Flinterman 1995, chaps. 1–2. I thank Grant Parker for discussing Apollonius with me. Philostratus investigated the evidence and popular interpretations of giant bones in *On Heroes*, discussed in chapter 3.

48. Dodson 1996, 225. Ronan 1973, 77. Plato *Laws* 6.782. Pausanias 9.21.4. The ancient Greek principle of natural change remained a viable alternative to the idea of immutable nature until the Middle Ages. Osborn 1894, 65–66. See Ovid *Metamorphoses* 15 ("The Teachings of Pythagoras"), esp. 177–429, for a wide-ranging discussion of ever-fluctuating nature. In Rudwick's view, advances in fossil interpretation were "delayed by the lack of any satisfactory explanation of geographical change," 1985, 39. But chapter 2 demonstrates ancient understandings of past and ongoing geophysical changes.

49. Concept of mythical and historical time in antiquity: Huxley 1975, 23–35. Difficulty of imagining immense time scales: Horner and Dobb 1997, 34–35, 37,

167–69, 197–98, quotes 35 and 49–50. Phillips 1964, 177. That fossils are organic remains of long-past ages is not easily recognized, even with signs of decay or petrifaction: Rudwick 1985, 50.

50. Science advances when "peaceful interludes" are interrupted by crises of contradictory evidence: Kuhn 1970, 52. Postclassical beliefs about fossils: Buffetaut 1987, 6 (fallen angels), chaps. 1 and 2; Wendt 1968, 9; Ley 1968, 192–93; Oakley 1975, 20; Bassett 1982; Kennedy 1976, 51, 53; Rudwick 1985; Leighton 1989. On the modern tension and distrust between popular culture and scientists, see Hofstadter 1998. Scientists confronting the "unknowable": Boxer 1998; and see Keller in Harding and Hintikka 1979, 190–91. Philip Fisher (1998) argues that intellectual curiosity about inexplicable wonders has been essential to philosophical-scientific thinking since Plato. But when it comes to the wonders of giant bones, it was outsider-thinkers like Apollonius of Tyana who seemed to approach Fisher's ideal. Greek contributions to science are disputed: see Lloyd 1979, 126, citing Karl Popper's assertion that the best ancient philosophical ideas had "nothing to do with observation," versus G. S. Kirk's belief that Greek thought was characterized by "common sense" and "respect for observation and experience." The mythical material served to "organize and categorize" the natural world: Greene 1992, 87–88. Eldredge 1991, 202.

Chapter 6

Centaur Bones: Paleontological Fictions

1. Pausanias 9.20–21, with Levi's n. 105. Aelian *On Animals* 13.21. Triton art motifs were popular in Boeotia, especially Tanagra: Shepard 1940, 14, 16, 17, fig. 12. Tritons: Gantz 1993, 1:62, 263, 364 (Pindar *Pythian* 4 on the Triton of Lake Tritonis in Libya), 405–6 (artistic portrayals of Tritons and other human–sea monsters), 16–17 (Nereids). Blanckenhagen 1987, 90–91. On Heracles clearing the seas of monsters, Boardman 1989b and 1987, 77 and n. 26; see, e.g., Euripides *Heracles* 400–422. Of the natural philosophers, Empedocles speculated that human-animal hybrids died out as unviable life-forms in the dim past: see chapter 5. The sea as a place of the unexpected and mysterious in antiquity: Buxton 1994, 100–103; King 1995, 159. In 1997, a scientific expedition led by a Smithsonian teuthologist and sponsored by National Geographic attempted (without success) to photograph giant squid; see Ellis 1998.

2. Pliny *Natural History* 9.9–11. Hansen 1996, 172–74; on 12–14, Hansen compares ancient paradoxography to modern tabloids. Deposits of Jurassic dinosaur remains occur near Lisbon (Norman 1985, 198–99). Perhaps some bizarre-

looking Jurassic fossils exposed in a sea cave were thought to resemble a Triton. Dio Cassius 54.21.2 reported a sea monster shaped like a woman except for the head, stranded on the beach in Gaul (ca. A.D. 164–229). Accounts of beached Nereids may have been inspired by garbled reports of single or mass strandings of cetaceans or seals on the Atlantic coast, or perhaps by observations of exposures of Miocene dugong fossils in the sands south of the mouth of the Gironde, or Mesozoic reptiles along the coast north of the Gironde. Buffetaut, personal communications, March 9 and September 13, 1998; and see chapter 3.

3. Paleozoologist Herbert Wendt, noting that dugongs inhabited the Mediterranean, the Red Sea, and the Indian Ocean in early antiquity, suggests that sailors' observations of human-seeming dugong characteristics influenced Greek lore about half-human marine creatures. Wendt 1959, 210–12. Levi n. 105 at Pausanias 9.20. The births of deformed animals and humans evoked intense interest in the Roman era; some were interpreted as avatars of long-vanished mythological creatures. Hansen 1996, 148–59, and Friedländer 1979, 1:6–7. Cf. Thompson 1968, 106–18 (medieval and modern mer-people sightings and teratogenic humans exhibited as fish-people). Starunia steep: Abel 1939, 88–89; Lister and Bahn 1994, 56–57; Savage and Long 1986, 3.

4. Thompson 1968, 110–15. Jenny Hanivers: Ley 1948 gives detailed descriptions of their manufacture in chap. 4, illustrations, 67. See also Purcell 1997, 78–79 (photos), 81–38; King 1995, 162 (with photo); Carrington 1957, 69–70. Examples of faked mer-people are kept in the British Museum and in the Peabody Museum, Harvard University. "Jenny Haniver" probably refers to their traditional production in Genoa (Jenny in nautical slang) and Antwerp (Anvers). Marco Polo revealed the method for creating mummified miniature men from Java that were exhibited in Italy: the recipe called for removing the hair from a tiny monkey and drying the reconfigured body with camphor. In the American West today, whimsical taxidermists still create trophy heads of legendary horned Jackalopes (large jackrabbits with horns attached) to dupe tourists.

5. Ambiguous or freakish evidence can fool modern scientists into dismissing genuine evidence as a hoax: the *Archaeopteryx* discovered in 1860, for example, was thought to be a crass forgery but turned out to be a feathered bird from the late Jurassic (140 million years ago), as demonstrated conclusively by the British Museum (Natural History) in 1986. Modern and medieval hoaxes are thoroughly studied, but ancient zoological hoaxes have not received serious attention. Vitaliano 1973, 6–7, 275–77; Ley 1948, 56–58, 61–69, 107–11; Krishtalka 1989, 67–71, 95–102; Tassy 1993, 31–32. The Empedocles misinformation originated by O. Abel and perpetuated by modern historians, as discussed in the introduction, is an example of how paleontological "hoaxes" supply missing information that somehow "deserves" to be true.

6. Diodorus of Sicily 4.8.2–5; cf. 1.21 (realistic waxen models of the corpse or body parts of the slain god Osiris fooled priests in ancient Egypt). Manilius *Astronomy* 4.101. Pliny *Natural History* 7.35, 10.5; and see bks. 33–35 for examples of artistic deceptions, fraudulent products, and counterfeits created by various techniques. Phoenix: Friedländer 1979, 1:7–9; Hansen 1996, 171. I thank Larry Kim for the reference to the human-headed snake in Lucian *Alexander* 12–16, 25. Pausanias's passage on the Triton (9.20–21) continues in a cryptozoological vein, discussing his knowledge of the African rhinoceros, buffalo, the Celtic elk, the Indian camel and tiger. "If you searched the furthest reaches of Libya or India or Arabia for the wild beasts we know in Greece, you would never find some of them at all, and others would look quite different. . . . animals take different forms in different climates and places." Triton: Aelian *On Animals* 13.21, 17.9. In antiquity, artifacts and natural wonders were displayed together; the distinctions among art, technology, and natural history are modern: Weschler 1995, 61.

7. Gantz 1993, 1:135–39, 146 (Centaur as offspring of Silenos and a Nymph); and Apollodorus 2.5.4.

8. Pausanias 2.21.5–6, 1.23.7. Xenophon *Anabasis* 1.2.8. Plutarch *Isis and Osiris* 356; cf. Herodotus 2.91, 156, and Diodorus of Sicily 1.18.2. Pliny *Natural History* 7.24. Strabo 7.316 described the burning bituminous soil at Dyrrhachium, in a meadow called Nymphaeum; see Forbes 1936, 29. That detail associates the meadow with the burning battlegrounds of the mythical Gigantomachy and reinforces the idea that the Satyr somehow survived extinction. Plutarch *Sulla* 27; cf. Herodotus 8.138 (Satyr captured by King Midas), 7.26 (Marsyas hide). Jerome cited in Friedländer 1979, 1:9–10; Alexander the Great dreamed of capturing a Satyr at Tyre, 376. Saint Augustine thought he saw a Satyr, according to Cuvier. Rudwick 1997, 213.

9. Pausanias 2.21.5–6, 1.23.7. Plutarch *Isis and Osiris* 356, *Sulla* 27. Herodotus 2.91 and 156, 8.138. Diodorus of Sicily 1.18.2. Seiterle 1988. See, e.g., Osborn 1921, 406, for an artist's conception of prehistoric man.

10. Gantz 1993, 1:144–47, 390–91; King 1995, 141–42. Heracles destroys Centaurs in Arcadia and Thessaly: Euripides *Heracles* 360–75. Blanckenhagen 1987, 87–89. Guthrie 1993. In his discussion of ancient zoological knowledge and fantasy, Cuvier found the imaginary creatures of Greek myth "repugnant to reason" but remarked on their "graceful mysticism." Rudwick 1997, 212, 208–15.

11. Aelian *On Animals* 17.9. The wording is obscure; it appears in the context of his describing a weird half-human, half-equine creature of exotic climes, an animal called an Onocentaur. It was thought to resemble the Centaur but was probably a garbled account of a chimpanzee or gorilla. See chapter 5 for philosophical reactions to the notion of Centaur-type hybrids.

12. Phlegon of Tralles *Book of Marvels* 34–35; Hansen 1996, 49, 170–71. Pliny *Natural History* 7.35. The location of Saune is unknown. The antibacterial, anaerobic qualities of honey were well known in antiquity. The corpse of Alexander the Great was supposedly preserved in a honey-filled glass coffin: Statius *Silvae* 3.2.117. The Centaur of Saune, sent to Rome after being embalmed in Egypt, was most likely an ancient Jenny Haniver assembled from mummified human and pony parts. According to Lucian (*Alexander* 15–18), the human-headed snake illusion "was shown in a dark little room to a wonderstruck crowd" primed to see a miracle. They were allowed to touch the snake's body but "herded through quickly and forced out the exit before they could look closely."

13. Pliny *Natural History* 7.35. The birth of such a throwback was considered a bad omen. Plutarch's "Feast of the Seven Sages," *Moralia* 149, seems to place him alongside other writers who noticed the philosophers' silence on anomalous evidence.

14. Williams 1994; Behrens 1997; Weschler 1995, 71–75, 157, and, on the "shimmering" sensation historically evoked by cabinets of curiosities, 60.

15. Willers, artistic statement and personal communications, October 1998. There is a Centaur of Volos Website on the Internet, with links to other hoaxes.

16. Willers, personal communications, October 1998. The phrase "paleontological fictions" is adapted from artist Beauvais Lyons, who specializes in creating "archaeological fictions." Hutcheon 1994, 171. The Greek myths that were used to explain the observed facts of large fossils were also a type of paleontological fiction. Cf. Eldredge 1991, 202, on the modern theory of evolution as a "picture" that explains fossils.

17. For statues of Tritons, Nereids, and a sea monster at the Temple of Poseidon at the Isthmus of Corinth, see Pausanias 2.1.7–8. At Sikyon, he saw a display of an "enormous sea-monster's skull, with a statue of the god of dreams behind it," 2.10.2. Diodorus of Sicily 4.8.2–5.

18. Lyons: Behrens 1997; Hutcheon 1994, 166–74; and Williams 1994, B4–5. Beauvais Lyons, personal communication, October 1998. Lyons's Hokes Archives Website has links to the Centaur Website. Another practitioner of the trickster/cabinet of curiosities genre, David Wilson, curator of the Museum of Jurassic Technology, Los Angeles, is the subject of Weschler 1995. Other artists working in the genre include Joan Fontcuberta (Helga de Alvear Gallery, Madrid), who creates blurred photographs documenting fantasy fauna with scientific labels (such as *Cercopithecus icarocornu* for a winged, horned monkey), alleged to come from mysterious German archives, and Jane Prophet's Website installation *Technosphere* and the CD-ROM *The Imaginary Internal Organs of a Cyborg* (1997). See Fisher 1998, 21–31, on the intellectual benefits of experiencing

momentary wonder. The revival of imaginative narrative in various disciplines that began in the 1990s may be allied to the realization that stories "are a fundamental unit of knowledge," in the words of Bill Buford 1996, 12.

19. Ruth Hubbard, in Harding and Hintikka 1979, 47, writes, "In trying to construct a coherent, self-consistent picture of the world, scientists" of each era "come up with questions and answers that depend on their perceptions of what has been, will be, and can be." Their beliefs about the possible often conflict with popular beliefs. "Absolutely nothing to do with science": Shermer 1997, 130. The limits of scientific knowability are controversial among scientists. In 1997, the Sloan Foundation gave $1.5 million in grants to establish the boundaries of what is unknown in various scientific disciplines: Boxer 1998.

20. Feynman, Einstein, Greenblatt, Eisner quoted in Weschler 1995, 90, 127, 78, 66. Fisher 1998, 61. Keller in Harding and Hintikka 1979, 194. Hofstadter 1998, 512.

21. Horner and Dobb 1997, 61. Kuhn 1970. I borrowed the phrase "a hoax is a hypothesis" from biomythologist Neil Greenberg, University of Tennessee, Knoxville.

22. Dodson 1996, 281. Horner and Dobb, 1997, 219, 232 (my emphasis), 10–11; but see 242 for their hypothetical illustration of the deleterious effects that fabricated hybrid remains could have on scientific knowledge if they are not recognized as false. See Stewart 1984, for a literary approach to the longing for monumental or miniature realms and fantastic creatures of the past. Nostalgia seeks a past that "never existed except as narrative," as a way of restoring "a lyric quality" to life, an "erasure of the gap between nature and culture, and hence a return to the utopia of biology and symbol" (21–23). Hollywood films featuring computer-animated dinosaurs, such as *Jurassic Park* (1993), *Lost World* (1997), and the IMAX film *T-rex: Back to the Cretaceous* (1998), fulfill our desire to magically transport animals of the deep past forward into our own time, to marvel at how they really looked and behaved. History of dinosaur casts and models: Krishtalka 1989, 253–63. Extinct animals and fossils in popular culture around the world: Thenius and Vávra 1996.

23. In 1988, a lifelike woolly mammoth robot constructed by French paleontologists breathed air through its trunk, stamped its feet, and trumpeted. In 1993, on a small Arctic island (Wrangel Island), Russian researchers reported the remains of dwarfed woolly mammoths that somehow survived extinction by more than six thousand years. Lister and Bahn 1994, 9, 137–39 (on recovering DNA from extinct mammoths). "Expedition News," *Explorers Journal* 74 (Fall 1996): 7. Adrian Lister, personal communication, October 29, 1998. Blashford-Snell and Lenska 1996, esp. 193, 225–26. The children's book by French artist-writer

François Place (*The Last Giants*, 1993) pursues the theme of nostalgia: an explorer follows a map inscribed on a huge human tooth and discovers a relict race of giants in remote Asian mountains.

24. Cryptozoologists search for prehistoric creatures that somehow, somewhere escaped extinction. Such pursuits have engendered their share of hoaxes and missteps, but several unknown animals and creatures thought to be long extinct have been discovered alive in the twentieth century. Coelacanth fish, for example, supposedly died out 80 million years ago, but specimens were caught off South Africa in 1938 and in Indonesia in 1998. *Science News* 154 (September 26, 1998): 196. Cases of other unknown creatures are unresolved. For example, paleozoologists note that mysterious "Nandi bear" sightings in Kenya correspond to the skeletal fossils of *Chalicotherium*, the bizarre Miocene creature that supposedly died out in the Pleistocene. Savage and Long 1986, 198–99. The hopes and successes of cryptozoologists are chronicled in *Cryptozoology: Interdisciplinary Journal of the International Society of Cryptozoology* (1982–); see also Mitchell and Rickard 1982, 32–67.

25. Taquet 1998, 127, 146, describes Efremov's novel *Shadow of the Past* (1953). William A. S. Sarjeant is another paleontologist whose writings (under the name Antony Swithin) "incorporate the concept of Tertiary mammals surviving into modern times": Sarjeant 1994, 324, 327. For Sarjeant's survey of fictional works that feature living dinosaurs, mammoths, and other extinct creatures surviving in the modern world, see 321–24. Connie Barlow, co-organizer with Paul Martin of the mammoth memorial event, personal communication, February 26, 1999. The idea is to reintroduce the closest living relatives of Pleistocene megafaunas to the American West. Martin 1999.

26. Dinosauroid hominid: Russell and Séguin 1982; Norman 1985, 54–55; Savage and Long 1986, 248; Horner and Dobb 1997, 100–102.

27. Members of the Vertebrate Paleontology Internet list helped me track the popular history of Russell's model. Dan Chure provided items from the *Weekly World News* ("Startling Discovery . . . Stuns Scientific World: Dinosaur People"; "Terrified Gardener [Meets] Space Alien") and *UFO Universe* (dinosaurs pilot UFOs). Bruce Townsley sent me a copy of the Topps collector card. Thanks also to Darren Naish for the quote from O'Neill 1989, 24, and to Darren Tanke for information about the dinosauroid museum mascot and its appearance in *Heavy Metal*. The live dinosauroid appeared in the last episode of the Granada TV (U.K.) series *Dinosaur!* See Norman 1991, 187; other animated robotic models, molds, and casts, 183–87.

28. Horner and Dobb 1997, 218. See also Krishtalka 1989, 7.

Appendix 1

Large Vertebrate Fossil Species in the Ancient World

1. Woodward 1901; Bodenheimer 1960; Gaudry 1862; Osborn 1921; Dermitzakis and Theodorou 1980; Symeonidis and Tataris 1982; Melentis 1974; Shoshani and Tassy 1996; Koufos and Syrides 1997, Evangelia Tsoukala, personal communication; Dermitzakis and Sondaar 1978.

2. Tobien 1980; Dermitzakis and Sondaar 1978; Sevket Sen, personal communication; Shoshani and Tassy 1996; Tsoukala and Melentis 1994; Major 1887, 1891, 1894; Koufos and Syrides 1997 and George Koufos, personal communication; Osborn 1936–42; Bernor, Fahlbusch, and Mittman 1996; Melentis and Psilovikos 1982; Symeonidis and Tataris 1982; Brown 1926, 1927; Reese 1996.

3. Lister and Bahn 1994; Shoshani and Tassy 1996; Adam 1988; Reese, "Elephant Remains from Turkey," n.d.; Osborn 1936–42; Savage and Long 1986; Ann Forsten, personal communication; Alexey Tesakov, personal communication.

4. Harris 1978; Shoshani and Tassy 1996; Gates 1996; Dermitzakis and Sondaar 1978; Sevket Sen, personal communication.

5. Dodson 1996; Buffetaut 1987; Shoshani and Tassy 1996, Savage and Long 1986; Ghosh 1989; Murchison 1868; Osborn 1921; Osborn 1936–42; Anna K. Behrensmeyer, personal communication.

6. Bodenheimer 1960; Reese 1985a; Adam 1988; Shoshani and Tassy 1996; Maglio 1973; Scullard 1974.

7. Bodenheimer 1960; Harris 1978; Reese 1985a and 1985b; Lister and Bahn 1994; Simon Davis, personal communication; Shoshani and Tassy 1996.

8. Maglio and Cooke 1978; Bodenheimer 1960; Fortau 1920; Sevket Sen, personal communication; Eric Buffetaut, personal communication; Elwyn Simons, personal communication.

9. Osborn 1936–42; Maglio and Cooke 1978; Shoshani and Tassy 1996; Harris 1978 and personal communication; Carrington 1971.

10. Maglio and Cooke 1978; Osborn 1936–42.

11. Shoshani and Tassy 1996; Lister and Bahn 1994.

12. Lister and Bahn 1994; Eric Buffetaut, personal communication.

13. Maiuri 1987; Federico 1993; Lister and Bahn 1994; Shoshani and Tassy 1996; de Angelis d'Ossat 1942; Melentis 1974; Osborn 1936–42; Carrington 1958; Filippo Barattolo, personal communication.

Appendix 2

Ancient Testimonia

1. The location of Zabirna (in the region of Siwa) is unknown, but the Miocene fossil deposits of Wadi Moghara, northeast of Siwa (and west of Wadi Natrun), contain large mammal skeletons, and Libya has similar remains. Fortau 1920. Ammon has been associated with fossil ammonites. The area around Siwa has abundant marine fossils.

2. Translation adapted from Jeffrey Rusten's translation in progress.

3. Dalmatia: the Cave of Artemis is thought to be on the coast near Split. Coastal caves in that region contain large fossil bones, as noted by Cuvier in Rudwick 1997, 56. Translations from William Hansen, *Phlegon of Tralles' Book of Marvels* (Exeter: University of Exeter Press, 1996). Used with permission.

4. Pliny implies that the huts were constructed of the jawbones and ribs of whales. But would enough whales beach on that coast to make possible the construction of a village? It is interesting to note that the Arabis (Purali or Habb) River flows to the Arabian Sea through Baluchistan, where *Indricotherium* and other Tertiary fossils are abundant. Indricotheres, the most stupendous land mammal that ever existed, stood over 17 feet at the shoulder (5.4 m), and the skull measures over 4 feet long (1.3 m). Savage and Long 1986, 193. People might have collected these and other fossil bones from the riverbanks to build dwellings, just as mammoth jaws and ribs were used to make complex houses in eastern Europe, Ukraine, and southern Russia 15,000 years ago. More than seventy such dwellings have been discovered by archaeologists: Lister and Bahn 1994, 104–9. Cf. Arrian, *Indica* 30.1–9.

5. See Bassett 1982 for postclassical European fossil lore about bull's hearts (bucardia), ammonites, belemnites, and crinoid stems.

6. Excavations in 1988–90 at Yarimburgaz Cave, western Istanbul, turned up thousands of Pleistocene cave bear bones. In an interesting coincidence, a chapel was carved into this fossiliferous cavern during the Byzantine era. The builders of the Byzantine church of Saint Menas may well have found a similar deposit of fossils. Thanks to Sinan Kilic for referring me to Stiner, Arsebük, and Howell 1996; see 280 on the chapel.

7. See Bromehead 1945, 102–3; Guthrie 1962, 387 and nn. 2–4; Kirk, Raven, and Schofield 1983, 176–77; Lloyd 1979, 143 and n. 92.

WORKS CITED

Abel, Othenio. 1914. *Die Tiere der Vorwelt.* Berlin: Teubner. Rpt. 1914 in *Abstammungslehre Systematik Paläontologie Biogeographie*, edited by R. Hertwig and R. von Wettstein. Berlin: Teubner.

———. 1939. *Das Reich der Tiere: Tiere der Vorzeit in ihrem Lebensraum.* Berlin: Deutscher Verlag.

Adam, Karl Dietrich. 1988. "On Finds of Pleistocene Elephants in the Environment of Erzurum in East Anatolia." [Summary]. *Stuttgarter Beiträge zur Naturkunde*, ser. B, 146:1–89.

Ager, Derek V. 1980. *The Geology of Europe.* New York: John Wiley and Sons.

Amyx, D. A. 1988. *Corinthian Vase-Painting of the Archaic Period.* 3 vols. Berkeley and Los Angeles: University of California Press.

Andrews, Roy Chapman. 1926. *On the Trail of Ancient Man.* New York: Garden City Publishing Co.

Antonaccio, Carla. 1995. *An Archaeology of Ancestors: Tomb Cult and Hero Cult in Early Greece.* Lanham, MD: Rowan and Littlefield.

Armour, Peter. 1995. "Griffins." In *Mythical Beasts*, edited by John Cherry, 72–103. London: British Museum Press.

Ascherson, Neal. 1996. *Black Sea.* New York: Hill and Wang.

Attenborough, David. 1987. *The First Eden: The Mediterranean World and Man.* London: Collins.

Bahn, Paul, ed. 1996. *Tombs, Graves and Mummies: Fifty Discoveries in World Archaeology.* London: Weidenfeld and Nicolson.

Barnett, Richard D. 1982. *Ancient Ivories in the Middle East.* Jerusalem: Institute of Archaeology, Hebrew University.

Bascom, William. 1965. "The Forms of Folklore: Prose Narratives." *Journal of American Folklore* 78:3–20.

Bass, George F. 1986. "A Bronze Age Shipwreck at Ulu Burun (Kas): 1984 Campaign." *American Journal of Archaeology* 90 (3): 269–96.

Bassett, Michael G. 1982. *Formed Stones, Folklore and Fossils.* Geological Series no. 1. Cardiff: National Museum of Wales.

Bate, Dorothea. 1913. "Notes and Observations [fossil remains in Crete]." In *Camping in Crete*, by Aubyn Trevor-Battye, 239–50. London: Witherby.

Bather, A. G., and V. W. Yorke. 1892–93. "The Probable Sites of Basilis and Bathos." *Journal of Hellenic Studies* 13:227–31.

Behrens, Roy R. 1997. "History in the Mocking." *Print*, May–June, 70–77.

Bentley, J. 1985. *Restless Bones: The Story of Relics.* London: Constable.

Bergman, Charles. 1997. *Orion's Legacy: A Cultural History of Man as Hunter.* New York: Penguin.

Bernor, Raymond L., Volker Fahlbusch, and Hans-Walter Mittman, eds. 1996. *The Evolution of Western Eurasian Neogene Mammal Faunas.* New York: Columbia University Press.

Bird, Eric. C. F., and Maurice L. Schwartz, eds. 1985. *The World's Coastline.* New York: Van Nostrand.

Blanckenhagen, Peter H. von. 1987. "Easy Monsters." In *Monsters and Demons in the Ancient and Medieval Worlds*, edited by Ann E. Farkas et al., 85–96. Mainz: Philipp von Zabern.

Blashford-Snell, John, and Rula Lenska. 1996. *Mammoth Hunt: In Search of the Giant Elephants of Nepal.* London: HarperCollins.

Blundell, Sue. 1986. *The Origins of Civilization in Greek and Roman Thought.* London: Croom Helm.

Boardman, John. 1987. "Very Like a Whale—Classical Sea Monsters." In *Monsters and Demons in the Ancient and Medieval Worlds*, edited by Ann E. Farkas et al., 73–84. Mainz: Philipp von Zabern.

———. 1989a. *Athenian Red-Figure Vases: The Classical Period.* London: Thames and Hudson.

———. 1989b. "Herakles at Sea." In *Festschrift für Nikolaus Himmelmann*, edited by Hans-Ulrich Cain et al., 191–95. Mainz: Philipp von Zabern.

Bodenheimer, F. S. 1960. *Animals and Man in Bible Lands.* Leiden: Brill.

Boedeker, Deborah. 1993. "Hero Cult and Politics in Herodotus: The Bones of Orestes." In *Cultural Poetics in Archaic Greece*, edited by Carol Dougherty and Leslie Kurke, 164–77. Cambridge: Cambridge University Press.

Boessneck, Joachim, and Angela von den Driesch. 1981. "Reste Exotischer Tiere aus dem Heraion auf Samos" (Exotic animal remains at the Heraion temple). *Mitteilungen des Deutschen Archäologischen Instituts* (Berlin) 96:245–48.

———. 1983. "Weitere Reste Exotischer Tiere aus dem Heraion auf Samos." *Mitteilungen des Deutschen Archäologischen Instituts* (Berlin) 98:21–24.

Bolton, J.D.P. 1962. *Aristeas of Proconnesus.* Oxford: Clarendon Press.

Boxer, Sarah. 1998. "Science Confronts the Unknowable." *New York Times*, January 24, B7, 9.

Brice, William C., ed. 1978. *The Environmental History of the Near and Middle East since the Last Ice Age.* London: Academic Press.

Brinkmann, R. 1976. *The Geology of Turkey.* Stuttgart: Ferdinand Enke Verlag.

Bromehead, C.E.N. 1943. "The Forgotten Uses of Selenite." *Mineralogical Magazine* 26:325–33.

———. 1945. "Geology in Embryo (up to 1600 A.D.)." *Proceedings of the Geologists Association* 56:89–134.

Brown, Barnum. 1926. "Is This the Earliest Known Fossil Collected by Man?" *Natural History* 26 (5): 535.

———. 1927. "Samos—Romantic Isle of the Aegean." *Natural History* 27 (1) (January–February): 19–32.

Browne, Malcolm W. 1997. "Huge and Rich Bed of Dinosaur Fossils Found in Remote China." *International Herald Tribune*, April 26, 1 and 7.

Brunton, Guy. 1927–30. *Qau and Badari.* Vols. 1 and 3. London: British School of Archaeology in Egypt.

Buffetaut, Eric. 1987. *A Short History of Vertebrate Paleontology.* London: Croom Helm.

———. 1991. *Fossiles et des Hommes.* Paris: Laffont.

Buffetaut, Eric, Gilles Cuny, and Jean Le Loeff. 1995. "The Discovery of French Dinosaurs." In *Vertebrate Fossils and the Evolution of Scientific Concepts*, edited by William Sarjeant, 159–80. London: Gordon and Breach.

Buford, Bill. 1996. "The Seductions of Storytelling." *New Yorker*, June 24–July 1, 11–12.

Burgio, Enzo. 1989. "Mammiferi del Pleistocene della Sicilia: Leggende e Realtà." In *Ippopotami di Sicilia: Paleontologia e Archaeologia nel territorio di Acquedolci.* Messina: EDAS.

Burkert, Walter. 1983. *Homo Necans: An Anthropology of Ancient Greek Sacrificial Ritual and Myth.* Translated by Peter Bing. Berkeley and Los Angeles: University of California Press.

Burnet, John. 1948 [1892]. *Early Greek Philosophy*, 4th ed. London: Adam and Charles Black.

Buxton, Richard. 1994. *Imaginary Greece: The Contexts of Mythology.* Cambridge: Cambridge University Press.

Canby, Sheila R. 1995. "Dragons." In *Mythical Beasts*, edited by John Cherry, 14–43. London: British Museum Press.

Carrington, Richard. 1957. *Mermaids and Mastodons.* London: Chatto and Windus.

———. 1958. *Elephants.* London: Chatto and Windus.

———. 1971. *The Mediterranean.* New York: Viking.

Casson, Lionel. 1974. *Travel in the Ancient World.* London: George Allen and Unwin.

Christopoulos, George A., and John C. Bastias, eds. 1974. *Prehistory and Protohistory.* Vol. 1, *History of the Hellenic World.* Translated by Philip Sherrard. Athens: Ekdotike Athenon; London: Heinemann.

Clutton-Brock, Juliet. 1987. *The Natural History of Domesticated Animals.* London: British Museum of Natural History.

Cohn, Norman. 1996. *Noah's Flood: The Genesis Story in Western Thought.* New Haven: Yale University Press.

Crump, James, and Irving Crump. 1963. *Dragon Bones in the Yellow Earth.* New York: Dodd, Mead.

Cunningham, Alexander. 1963 [1871]. *The Ancient Geography of India.* Varanasi: Indological Book House.

Currie, Philip, and Kevin Padian, eds. 1997. *Encyclopedia of Dinosaurs.* San Diego: Academic Press.

Cuvier, Georges. 1806. "Sur les éléphans vivans et fossiles." *Annales du Muséum d'Histoire Naturelle* (Paris) 8:1–58, 93–155, 249–69. Rpt. New York: Arno Press, 1980.

———. 1826. *Discours sur les révolutions de la surface du globe, et sur les changements qu'elles ont produit dans le règne animal.* Paris: n.p. Translation in Rudwick 1997, text 19.

Davis, Simon. 1987. *The Archaeology of Animals.* New Haven: Yale University Press.

de Angelis d'Ossat, Gioacchino. 1942. "Elefanti nella regione romana." *L'Urbe* 7 (8): 1–8.

DeLoach, Charles. 1995. *Giants: A Reference Guide from History, the Bible, and Recorded Legend* Metuchen, NJ: Scarecrow Press.

Dermitzakis, M. D., and P. Y. Sondaar. 1978. "The Importance of Fossil Mammals in Reconstructing Paleogeography with Special Reference to the Pleistocene Aegean Archipelago." *Annales Géologiques des Pays Helléniques.* Athens: Laboratoire de Géologie de l'Université.

Dermitzakis, M., and G. Theodorou. 1980. "Map of the Main Fossiliferous Localities of Proboscidean[s] in Aegean Area." Athens: n.p.

"Digging into Natural World Insights." 1996. *Science News* 150 (November 16): 308

Dodson, Peter. 1996. *The Horned Dinosaurs: A Natural History.* Princeton: Princeton University Press.

Drew, John. 1987. *India and the Romantic Imagination.* Oxford: Oxford University Press.

Dubs, Homer H. 1941. "Ancient Military Contact between Romans and Chinese." *American Journal of Philology* 62 (July): 322–30.

Edwards, W. N. 1967. *The Early History of Paleontology.* London: British Museum (Natural History).

Eichholz, D. E., ed. and trans. 1965. *Theophrastus De Lapidibus.* Oxford: Clarendon Press.

Eldredge, Niles. 1991. *Fossils: The Evolution and Extinction of Species.* New York: Abrams.

Ellis, Richard. 1998. *The Search for the Giant Squid.* New York: Lyons Press.

Embleton, Clifford, ed. 1984. *Geomorphology of Europe.* New York: John Wiley and Sons.

Erman, Georg Adolph. 1834. *Fragmens sur Herodote et sur la Sibérie.* Berlin, n.p.

———. 1848. *Travels in Siberia.* Translated by W. D. Cooley. 2 vols. London.

Erol, Oguz. 1985. "Turkey and Cyprus." In Bird and Schwartz 1985, 491–98.

"Expedition News." 1996. *Explorers Journal* 74 (Fall): 7.

Farlow, James O., and M. K. Brett-Surman, eds. 1997. *The Complete Dinosaur.* Bloomington: Indiana University Press.

Federico, E. 1993. "Ossa di giganti ed armi di eroi—Sugli ornamenti delle ville augustee di Capri (Svetonio, Aug. 72)." *Civiltà del Mediterraneo* 1:7–19.

Felton, Debbie. 1990. "A Survey and Analysis of Metamorphosis to Stone." Master's thesis, University of North Carolina, Chapel Hill.

Fischer, Henry. 1987. "The Ancient Egyptian Attitude towards the Monstrous." *In Monsters and Demons in the Ancient and Medieval Worlds,* edited by Ann E. Farkas et al., 13–26. Mainz: Philipp von Zabern.

Fisher, Philip. 1998. *Wonder, the Rainbow, and the Aesthetics of Rare Experiences.* Cambridge: Harvard University Press.

Flinterman, Jaap-Jan. 1995. *Power,* Paideia *and Pythagoreanism: Greek Identity, Conceptions of the Relationship between Philosophers and Monarchs and Political Ideas in Philostratus'* Life of Apollonius. Amsterdam: Gieben.

Forbes, R. J. 1936. *Bitumen and Petroleum in Antiquity.* Leiden: Brill.

Forbes Irving, P.M.C. 1990. *Metamorphosis in Greek Myths.* Oxford: Oxford University Press.

Fortau, R. 1920. *Contribution à l'étude des vertébrés miocènes de l'Egypte.* Cairo: Ministry of Finance, Egypt, Survey Department, Government Press.

Frazer, James G., trans. and comm. 1898. *Pausanias's Description of Greece.* 6 vols. London: Macmillan.

Friedländer, Ludvig. 1979 [1913]. *Roman Life and Manners under the Early Empire.* 4 vols. Translated by A. B. Gough. London. Rpt. translated by Leonard Magnus, 1:367–94, 4:6–11. New York: Arno Press.

Gantz, Timothy. 1993. *Early Greek Myth: A Guide to Literary and Artistic Sources.* 2 vols. Baltimore: Johns Hopkins University Press.

Gardner, E. A., et al. 1892. *Excavations at Megalopolis, 1890–91.* London: Macmillan.

Gates, Marie-Henriette. 1996. "Archaeology in Turkey." *American Journal of Archaeology* 100 (2): 280–81.

Gaudry, Albert. 1862–67. *Animaux fossiles et géologie de l'Attique* and *Atlas.* 2 vols. Paris: Libraire de la Société Géologique de France.

Ghosh, A., ed. 1989. *An Encyclopedia of Indian Archaeology*. New Delhi: Indian Council of Historical Research.

Gill, D.W.J. 1992. "The Ivory Trade." In *Ivory in Greece*, edited by J. Lesley Fitton, 233–38. Occasional Paper 85. London: British Museum.

Ginsburg, Léonard. 1984. "Nouvelles lumières sur les ossements autrefois attribués au géant Theutobochus." *Annales de Paléontologie* 70:181–219.

Glenn, Justin. 1978. "The Polyphemus Myth: Its Origin and Interpretation." *Greece and Rome* 25:141–55.

Glut, Donald. 1997. *Dinosaurs: The Encyclopedia*. Jefferson, NC: McFarland.

Goodstein, Laurie. 1997. "New Light for Creationism." *New York Times*, December 21, sec. 4, pp. 1, 4.

Gore, Rick. 1982. "The Mediterranean—Sea of Man's Fate." *National Geographic* 162:694–737.

Greene, Mott T. 1992. *Natural Knowledge in Preclassical Antiquity*. Baltimore: Johns Hopkins University Press.

Grene, David, trans. 1987. *Herodotus, The History*. Chicago: University of Chicago Press.

Griffith, Mark, ed. and comm. 1983. *Aeschylus, Prometheus Bound*. Cambridge: Cambridge University Press.

Guthrie, Stewart. 1993. *Faces in the Clouds: A New Theory of Religion*. New York: Oxford University Press.

Guthrie, W.K.C. 1962. *A History of Greek Philosophy*. Vol. 1, *The Earlier Presocratics and the Pythagoreans*. Cambridge: Cambridge University Press.

———. 1965. *A History of Greek Philosophy*. Vol. 2, *The Presocratic Tradition from Parmenides to Democritus*. Cambridge: Cambridge University Press.

Halliday, W. R. 1928. *The Greek Questions of Plutarch, Translation and Commentary*. Oxford: Clarendon Press.

Hansen, William, trans. and comm. 1996. *Phlegon of Tralles' Book of Marvels*. Exeter: University of Exeter Press.

Harding, Sandra, and Merrill B. Hintikka, eds. 1979. *Discovering Reality*. Rochester, VT: Schenkman Publishing Company.

Harris, H. A. 1964. *Greek Athletes and Athletics*. London: Hutchinson.

Harris, John M. 1978. "Deinotherioidea and Barytherioidea." In *Evolution of African Mammals*, edited by Vincent J. Maglio and H.B.S. Cooke, 315–32. Cambridge: Harvard University Press.

Haynes, Gary. 1991. *Mammoths, Mastodonts, and Elephants: Biology, Behavior, and the Fossil Record*. Cambridge: Cambridge University Press.

Hayward, Lorna G., and David S. Reese. *Elephants in Ancient Western Asia*. In preparation.

Hecht, Jeff. 1997. "China Unveils First Bird's Feathered Cousin." *New Scientist*, April 19, 6.

Higgins, Michael D., and Reynold Higgins. 1996. *A Geological Companion to Greece and the Aegean*. Ithaca: Cornell University Press.

Hofstadter, Douglas R. 1998. "Popular Culture and the Threat to Rational Inquiry." *Science* 281 (24 July): 512–13.

"Holy Dacian." 1979. *Society of Vertebrate Paleontology News Bulletin* 116 (June): 71.

Horner, John R., and Edwin Dobb. 1997. *Dinosaur Lives: Unearthing an Evolutionary Saga*. New York: HarperCollins.

How, W. W., and J. Wells. 1964–67. *A Commentary on Herodotus*. 2 vols. Oxford: Clarendon Press.

Hughes, J. Donald. 1996. *Pan's Travail: Environmental Problems of the Ancient Greeks and Romans*. Baltimore: Johns Hopkins University Press.

Humphreys, S. C. 1997. "Fragments, Fetishes, and Philosophies: Towards a History of Greek Historiography after Thucydides." In *Collecting Fragments, Fragmente sammeln*, edited by Glenn W. Most, 207–24. Göttingen: Vandenhoeck & Ruprecht.

Hutcheon, Linda. 1994. *Irony's Edge*. London: Routledge.

Huxley, George. 1973. "Aristotle as Antiquary." *Greek, Roman and Byzantine Studies* 14:271–86.

———. 1975. *Pindar's Vision of the Past*. Belfast: Queen's. Rpt. 1982.

———. 1979. "Bones for Orestes." *Greek, Roman and Byzantine Studies* 20 (2) (Summer): 145–48.

Inwood, Brad. 1992. *The Poem of Empedocles*. Toronto: University of Toronto Press.

Jameson, Michael, Curtis Runnels, and Tjeerd van Andel. 1994. *A Greek Countryside: The Southern Argolid from Prehistory to the Present Day*. Stanford: Stanford University Press.

Kees, Hermann. 1961. *Ancient Egypt: A Cultural Topography*. Translated by I. Morrow. Chicago: University of Chicago Press.

Kennedy, Chester. 1976. "A Fossil for What Ails You: The Remarkable History of Fossil Medicine." *Fossils* 1:42–57.

Kindle, E. M. 1935. "American Indian Discoveries of Vertebrate Fossils." *Journal of Paleontology* 9 (5) (July): 449–52.

King, Helen. 1995. "Half-Human Creatures." In *Mythical Beasts*, edited by John Cherry, 138–67. London: British Museum Press.

Kirk, G. S., J. E. Raven, and M. Schofield. 1983. *The Presocratic Philosophers*. 2d ed. Cambridge: Cambridge University Press.

Klein, Richard, and Kathryn Cruz-Uribe. 1984. *The Analysis of Animal Bones from Archaeological Sites*. Chicago: University of Chicago Press.

Konrad, C. F. 1994. *Plutarch's Sertorius: A Historical Commentary*. Chapel Hill: University of North Carolina Press.

Koufos, George D., and George E. Syrides. 1997. "A New Early/Middle Miocene Mammal Locality from Macedonia, Greece. *Comptes Rendus de L'Académie des Sciences* 325 (7) (October): 511–16.

Krishtalka, Leonard. 1989. *Dinosaur Plots and Other Intrigues in Natural History*. New York: Morrow.

Krzyszkowska, Olga. 1990. *Ivory and Related Materials: An Illustrated Guide*. Classical Handbook 3, Bulletin suppl. 59. London: Institute of Classical Studies.

Kuhn, Thomas S. 1970. *The Structure of Scientific Revolutions*. 2d rev. ed. Chicago: University of Chicago Press.

Kyrieleis, Helmut. 1988. "Offerings of the 'Common Man' in the Heraion at Samos." In *Early Greek Cult Practice*, edited by R. Hägg, N. Marinatos, and G. C. Nordquist, 215–21. Athens: Swedish Institute.

Lanham, Url. 1973. *The Bone Hunters*. New York: Columbia University Press.

Lapatin, Kenneth. 1997. "Pheidias Elephantourgos." *American Journal of Archaeology* 101:663–82.

———. 2001. *Chryselephantine Statuary in the Ancient Mediterranean World*. Oxford: Oxford University Press.

Leighton, Robert. 1989. "Antiquarianism and Prehistory in Western Mediterranean Islands." *Antiquaries Journal* 69:183–204.

Lemprière's Classical Dictionary of Proper Names Mentioned in Ancient Authors. 1978. Rev. ed. by F. A. Wright. London: Routledge & Kegan Paul.

Levi, Peter, trans. 1979. *Pausanias, Guide to Greece.* 2 vols. Harmondsworth: Penguin.

Levin, David. 1988. "Giants in the Earth: Science and the Occult in Cotton Mather's Letters to the Royal Society." *William and Mary Quarterly,* 3d ser., 45:751–70.

Ley, Willy. 1948. *The Lungfish, the Dodo, and the Unicorn: An Excursion into Romantic Zoology.* New York: Viking.

———. 1968. *Dawn of Zoology.* Englewood Cliffs, NJ: Prentice Hall.

Lister, Adrian, and Paul Bahn. 1994. *Mammoths.* New York: Macmillan.

Lloyd, Alan B. 1976. *Herodotus, Book II Commentary.* 2 vols. Leiden: Brill.

Lloyd, G.E.R. 1979. *Magic, Reason, and Experience.* Cambridge: Cambridge University Press.

———. 1983. *Science, Folklore and Ideology.* Cambridge: Cambridge University Press.

———. 1996. *Aristotelian Explorations.* Cambridge: Cambridge University Press.

Loenen, J. H. 1954. "Was Anaximander an Evolutionist?" *Mnemosyne,* ser. 4, 7:214–32.

Maglio, Vincent J. 1973. "Origin and Evolution of the Elephantidae." *Transactions of the American Philosophical Society,* n.s., 63 (pt. 3): 1–149.

Maglio, Vincent J., and H.B.S. Cooke, eds. 1978. *Evolution of African Mammals.* Cambridge: Harvard University Press.

Maiuri, Amadeo. 1987 [1956]. *Capri: Its History and Its Monuments.* Rome: English language rpt.

Major, C. I. Forsyth. 1887. "Sur un gisement d'ossements fossiles dans l'ile d Samos, contemporains de l'âge de Pikermi." *Comptes Rendus de L'Académie des Sciences,* 4.

———. 1891. "Le gisement ossifère de Mytilini." In *Samos: Etude géologique, paléontologique et botanique,* by C. de Stefani, C. I. Forsyth Major, and W. Barbey, 85–99. Lausanne: n.p.

———. 1894. *Le Gisement ossifère de Mytilini et catalogue d'ossements fossiles . . . à Lausanne.* Lausanne: Bridel.

Marangou-Lerat, Antigone. 1995. *Le Vin et les amphores de Crète de l'époque classique à l'époque impériale.* Athens: French School of Athens.

Marinatos, Nanno, and Robin Hägg. 1993. *Greek Sanctuaries: New Approaches*. London: Routledge.

Martin, Paul. 1999. "Bring Back the Elephants." *Wild Earth* (Spring): 57–64.

Mathews, W. H. 1962. *Fossils*. New York: Barnes and Noble.

Mayor, Adrienne. 1989. "Paleocryptozoology: A Call for Collaboration between Classicists and Cryptozoologists." *Cryptozoology* 8:19–21.

———. 1990. "Hunting Griffins." *Southeastern Review* (Athens) 1 (2): 193–206.

———. 1991. "Griffin Bones: Ancient Folklore and Paleontology." *Cryptozoology* 10:16–41.

———. 1994. "Guardians of the Gold." *Archaeology*, November–December, 53–58.

———. 2000. "The 'Monster of Troy' Vase: The Earliest Artistic Record of a Vertebrate Fossil Discovery?" *Oxford Journal of Archaeology* 19 (1) (February).

Mayor, Adrienne, and Michael Heaney. 1993. "Griffins and Arimaspeans." *Folklore* (London) 104:40–66.

Melentis, John K. 1974. "The Natural Setting: The Sea and the Land." In Christopoulos and Bastias 1974, 18–25.

Melentis, John, and A. Psilovikos. 1982. *The Paleontological Treasures of Mytilini, Samos, Greece*. Booklet. Mytilini, Samos: Paleontological Museum.

Mitchell, John, and Robert Rickard. 1982. *Living Wonders: Mysteries and Curiosities of the Animal World*. London: Thames and Hudson.

Monastersky, Richard. 1998. "Mongolian Dinosaurs Give Up Sandy Secrets." *Science News* 153 (January 3): 6.

Murchison, Charles, ed. 1868. *Palaeontological Memoirs and Notes of the Late Hugh Falconer*. Vol. 1, *Fauna Antiqua Sivalensis*. London: Robert Hardwicke.

Norman, David. 1985. *The Illustrated Encyclopedia of Dinosaurs*. New York: Crown.

———. 1991. *Dinosaur!* New York: Prentice Hall.

Novacek, Michael. 1996. *Dinosaurs of the Flaming Cliffs*. New York: Doubleday.

Oakley, John. 1997. "Hesione." *Lexicon Iconographicum Mythologiae Classicae*, Supplementum 8, pt. 1. Zurich: Artemis.

Oakley, Kenneth P. 1965. "Folklore of Fossils, Parts I and II." *Antiquity* 39 (March–June): 9–17, 117–24.

———. 1975. *Decorative and Symbolic Uses of Vertebrate Fossils.* Occasional Papers on Technology 12. Oxford: Pitt Rivers Museum, University of Oxford.

O'Neill, Mary. 1989. *Dinosaur Mysteries.* London: Hamlyn.

Osborn, Henry Fairfield. 1894. *From the Greeks to Darwin.* New York: Macmillan.

———. 1921. *The Age of Mammals in Europe, Asia, and North America.* New York: Macmillan.

———. 1936–42. *Proboscidea: The Discovery, Evolution, Migration, and Extinction of the Mastodonts and Elephants of the World.* 2 vols. New York: American Museum of Natural History.

Papadopoulos, John, and Deborah Ruscillo. "A Ketos in Early Athens." In preparation.

Parke, H. W., and Wormell, D.E.W. 1956. *The Delphic Oracle.* 2d ed. 2 vols. Oxford: Oxford University Press.

Petrie, W. M. Flinders. 1925. "Early Man in Egypt." *Man* 25 (78): 130.

Pfister, Friedrich. 1909–12. *Der Reliquienkult im Altertum.* 2 vols. Giessen: Topelmann.

Phillips, E. D. 1955. "The Legend of Aristeas: Fact and Fancy in Early Greek Notions of East Russia, Siberia, and Inner Asia." *Artibus Asiae* 18:161–77.

———. 1964. "The Greek Vision of Prehistory." *Antiquity* 38:171–78.

Poplin, François. 1995. "Sur le polissage des oeufs d'autruche en archéologie." In *Archaeozoology of the Near East II*, edited by H. Buitenhuis and H.-P. Uerpmann. Leiden: Backhuys.

Pritchett, W. K. 1982. *Studies in Ancient Greek Topography.* Part 4, *Passes.* Berkeley and Los Angeles: University of California Press.

Purcell, Rosamond. 1997. *Special Cases: Natural Anomalies and Historical Monsters.* San Francisco: Chronicle Books.

Rackham, James. 1994. *Animal Bones.* London: British Museum.

Rapp, George, Jr., and S. E. Aschenbrenner, eds. 1978. *Excavations at Nichoria in Southwestern Greece.* Vol. 1, *Sites, Environs, and Techniques.* Minneapolis: University of Minnesota Press.

Ray, John D. 1998. [Opening Address]. *Proceedings of the Seventh Inter-*

national Congress of Egyptologists, edited by C. J. Eyre. Leuven: Uitgeverij Peeters.

Reese, David S. 1976. "Men, Saints, or Dragons?" *Folklore* 87:89–95.

———. 1984. "Shark and Ray Remains in Aegean and Cypriote Archaeology." *Opuscula Atheniensia* 15:188–92.

———. 1985a. "Appendix VIII(D)." In *Excavations at Kition* 5, pt. 2, by V. Karageorghis. Nicosia, Cyprus: Department of Antiquities.

———. 1985b. "Fossils and Mediterranean Archaeology." Abstract. *American Journal of Archaeology* 89:347–48.

———. 1992. "Appendix I: Recent and Fossil Invertebrates." In *Excavations at Nichoria in Southwest Greece* II, *The Bronze Age Occupation*, edited by W. McDonald and N. Wilkie. Minneapolis: University of Minnesota Press.

———. N.d. "Elephant Remains from Turkey." Unpublished paper, Dept. of Anthropology, Field Museum of Natural History, Chicago.

———, ed. 1996. *Pleistocene and Holocene Fauna of Crete and Its First Settlers.* Monographs in World Archaeology 28. Philadelphia: Prehistory Press, University Museum Publications.

Reese, David S., and Olga Krzyszkowska. 1996. "Elephant Ivory at Minoan Kommos." In *Kommos: An Excavation on the South Coast of Crete. I/2 The Kommos Region and Houses of the Minoan Town*, edited by J. Shaw and M. Shaw. Princeton: Princeton University Press.

Rolle, Renate. 1989. *The World of the Scythians.* Translated by F. G. Walls. Berkeley and Los Angeles: University of California Press.

Ronan, Colin. 1973. *Lost Discoveries: The Forgotten Science of the Ancient World.* New York: McGraw-Hill.

Rose, H. J. 1959. *A Handbook of Greek Mythology.* 6th ed. New York: E. P. Dutton.

Rothenberg, B. 1988. *The Egyptian Mining Temple at Timna.* London: Institute for Archaeo-Metallurgical Studies.

Rouse, W.H.D. 1902. *Greek Votive Offerings.* Cambridge: Cambridge University Press.

Rudkin, David, and Robert Barnett. 1979. "Magic and Myth: Fossils in Folklore." *Rotunda* 12 (Summer): 13–18.

Rudwick, Martin J. S. 1985 [1972]. *The Meaning of Fossils: Episodes in the History of Paleontology.* 2d rev. ed. Chicago: University of Chicago Press.

———. 1997. *Georges Cuvier, Fossil Bones, and Geological Catastrophes: New Translations and Interpretations of the Primary Texts.* Chicago: University of Chicago Press.

Russell, Dale, and R. Séguin. 1982. "Reconstructions of the Small Cretaceous Theropod *Stenonychosaurus inequalis* and a Hypothetical Dinosauroid." *Syllogeus* (National Museums of Canada) 37:1–43.

Russell, W.M.S. 1981. "An Iguanodon Proper: The Fascination of Fossils." *Social Biology and Human Affairs* 45 (2): 75–87.

Sandford, K. S. 1929. "The Pliocene and Pleistocene Deposits of Wadi Qena and of the Nile Valley between Luxor and Assiut (Qau)." *Quarterly Journal of the Geological Society of London* 85 (4): 493–548.

Sarjeant, William A. S. 1994. "Geology in Fiction." In *Useful and Curious Geological Enquiries beyond the World*, edited by D. F. Branagan and G. H. McNally, 318–37. Sydney: International Commission on the History of Geological Sciences.

Sarton, George. 1964 [1952]. *A History of Science: Ancient Science through the Golden Age of Greece.* New York: John Wiley.

Savage, Donald E., and Donald E. Russell. 1983. *Mammalian Paleofaunas of the World.* Reading, MA: Addison-Wesley.

Savage, R.J.G., and M. R. Long. 1986. *Mammal Evolution: An Illustrated Guide.* New York: Facts on File.

Schefold, Karl. 1992. *Gods and Heroes in Late Archaic Greek Art.* Translated by Alan Griffiths. Cambridge: Cambridge University Press.

Schliemann, Heinrich. 1880. *Ilios: The City and Country of the Trojans.* London: John Murray.

Schmidt, Leopold. 1951. "Pelops und die Haselhexe: Ein sagenkartographischer Versuch." *Laos* 1:67–78.

Scullard, H. H. 1974. *The Elephant in the Greek and Roman World.* Ithaca: Cornell University Press.

Seiterle, Gérard. 1988. "Maske, Ziegenbock und Satyr." *Antike Welt* 19:2–14.

Shepard, Katharine. 1940. *The Fishtailed Monster in Greek and Etruscan Art.* New York: privately printed.

Shermer, Michael. 1997. *Why People Believe Weird Things: Superstitions and Other Confusions of Our Time.* New York: W. H. Freeman.

Shipley, Graham. 1987. *A History of Samos.* Oxford: Clarendon Press.

Shipman, Pat. 1998. *Taking Wing: Archaeopteryx and the Evolution of Bird Flight.* New York: Simon and Schuster.

Shoshani, Jeheskel, and Pascal Tassy, eds. 1996. *The Proboscidea: Evolution and Palaeoecology of Elephants and Their Relatives.* Oxford: Oxford University Press.

Simons, Marlise. 1996. "The Dutch Want Back the Fossil Napoleon Took Away." *New York Times*, June 7.

Simpson, George G. 1942. "The Beginnings of Vertebrate Paleontology in North America." *Proceedings of the American Philosophical Society* 86:130–88.

Snodgrass, Anthony. 1998. *Homer and the Artists: Text and Picture in Early Greek Art.* Cambridge: Cambridge University Press.

Solmsen, F. 1949. *Hesiod and Aeschylus.* Ithaca: Cornell University Press.

Solounias, Nikos. 1981a. *The Turolian Fauna from the Island of Samos, Greece.* Basel: S. Karger.

———. 1981b. "Mammalian Fossils of Samos and Pikermi. Part 2. Resurrection of a Classic Turolian Fauna." *Annals of Carnegie Museum* 50:231–70.

Solounias, Nikos, and Adrienne Mayor. "The Earliest Documented Discovery of Vertebrate Fossils: Samos, Greece." In preparation.

Sondaar, P. Y. 1971. "The Samos Hipparion I." *Proceedings of the Koninkljke Nederlandse Akademie van Wetenschappen.* ser. B, 74. (4): 417–41.

Spalding, David. 1993. *Dinosaur Hunters.* Toronto: Key Porter.

Stanley, Steven. 1989. *Earth and Life through Time.* 2d ed. New York: W. H. Freeman.

Stephens, Walter. 1989. *Giants in Those Days: Folklore, Ancient History and Nationalism.* Lincoln: University of Nebraska Press.

Stern, Jacob, trans. and comm. 1996. *Palaephatus On Unbelievable Tales.* Wauconda, IL: Bolchazy-Carducci.

Stewart, Susan. 1984. *On Longing: Narratives of the Miniature, the Gigantic, the Souvenir, and the Collection.* Baltimore: Johns Hopkins University Press.

Stiner, Mary C., Güven Arsebük, and F. Clark Howell. 1996. "Cave Bears and Paleolithic Artifacts in Yarimburgaz Cave, Turkey." *Geoarchaeology* 11:279–327.

Sutcliffe, A. 1985. *On the Track of Ice Age Mammals.* London: British Museum.

Swinton, William E. 1966. *Giants Past and Present.* London: Robert Hale.

Swiny, Stuart. 1995 "Giants, Dwarfs, Saints, or Humans, Who First Reached Cyprus?" In *Visitors, Immigrants, and Invaders in Cyprus*, edited by Paul Wallace, 1–19. Albany: SUNY Press.

Symeonidis, N., and A. Tataris. 1982. "The First Results of the Geological and Paleontological Study of the Sesklo Basin and Its Broader Environment (Eastern Thessaly—Greece)." *Annales Géologiques des Pays Helleniques*, 146–90.

Symeonidis, N., and G. Theodorou. 1990. "Introducing the Last Elephants of Europe." Press release, Dwarf and Giant Elephants exhibit, Goulandris Natural History Museum, Kifissia, Greece.

Taquet, Phillipe. 1998. *Dinosaur Impressions.* Translated by Kevin Padian. Cambridge: Cambridge University Press.

Tassy, Pascal. 1993. *The Message of Fossils.* Translated by Nicholas Hartmann. New York: McGraw-Hill.

Thenius, Erich. 1973. *Fossils and the Life of the Past.* Translated by B. Crook. New York: Springer-Verlag.

Thenius, Erich, and Norbert Vávra. 1996. *Fossilien im Volksglauben und im Alltag.* Senckenberg-Buch 71. Frankfurt am Main: Kramer.

Theodorou, George E. 1990. "The Dwarf Elephants of Tilos." *Athenian* (Athens), May, 17–19.

Thompson, C.J.S. 1968. *The Mystery and Lore of Monsters.* New York: Bell.

Tobien, Heinz. 1980. "A Note on the Skull and Mandible of a New Choerolophodont Mastodont from the Middle Miocene of Chios (Greece)." In *Aspects of Vertebrate History: Essays in Honor of Edwin Colbert*, edited by Louis Jacobs, 299–307. Flagstaff, AZ: Museum of Northern Arizona Press.

Tsoukala, Evangelia, and John Melentis. 1994. "Deinotherium giganteum KAUP (Proboscidea) from Kassandra Peninsula (Chalkidiki, Macedonia, Greece)." *Geobios* 27:633–64.

Tylor, Edward Burnet. 1964 [1865]. *Researches into the Early History of Mankind.* London. Rpt. Chicago: University of Chicago Press.

Tziavos, Christos, and John C. Kraft. 1985. "Greece." In Bird and Schwartz 1985, sec. 61.

Ubelaker, Douglas, and Henry Scammell. 1992. *Bones: A Forensic Detective's Casebook.* New York: HarperCollins.

Van Couvering, J. A., and J. A. Miller. 1971. "Late Miocene Marine and Non-marine Time Scale in Europe [Samos]." *Nature* 230:559–63.

Vermeule, Cornelius. 1963. [Recent acquisitions]. *Museum of Fine Arts, Boston Bulletin* 61:159–64, figs. 10–12.

Vian, Francis. 1952. *La Guerre des géants.* Paris: Klincksieck.

Vigne, Jean-Denis. 1996. "Did Man Provoke Extinctions of Endemic Large Mammals on the Mediterranean Islands? The View from Corsica." *Journal of Mediterranean Archaeology* 9 (1): 117–20.

Vitaliano, Charles J. 1987. "Geological History." In *Landscape and People of the Franchthi Region*, edited by Tjeerd van Andel and Susan Sutton, 12–16. Bloomington: Indiana University Press.

Vitaliano, Dorothy. 1973. *Legends of the Earth.* Bloomington: Indiana University Press.

Warmington, B. H. 1964. *Carthage.* Baltimore: Penguin.

Weidmann, M., et al. 1984. "Neogene Stratigraphy of the Eastern Basin, Samos Island, Greece." *Geobios* 17:477–90.

Wendt, Herbert. 1959. *Out of Noah's Ark: The Story of Man's Discovery of the Animal Kingdom.* Translated by Michael Bullock. Boston: Houghton Mifflin.

———. 1968 [1965]. *Before the Deluge.* Translated by R. Winston and C. Winston. London: Gollancz.

Weschler, Lawrence. 1995. *Mr. Wilson's Cabinet of Wonder.* New York: Vintage.

Wilford, John Noble. 1998. "New Picture Emerges on How Dinosaurs of Gobi Were Killed." *New York Times*, January 20, F3.

Williams, Don. 1994. "Do You Believe in Centaurs?" *Knoxville News-Sentinel* (Tennessee), October 11, B4–5.

Wood, Edward J. 1868. *Giants and Dwarves.* London: Richard Bentley.

Woodward, A. Smith. 1901. "On the Bone Beds of Pikermi, Attica, and on Similar Deposits in Northern Euboea." *Geological Magazine*, n.s., 4, 8:481–86.

Wright, M. R., ed. 1995 [1981]. *Empedocles: The Extant Fragments.* Bristol: Classical Press.

Zammit-Maempel, George. 1989. "The Folklore of Maltese Fossils." *Papers in Mediterranean Social Studies* 1:1–29.

INDEX

DATE
AUG 1 4 2000
JAN 0 5 2003
JAN 0 5 2003
GAYLORD